Recent Trends in Roughness Measurement and Data Analysis of Machined Surfaces

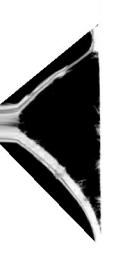

Recent Trends in Roughness Measurement and Data Analysis of Machined Surfaces

Guest Editor

Przemysław Podulka

Basel • Beijing • Wuhan • Barcelona • Belgrade • Novi Sad • Cluj • Manchester

Guest Editor
Przemysław Podulka
Faculty of Mechanical
Engineering and Aeronautics
Rzeszów University of Technology
Rzeszów
Poland

Editorial Office
MDPI AG
Grosspeteranlage 5
4052 Basel, Switzerland

This is a reprint of the Special Issue, published open access by the journal *Materials* (ISSN 1996-1944), freely accessible at: www.mdpi.com/journal/materials/special_issues/Roughness_Surfaces.

For citation purposes, cite each article independently as indicated on the article page online and using the guide below:

Lastname, A.A.; Lastname, B.B. Article Title. *Journal Name* **Year**, *Volume Number*, Page Range.

ISBN 978-3-7258-3422-8 (Hbk)
ISBN 978-3-7258-3421-1 (PDF)
https://doi.org/10.3390/books978-3-7258-3421-1

© 2025 by the authors. Articles in this book are Open Access and distributed under the Creative Commons Attribution (CC BY) license. The book as a whole is distributed by MDPI under the terms and conditions of the Creative Commons Attribution-NonCommercial-NoDerivs (CC BY-NC-ND) license (https://creativecommons.org/licenses/by-nc-nd/4.0/).

Contents

About the Editor vii

Preface ix

Przemysław Podulka
Thresholding Methods for Reduction in Data Processing Errors in the Laser-Textured Surface Topography Measurements
Reprinted from: *Materials* **2022**, *15*, 5137, https://doi.org/10.3390/ma15155137 1

Elżbieta Doluk, Anna Rudawska and Izabela Miturska-Barańska
Investigation of the Surface Roughness and Surface Uniformity of a Hybrid Sandwich Structure after Machining
Reprinted from: *Materials* **2022**, *15*, 7299, https://doi.org/10.3390/ma15207299 22

Tomasz Trzepieciński, Ján Slota, Ľuboš Kaščák, Ivan Gajdoš and Marek Vojtko
Friction Behaviour of 6082-T6 Aluminium Alloy Sheets in a Strip Draw Tribological Test
Reprinted from: *Materials* **2023**, *16*, 2338, https://doi.org/10.3390/ma16062338 40

Krzysztof Żaba and Tomasz Trzepieciński
Cold Drawing of AISI 321 Stainless Steel Thin-Walled Seamless Tubes on a Floating Plug
Reprinted from: *Materials* **2023**, *16*, 5684, https://doi.org/10.3390/ma16165684 58

Kamil Czapla, Krzysztof Żaba, Marcin Kot, Ilona Nejman, Marcin Madej and Tomasz Trzepieciński
Tribological Performance of Anti-Wear Coatings on Tools for Forming Aluminium Alloy Sheets Used for Producing Pull-Off Caps
Reprinted from: *Materials* **2023**, *16*, 6465, https://doi.org/10.3390/ma16196465 77

Damian Gogolewski, Paweł Zmarzły, Tomasz Kozior and Thomas G. Mathia
Possibilities of a Hybrid Method for a Time-Scale-Frequency Analysis in the Aspect of Identifying Surface Topography Irregularities
Reprinted from: *Materials* **2023**, *16*, 1228, https://doi.org/10.3390/ma16031228 101

Paweł Zmarzły, Tomasz Kozior and Damian Gogolewski
The Effect of Non-Measured Points on the Accuracy of the Surface Topography Assessment of Elements 3D Printed Using Selected Additive Technologies
Reprinted from: *Materials* **2023**, *16*, 460, https://doi.org/10.3390/ma16010460 113

Damian Gogolewski, Paweł Zmarzły and Tomasz Kozior
Multiscale Analysis of Functional Surfaces Produced by L-PBF Additive Technology and Titanium Powder Ti6Al4V
Reprinted from: *Materials* **2023**, *16*, 3167, https://doi.org/10.3390/ma16083167 129

Przemysław Podulka, Wojciech Macek, Ricardo Branco and Reza Masoudi Nejad
Reduction in Errors in Roughness Evaluation with an Accurate Definition of the S-L Surface
Reprinted from: *Materials* **2023**, *16*, 1865, https://doi.org/10.3390/ma16051865 143

Monika Gwoździk, Mirosław Bramowicz and Sławomir Kulesza
Analysis on the Morphology and Interface of the Phosphate Coating Prepared on X39Cr13 and S355J2 Steels
Reprinted from: *Materials* **2024**, *17*, 2805, https://doi.org/10.3390/ma17122805 167

Lukasz Nowakowski, Slawomir Blasiak and Michal Skrzyniarz
Influence of the Relative Displacements and the Minimum Chip Thickness on the Surface Texture in Shoulder Milling
Reprinted from: *Materials* **2023**, *16*, 7661, https://doi.org/10.3390/ma16247661 179

Dana Stančeková, Filip Turian, Michal Šajgalík, Mário Drbúl, Nataša Náprstková and Anna Rudawská et al.
Identification of the Production of Small Holes and Threads Using Progressive Technologies in Austenite Stainless Steel 1.4301
Reprinted from: *Materials* **2023**, *16*, 6538, https://doi.org/10.3390/ma16196538 192

About the Editor

Przemysław Podulka

Przemysław Podulka studied his Eng. and M.Sc. in Technical and Informatic Sciences at the University of Rzeszów, Rzeszów, Poland. Before his graduation in 2009 and 2013, respectively, he received three scholarships from the Polish Minister of Higher Education in the 2006/2007, 2007/2008, and 2008/2009 academic years. In 2017, he received his Ph.D. in Mechanical Engineering from the Faculty of Mechanical Engineering and Aeronautics, Rzeszów University of Technology, Rzeszów. From 2017 till now, he has been affiliated with the Department of Manufacturing Processes and Production Engineering, Rzeszów University of Technology in Rzeszów, where he was a Research Associate Professor. He received three awards, being classified as 2% of authors from an updated science-wide author database of standardized citation indicators (Scopus) in 2022, 2023, and 2024.

His research activities have encompassed such fields as surface engineering, surface metrology, surface finishing, machining, quantitative characterization of surface topography, analysis of surface topography measurement noise of various rough materials, characterization of coating materials, and the relationship between structure and properties of various engineering materials.

Preface

A comprehensive study of the surface topography can provide valuable information on the material layers. Considering the nature of the surface, each topography can respond to the manufacturing processes during its creation. Surface texturing may be advantageous when layer-specific performances are required. Even when precise measuring methods are used, results may not be accurate if the data are collected and processed improperly. For this reason, there have been many studies on the errors resulting from the measurement process. As far as international requirements go, the standardization of areal techniques responded to the definition of the measurement noise. The time reduces when the repetitions improve the final accuracy of the surface topography results. This reprint will present the current advantages of methods for measuring, describing, and evaluating surface topography measurement results due to the errors that occur when both measurement and data processing actions are designated.

Przemysław Podulka
Guest Editor

Article

Thresholding Methods for Reduction in Data Processing Errors in the Laser-Textured Surface Topography Measurements

Przemysław Podulka

Department of Manufacturing and Production Engineering, Faculty of Mechanical Engineering and Aeronautics, Rzeszow University of Technology, Powstancow Warszawy 12 Str., PL-35959 Rzeszow, Poland; p.podulka@prz.edu.pl; Tel.: +48-17-743-2537

Abstract: There are many factors influencing the accuracy of surface topography measurement results: one of them is the vibrations caused by the high-frequency noise occurrence. It is extremely difficult to extract results defined as noise from the real measured data, especially the application of various methods requiring skilled users and, additionally, the improper use of software may cause errors in the data processing. Accordingly, various thresholding methods for the minimization of errors in the raw surface topography data processing were proposed and compared with commonly used (available in the commercial software) techniques. Applied procedures were used for the minimization of errors in the surface topography parameters (from ISO 25178 standard) calculation after the removal and reduction, respectively, of the high-frequency noise (S-filter). Methods were applied for analysis of the laser-textured surfaces with a comparison of many regular methods, proposed previously in the commercial measuring equipment. It was found that the application of commonly used algorithms can be suitable for the processing of the measured data when selected procedures are provided. Moreover, errors in both the measurement process and the data processing can be reduced when thresholding methods support regular algorithms and procedures. From applied, commonly used methods (regular Gaussian regression filter, robust Gaussian regression filter, spline filter and fast Fourier transform filter), the most encouraging results were obtained for high-frequency noise reduction in laser-textured details when the fast Fourier transform filter was supported by a thresholding approach.

Keywords: laser texturing; surface texture; surface topography measurement; data analysis; data processing errors; thresholding

Citation: Podulka, P. Thresholding Methods for Reduction in Data Processing Errors in the Laser-Textured Surface Topography Measurements. *Materials* **2022**, *15*, 5137. https://doi.org/10.3390/ma15155137

Academic Editors: Grzegorz Królczyk and Jingwei Zhao

Received: 30 June 2022
Accepted: 22 July 2022
Published: 24 July 2022

Publisher's Note: MDPI stays neutral with regard to jurisdictional claims in published maps and institutional affiliations.

Copyright: © 2022 by the author. Licensee MDPI, Basel, Switzerland. This article is an open access article distributed under the terms and conditions of the Creative Commons Attribution (CC BY) license (https://creativecommons.org/licenses/by/4.0/).

1. Introduction

Currently, the mechanical behaviour of machined parts is often improved by different manufacturing techniques. Much popular surface finishing uses laser-based methods, and laser surface texturing (LST) is a common example. LST is a surface engineering process used to improve tribological characteristics of materials, by creating patterned microstructures on the mechanical contact surface [1]. LST by dimpling has been shown analytically and experimentally to enhance the mixed, hydrodynamic, and hydrostatic lubrication of conformal sliding components to improve their load-carrying capacity, higher wear resistance, and lower friction coefficients; these were observed in LST mechanical seals and thrust bearings and presented in many scientific works previously [2].

In general, when tribological issues are considered, the laser-textured surfaces showed less friction than surfaces manufactured by conventional honing [3]. The introduction of laser texturing caused lower engine oil consumption compared to conventional honed structures [4]. Cylinder liners co-act with piston rings, and the benefits of applying LST to piston ring surfaces were demonstrated theoretically and experimentally [5]. The results of the work showed a reduction in friction force of about 30% by ring surface texturing in comparison to untextured rings under lubrication conditions [6]. The potential of LST was also evaluated regarding surface wettability [7], superhydrophobicity [8] or achieving extreme surface wetting behaviours [9]. The influences of preparation methods and texture density on the tribological properties of coatings were also investigated, and research results show that a thicker and denser coating has better adhesion with the textured steel substrate when fabricated by this combination technology, which results in excellent tribological properties [10]. A textured surface can improve anti-friction and anti-wear abilities [11].

In addition, laser-based additive manufacturing has attracted much attention as a promising 3D printing method for metallic components in recent years. The surface roughness of additive manufactured components has been considered a challenge to achieve high performance [12]. The tribological performance of laser-textured aluminium alloy was studied in unidirectional sliding tests under boundary lubrication; these showed that the tribological property of aluminium alloy is critical for its reliable operation in practical applications. It was found that the beneficial effects of LST are more pronounced at higher speeds and loads with higher viscosity oil [13]. Generally, the potential of a multi-dimple textured surface as a viable engineering surface for friction reduction and extending wear life was improved in many previous studies [14].

Precise studies of the LST surface topography measurement process can influence the tribological performance applications like friction [15], sealing [16], lubricant retention [17], wear [18] or wear resistance [19], corrosion [20], fatigue [21] or, generally, material contact [22] and material properties. In general, the surface topography analysis can be roughly divided into measurement and data analysis processes [23]. Each part of the surface topography study, simultaneously, can be fraught with many factors influencing the accuracy of the analysis. It was found that even when precise measurement techniques (device) were applied, the processes of raw measured data were selected inappropriately and the accuracy of the surface topography assessment was lost [24].

Generally, all of the measurement errors can be selected into those directly related to the measuring methods [25], caused by the digitisation [26] or data processing [27], software [28], object measuring [29] or other errors [30]. Furthermore, those errors found when the measurement process occurs are defined as noise [31], or in particular, measurement noise [32]. In general, measurement noise can be defined as the noise added to the output signal when the normal use of the measuring instrument occurs [33]. There are many types of measurement noise [34] in surface topography studies, considering both stylus and non-contact techniques. It was found that various types of environmental disturbances can introduce noise in different bandwidths [35]. One of the types of errors caused by the environment of the measuring system is high-frequency noise (HFN) [36]; this can be caused by instability of the mechanics with any influences from the environment or by internal electrical noise, but in most cases the HFN is the result of vibration [37] and, simultaneously, in real measurement, this can greatly affect the stability of slope estimation [38].

Some strategies were tried to reduce vibration noise by minimising vibration sources, isolating those sources or, correspondingly, isolating the instrument [39], optimising the mechanical structure of the instrument [40], and compensating for the vibrational effect, like a piezoelectric transducer [41]. Moreover, some extensive studies of environmental noise, such as thermal variation and vibration, in the definition of the accuracy of in-process measurement results, seem to be challenging to develop [42]. However, considering the measured data, it is extremely difficult to state that the received data is, in fact, raw [43].

Many research items have presented a comprehensive analysis of surface topography and errors received when studying its properties. Parametric description [44] found a wide range of applications in many studies, even if it requires mindful users. The power spectral density (PSD) function is very useful in surface metrology. Even though this technique contains plenty of limitations, many benefits belong to its application when dry and MCQL processes are considered; it was found possible to characterize turning regarding applied cooling methods, by qualitative and quantitative comparison of this function for the inspected surface [45]. PSD characterisation can be especially useful when a parametric description is ambiguous, like Rk (from ISO 13565-2) or Rq (ISO 13565-3) parameters, compared with considering their practical significance and sensitivity to measurement errors [46]. Alternatively, for areal details, the Sk group parameters analysis can be extremely useful for the description of the functional importance of surface topography consideration [47], like in honed cylinder liner surface texture studies.

From the above, even in many published papers on surface topography measurement and data analysis, there is a lack of full response on how to deal with the high-frequency measurement noise when LST details are considered. Moreover, general, commonly used algorithms and procedures (e.g., those from the commercial software) can be supported by various thresholding methods, so consequently, guidance on how they should be applied is also required.

For that matter, the main purpose of this paper is to select the appropriate, widely available (e.g., in commercial software) procedure to reduce the influence of data processing errors for surface topography analysis when high-frequency measurement noise is considered. Reduction in errors in received data studies can cause, correspondingly, an inaccurate analysis of the surface tribological features. Moreover, detailed studies of high-frequency errors were not previously provided for LST details.

2. Materials and Methods

2.1. Analysed Details

Reduction in the effect of data processing errors was considered for a laser-textured surface. Surfaces with a different angle of laser texturing process were studied, e.g., 30°, 60°, 90° and 120°. The average depth of the LST features was around 50 μm for each type of detail studied, and the distance between each of the LST features was around 0.5 mm, on average. Examples of analysed surfaces were presented in Figure 1, where contour map plots (a,d), isometric views (b,e) and material ratio curves (c,f) were introduced.

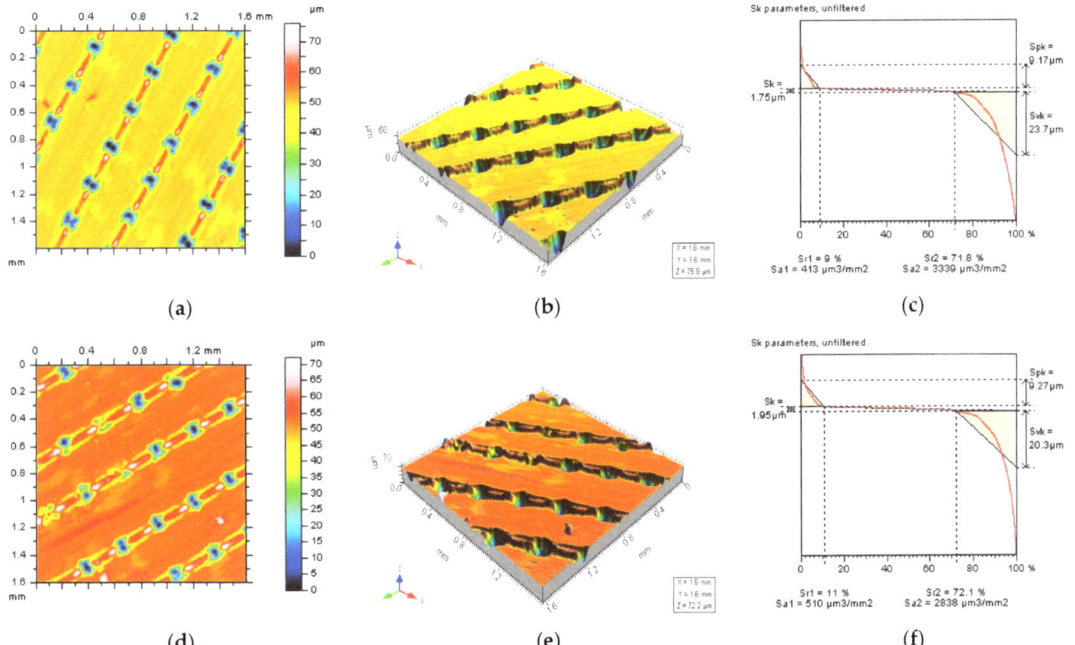

Figure 1. Contour map plots: (a,d), isometric views; (b,e) and their material ratio curve graphs; (c,f) of 30° (a–c) and 60° (d–f) LST; measured with a stylus equipment.

All of the studied details were provided with an areal form removal process as a preliminary data analysis. Types of procedures for the extraction of long-frequency components (definition of an L-surface) from the raw measured data were widely studied in previous papers [48], so they were not currently examined. The selection of a method for defining an appropriate reference plane was already widely studied and presented, by the author as well. Providing an areal form (shape and waviness) removal process caused the studied details to be generally flat. Nevertheless, the accuracy of form and waviness removal on the results obtained was not considered. If the surface contained other measurement errors, like non-measured points [49] or spikes (individual peaks [50]), they were extracted and, respectively, removed from the measured data.

More than 10 surfaces from each type of LST detail (with a different LST angle) were measured and studied, but only some of them were presented in detail. Furthermore, all the studies were substantially improved with modelled data and compared with those measured to define some general guidance.

2.2. Measurement Process

Analysed details were measured by different techniques, stylus and optical. The first instrument was Talyscan 150 with a nominal tip radius of 2 µm, containing a height resolution of about 10 nm. The measured area was 5 by 5 mm with, respectively, 1000×1000 measured points and the sampling interval equal to 5 µm. The measurement velocity was 1 mm/s and so its influence on the results presented was not studied; that was carefully considered in previous detailed studies and was not the subject of the current research.

The non-contact measurement device was a white light interferometer, Talysurf CCI Lite with a height resolution equal to 0.01 nm. The measured area was 3.35 by 3.35 mm with 1024×1024 measured points, proportionately, with a spacing equal to 3.27 µm. The effect of both sampling and spacing on values of areal surface texture parameters was not analysed in this paper.

2.3. Applied Methods

Generally, the approaches proposed in current studies can be divided according to their performance that the process of reduction in the influence of an HFN can be separated into processes of detection and reduction qualifications. From that fact, selected procedures can be classified as those relevant in the detection operations and, simultaneously, those providing accurate noise reduction results.

It was found in many previous studies that the application of both profile (2D) and an areal (3D) analysis may provide more accurate results considering noise detection than using them separately; profile estimation was more relevant for a honed cylinder liner analysis when an HFN was separated.

The general purpose of the current study was to use commonly used, available commercial software and techniques to detect and reduce the HFN and provide some guidance for a regular user. Very valuable in an HFN analysis are power spectral density (PSD) and autocorrelation function (ACF) techniques. PSD, in its two-dimensional form, has been designated as the preferred quantity for specifying surface roughness on a draft international drawing standard for surface texture [51]. Generally, the PSD scheme, which is based on Fourier analysis [52], was introduced to distinguish the scale-dependent smoothing effects, resulting in a novel qualitative and quantitative description of surface topography [53]. Secondly, the ACF assessment provides practical advice regarding the autocorrelation length and its properties as a function of surface irregularities. Comparatively, both ACF and PSD are related to the frequency or, simplifying, spectral analysis; nevertheless, ACF can be more relevant for the study of irregular textures and PSD for the analysis of periodic surfaces [54].

Often used, nevertheless, are approaches based on the thresholding operations. A typical example is a thresholding on heights which, due to their simplicity, become a common method to obtain a segmentation of the surface topography. However, simple thresholding is not a stable method when surfaces have a stochastic content and can produce many insignificant features that can cause problems for many characterisation parameters, such as the number of defects and the density of features [55]. Alternatively, morphological segmentation into hills or dales is the only partitioning operation currently endorsed by the ISO specification standards on surface texture metrology [56]. More sophisticated thresholding techniques may allow determination of valuable results of complex surfaces. A height thresholding operation can be used to isolate the topmost regions of the filtered topography, most likely belonging to protruded formations such as spatter and particles, using different threshold values depending on surface orientation [57]. A multilevel surface thresholding algorithm for enhancing the representation of topographic values by slicing the 3D surface topography into cumulative levels about the characteristics of the in-control surfaces was proposed previously, where the spatial and random properties of topographic values were quantified at each surface level through the proposed spatial randomness profile [58]. Some optical measurement errors, like spikes, can be also extracted and reduced with relevant thresholding applications [50].

The thresholding process can be applied to the reduction in the length of profiles or areas of the surface as well. It was found in previous studies that analysis of surface 3D or 2D data can be improved for HFN detection when data excluding the deep and (or) wide features are considered. This technique, shortly defined as out-of-dimple or, generally, out-of-feature characterisation, can be valuable for details where the number of features is relatively small, like plateau-honed cylinder liners with additionally burnished oil pockets or turned topographies containing dimples with various (usually huge) sizes (depth and width). LST can be classified as surfaces where the density of features (dimples) is relatively large, so consequently, out-of-feature techniques may not provide relevant results. Generally, when LST surfaces contain dimples, the detection of the HFN can be difficult and may not allow for measurement error characterisation. In Figure 2 examples of a 3D detection of an HFN are presented. It was found that when the surface contained dimples, the detection of an HFN with PSD and ACF methods did not allow for receiving differences or, correspondingly, they were negligible. Even though the ACF graph was studied with a 3D or 2D (for extraction in the centre part of the graph) performance, it did not give a valuable response if the HFN was present in the results of surface texture measurements.

As an alternative to the out-of-feature technique, the thresholding can be used in both directions, height (amplitude) or length (width) of the profile or areal data. In Figure 3 the example of the thresholding method of the LST profile is presented. It was found that if data (profile in particular) contained dimples (b–d), created in the LST process, the process of detection of the HFN was impossible with an application of commonly used procedures, like PSD, and ACF techniques. Therefore, the measured data (a) were thresholded into parts containing the plateau-part of profile and dimples (A1, A2, A3 and A4). In the next step, dimple parts were removed from the data, and correspondingly only the plateau parts were considered. When wide/deep dimples were removed (omitted) from the data, the HFN detection was improved with the application of regular (available in commercial software) methods, such as the PSD and ACF approaches.

The differences in PSD and ACF graphs were clearly visible in the example presented in Figure 4 where two various (profile from surface where an HFN was not defined, a, and the same surface where an HFN was observed, b) measurement results were introduced. When measured data contained an HFN and the thresholding method was applied (g–s) the differences in the PSD and ACF graphs were easily visible when the surface did not contain (g–i and m–o) HFN against results where the noise was found (j–l and p–r).

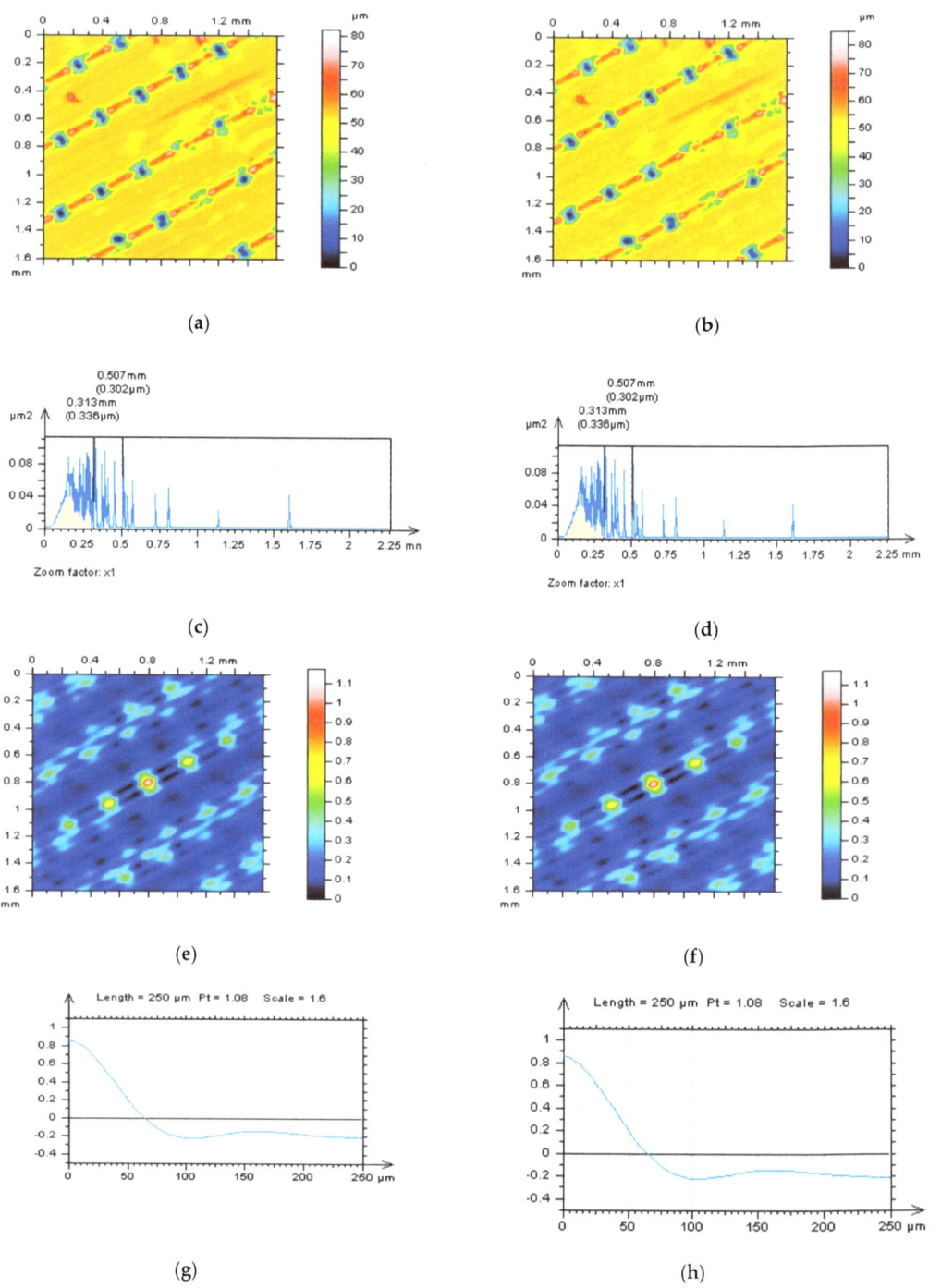

Figure 2. Laser-textured surface topography: contour map plots (**a**,**b**); their PSDs (**c**,**d**); ACFs (**e**,**f**); and centre-extracted ACF profiles (**g**,**h**); measured (**left** column) and containing a high-frequency measurement noise (**right** column).

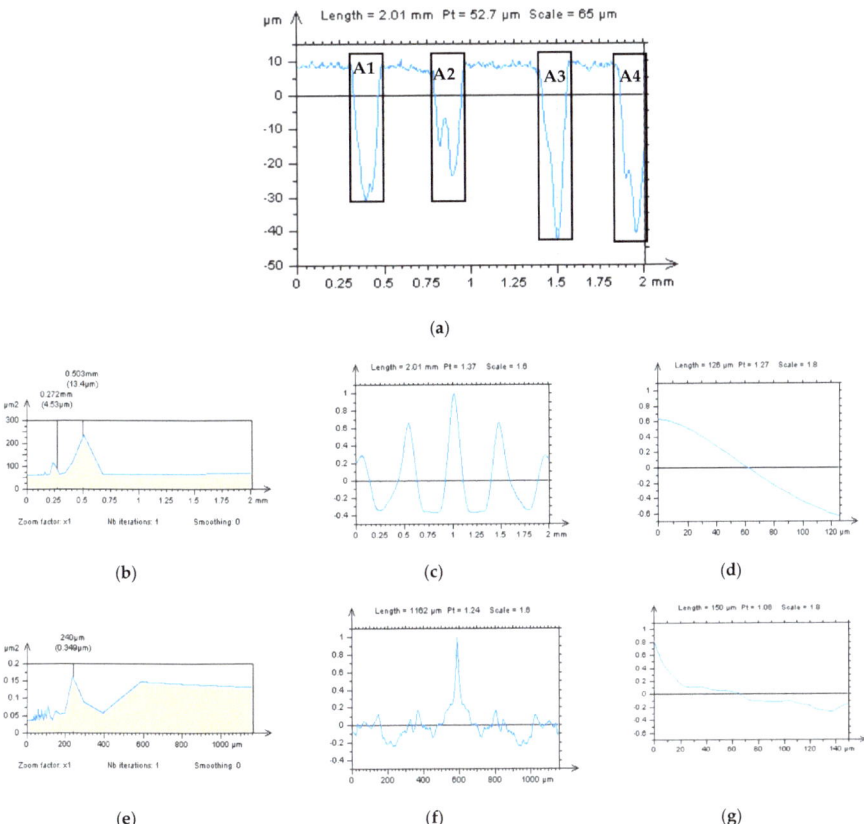

Figure 3. Profile (**a**), its PSD (**b**), ACF (**c**) and centre-part ACF (**d**), and, correspondingly, PSD (**e**), ACF (**f**) and its centre-part ACF (**g**) of profile with thresholded A1-A2-A3-A4 parts from primary profile from (**a**) sub-figure; profile extracted from 120-angle LST detail containing an HFN.

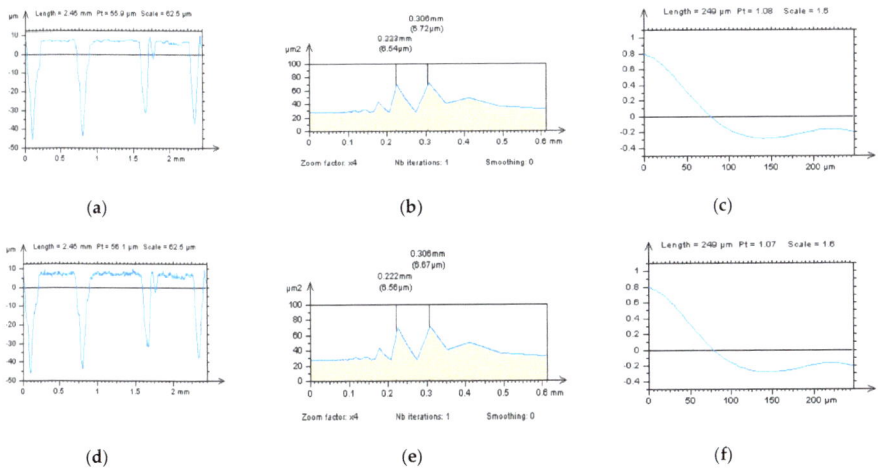

Figure 4. *Cont.*

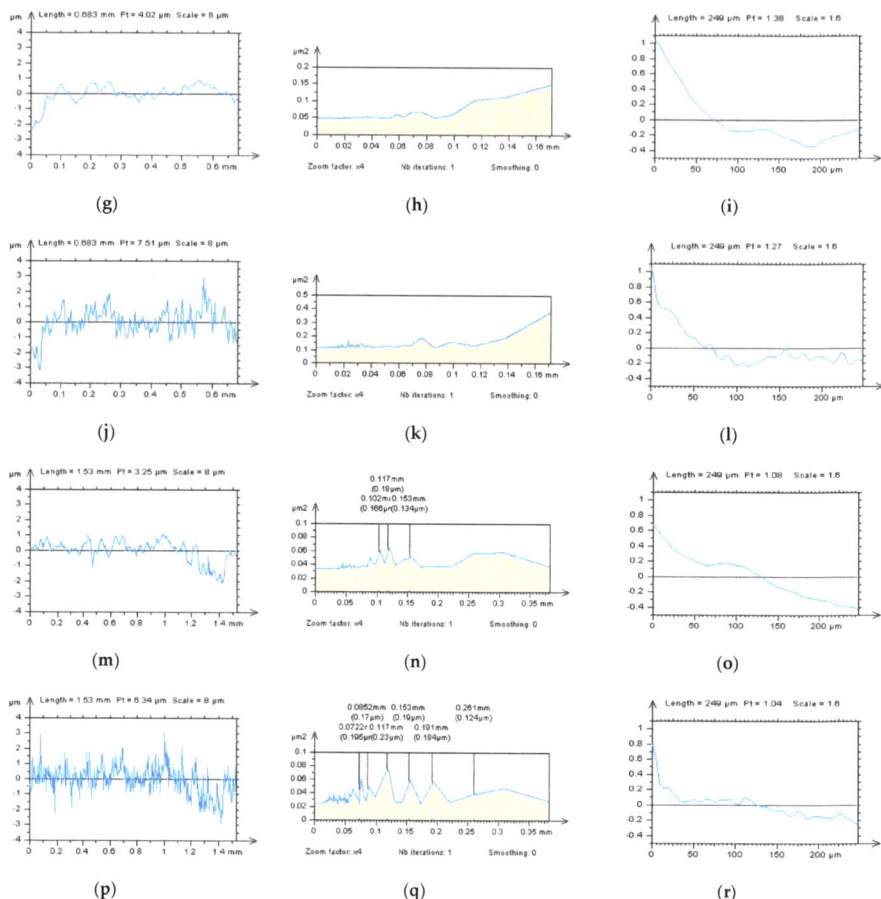

Figure 4. Extracted profiles (**left**), their PSDs (**middle**) and ACFs (**right** column) of the LST detail: measured (**a–c**) and with high-frequency errors (**d–f**) and, respectively, after an application of thresholding method (**g–r**).

Compared to the previously published results (papers), the thresholding methods can give the advantage to analyse larger (longer) 3D details (2D profiles) than the out-of-dimple technique. Moreover, the accuracy of the algorithm in selecting the data (points) as a 'dimple point', as it was proposed in the out-of-dimple method, does not influence the final results. Classification of the 'dimple point' as deep enough to be a dimple can be also fraught with errors. Furthermore, the thresholding techniques can be more intuitive than the out-of-dimples (or simplifying out-of-feature) technique. For that matter, even an inexperienced user can apply this method instead of other, more complicated approaches.

3. Results

Analysis of data processing errors was divided into three main subsections. In the first (Section 3.1) the problem in the definition (detection) of the high-frequency measurement noise was considered. Secondly, (in Section 3.2) various regular (commonly available in the commercial software) methods (filters) were compared with a specification of the high-frequency measurement errors reduction. Finally, in the last (Section 3.3) subsection, all of the proposed thresholding approaches were improved by analysis of the modelled data.

3.1. Detection of the High-Frequency Errors from the Results of LST Measurement with a Thresholding Approach

The process of detection (definition) of the HFN from the results of surface topography measurements was, firstly, provided for areal (3D) data. As in previous studies, it was found that the 3D surface topography HFN definition is difficult to provide in a conscious way that, for both commonly used (available in commercial software) functions, PSD and ACF, differences against results without an HFN did not exist or, at least, are negligible (Figure 5a–c). Moreover, when the surface contained some deep or wide dimples, like those after laser treatment, the HFN detection with a profile exploration is also not convincing (Figure 5d–f). Even where, in some cases, the PSD function could suggest the occurrence of the HFN, the ACF was not modified against the surface where the HFN was not found. Some encouraging results were received when a thresholding method was proposed where both the PSD and ACF functions were modified (Figure 5g–i). From that point of view, the 2D (profile) detection of HFN with a thresholding approach was used and, simultaneously, was suggested in further studies.

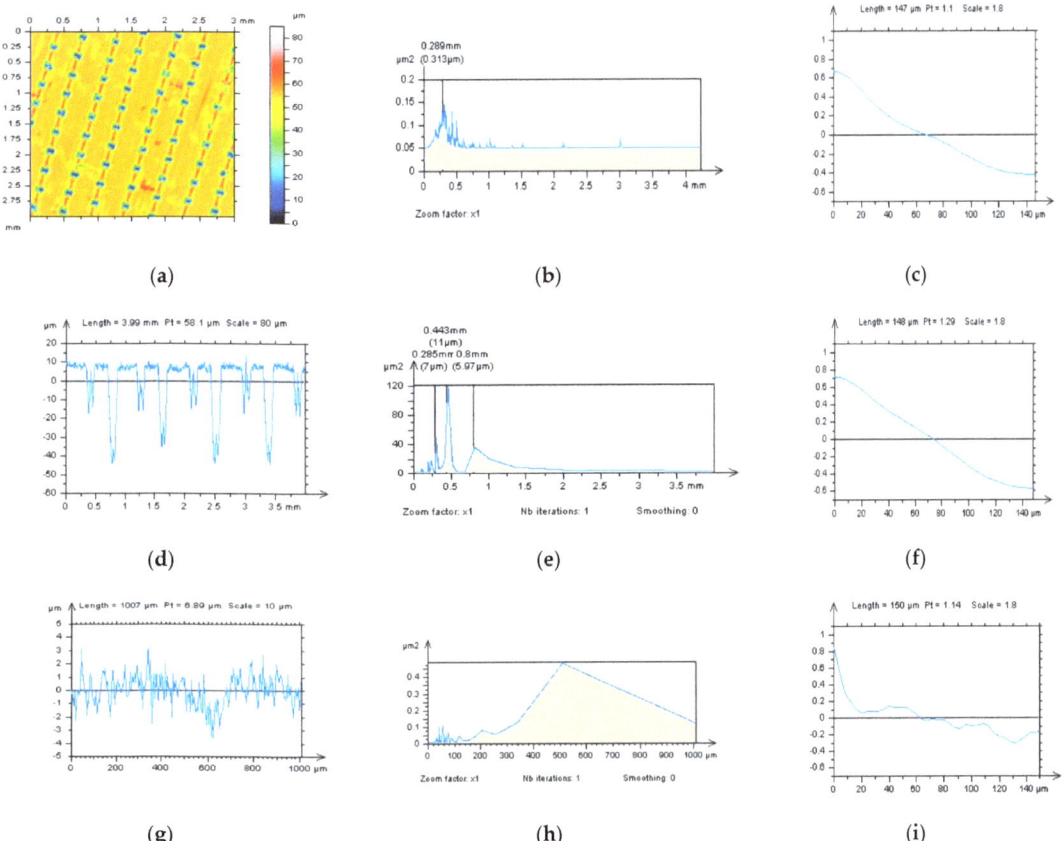

Figure 5. Contour map plot (**a**), its PSD (**b**) and ACF (**c**) graphs, extracted profile (**d**) with PSD (**e**) and ACF (**f**) graphs and, respectively, profile defined with a thresholding method (**g**) with PSD (**h**) and ACF (**i**) graphs, all defined for a laser-textured surface.

3.2. Comparison of Regular Filters for High-Frequency Measurement Noise Removal

For removal (reduction) of the HFN from the results of laser-textured surface topography measurements, the regular methods (available in commercial software), were proposed, like regular Gaussian regression (GF) or robust Gaussian regression (RGF) filters, regular isotropic spline method (SF) and the Fast Fourier Transform approach (FFTF). The selection of cut-off values was not provided in the paper that was studied previously by the author, and consequently, the 0.025 mm bandwidth was suggested and applied.

It was found in previous studies that analysis of the high-frequency measurement noise can be effectively performed by studies of the results of removed data, defined as 'noise surface' (NS). From the commonly available (in the commercial software) procedures and functions, except PSD and ACF, the texture direction (TD) graph can provide valuable information about the properties of the measured data. The NS was expected to be isotropic, or at least, not consist of one dominant direction. Generally, when NS was not isotropic and, simultaneously, contained a dominant direction, this direction was equal to the direction of dominant features like dimples or oil pockets. Usually, this property indicated that in the removed noise data (NS) some features, not defined in the noise domain, could be found. This would indicate that the used method (filter) for noise suppression was selected inaccurately, and caused a distortion of the results obtained, with the accuracy in the whole process of surface topography measurement lost as well.

From the above, LST details could be considered with the NS analysis method, where this type of surface topography contains some features with directional performance. As the NS should contain only high-frequency components, or at least, these components should be dominant, the PSD characterisation can be useful. For each of the filtering methods (GF, RGF, SF or FFTF), the high-frequency components were those dominant (Figure 6b,e,h,j); nevertheless, this dominance was greater for some methods, e.g., FFTF, in contrast to the other approaches.

A more confident result could be found when analyzing the TD graphs. For commonly used schemes, GF, RGF and SF, the dominant direction was recognized, and this indicated that NS was not isotropic (Figure 6c,f,i, respectively). Contrary to those three filters, the FFTF gave more confident results (Figure 6l). Furthermore, when analysing the contour map plots of NS for regular GF, RGF and SF methods, some non-noise data were visible, indicating the traces of the LST process (Figure 6a,d,g). However, when using an FFTF approach, those non-noise and non-required features on the NS were not observed, or at least, were negligible. From the above, the conclusion that the application of the FFTF method can be more valuable for an HFN reduction than regular GF, RGF or SF approaches, can be proposed.

When analyzing profiles received from the NS data (Figure 7a,d,g,j), even the PSD graphs present no different (or they are negligible), and some modifications can be defined with an analysis of the ACF method. It was introduced in previous studies that if measured data contain an HFN, the value of ACF increased more rapidly near the highest value (i.e. near the value '1'). This property was found when SF (Figure 7i) and FFTF (Figure 7l) were applied, contrary to the Gaussian filters, GF (Figure 7c) or RGF (Figure 7f).

All of the results from the above profile studies can be significantly improved by the analysis of the view of profiles (Figure 7a,d,g,j). Some non-noise features can be found after the application of Gaussian filters (marked in Figure 7a,d). Furthermore, when closely studying the profiles received after the application of SF, some non-noise components can be also found (marked in Figure 7g).

Similar to the PSD, ACF and TD analysis, the most encouraging results from all four methods were obtained when an FFTF approach was used. Reducing the occurrence of non-noise features on the NS data can, simultaneously, reduce the errors in an HFN removal and calculation of surface topography parameters and evaluation of its tribological performance.

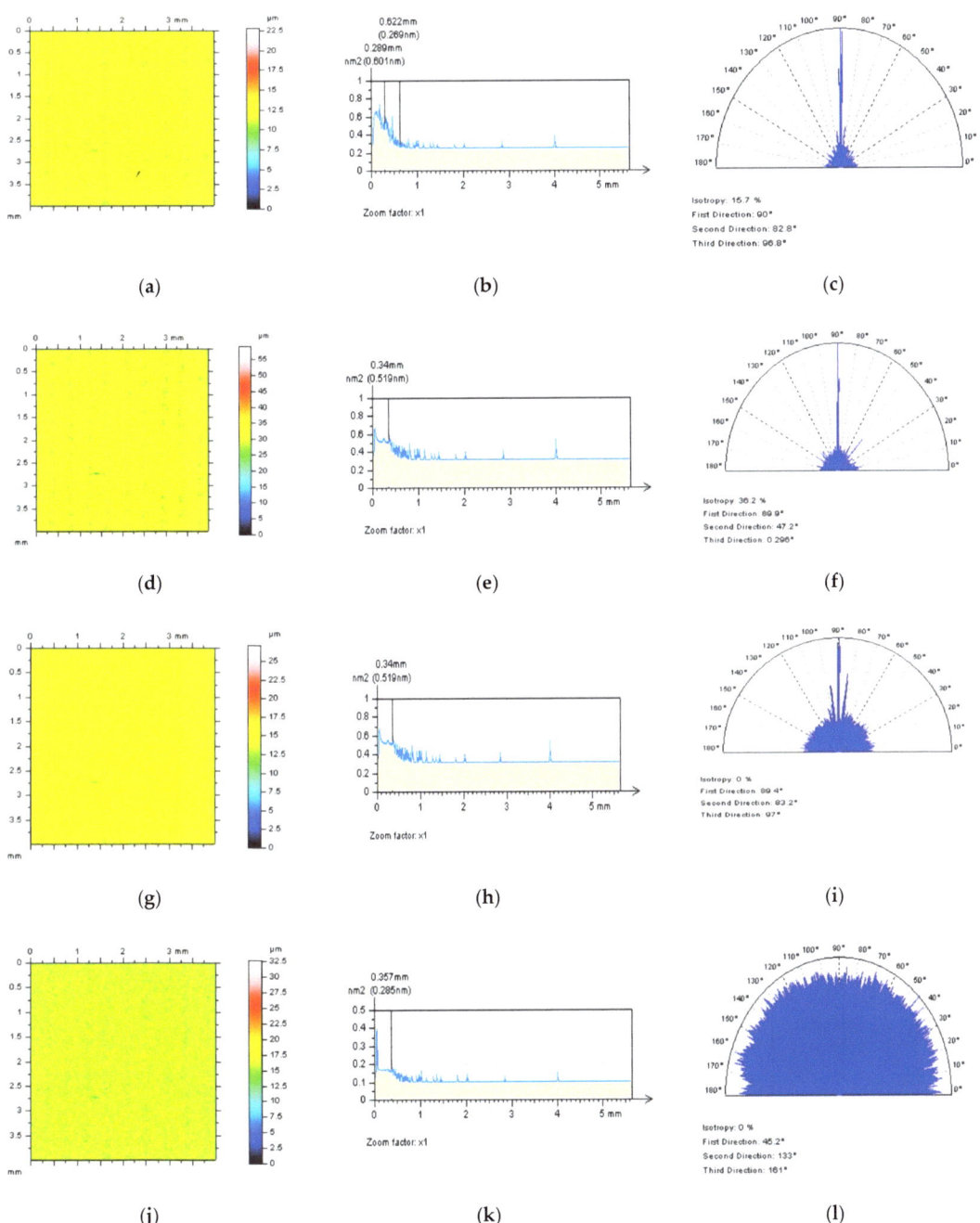

Figure 6. Contour map plots (**left** column), their PSDs (**middle**) and ACFs (**right** column), of a NS received by application of GF (**a–c**), RGF (**d–f**), SF (**g–i**) and FFTF (**j–l**) method with the cutoff = 0.025 mm.

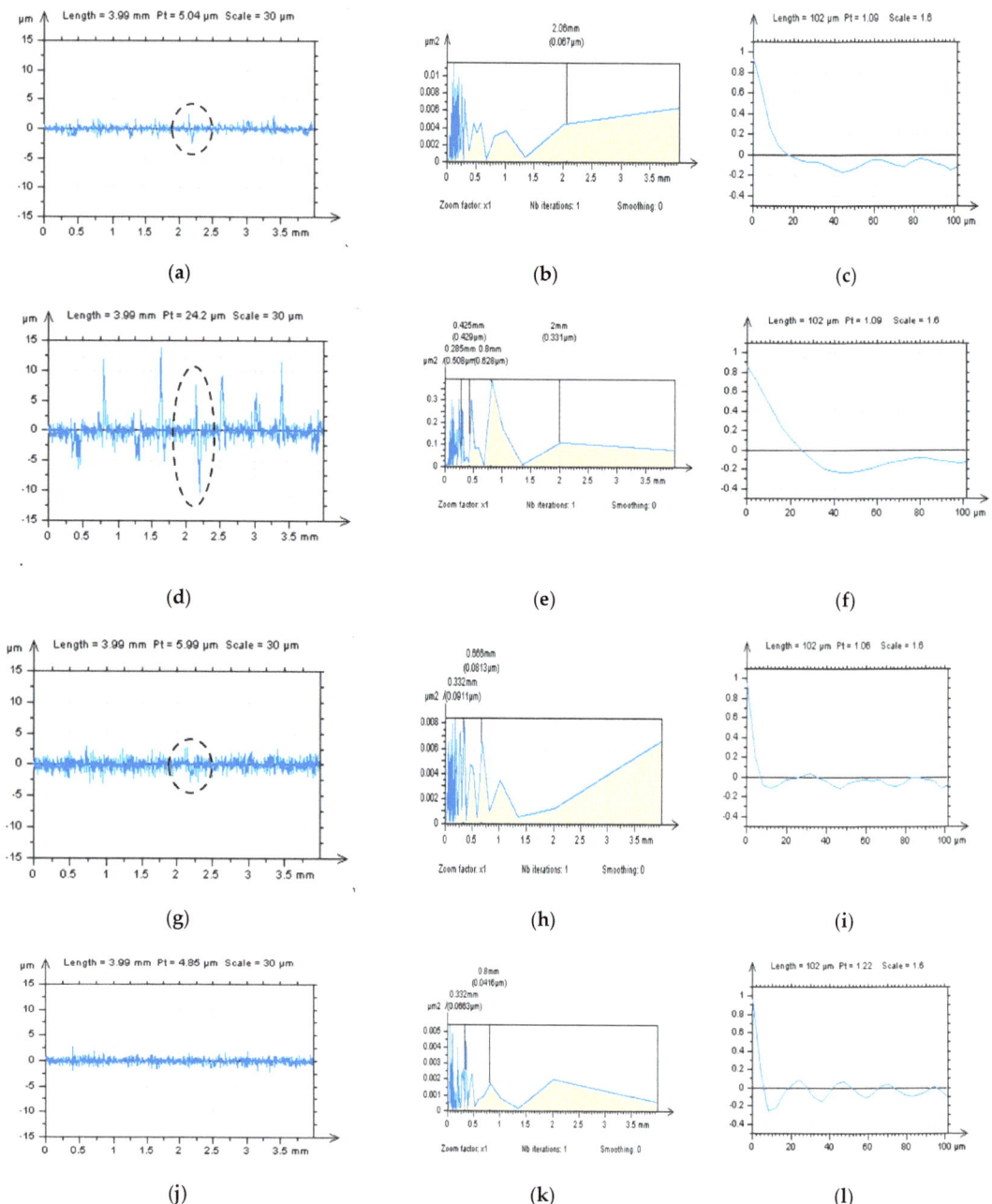

Figure 7. Profiles (**left** column), their PSDs (**middle**) and ACFs (**right** column) received from the NS created by application of GF (**a–c**), RGF (**d–f**), SF (**g–i**) and FFTF (**j–l**) method, cutoff = 0.025 mm.

3.3. Improving the Procedures for High-Frequency Measurement Noise Suppressions with Analysis of Modelled Data

The validation of the proposed methods can be proposed:. firstly, the difference in surface topography parameters can be considered; and secondly, the properties of an NS can be evaluated.

From the ISO 25178 standard, the following surface topography parameters were measured and analyzed: root mean square height Sq; skewness Ssk; kurtosis Sku; maximum peak height Sp, maximum valley depth Sv, the maximum height of surface Sz, arithmetic mean height Sa; auto-correlation length Sal; texture parameter Str; texture direction Std; root mean square gradient Sdq; developed interfacial areal ratio Sdr; peak density Spd; arithmetic mean peak curvature Spc; core roughness depth Sk; reduced summit height Spk; reduced valley depth Svk; surface bearing index Sbi; core fluid retention index Sci; and valley fluid retention index Svi.

For a 30° LST detail, it was found that the smallest differences between primary data and the data received after the noise removal process (filtration by various methods) were received after filtration by the FFTF method (Table 1). In some cases, parameters were similar (or differences were smaller than 5%), like Ssk, Sku, Str, Std, Spk, Sbi, Svi or, what is especially crucial, Sz, Sa, Sal and Spd. Generally, in some cases, when 60° LST (Table 2), 90° LST or 120° LST (Table 3) were considered, an effective alternative to the fast Fourier (FFTF) method can be spline (SF) filtration. However, when applying filters for HFN removal, the differences in roughness parameters calculation, and consequently, evaluation, must be provided in a required manner.

When studying the NS properties, for PSD and TD properties (dominant component in a high-frequency domain and isotropic, respectively) consideration, the FFTF method gave the most encouraging results (Figure 8). For studies of contour map plots, non-noise features were not found for NS created by SF and FFTF filtrations. Analysis of ACF (Figure 9) improved both mentioned approaches, SF and FFTF, that present the spline techniques as an acceptable alternative.

Table 1. Values of surface topography parameters, from the ISO 25178 standard, defined for a primary surface data (start data), containing an HFN (noise data) and after reduction in HFN by various methods: GF, RGF, SF and FFTF, cut-off = 0.025 mm.

	30° LST Detail Analysis after an HFN Removal by Various Methods					
	Start Data	Noise Data	GF	RGF	SF	FFTF
Sq, μm	9.32	9.36	9.07	8.81	9.27	9.27
Ssk	2.65	−2.61	−2.71	−2.91	−2.67	−2.66
Sku	11.5	11.3	11.6	12.2	11.6	11.5
Sp, μm	33.5	34.2	30.4	28.4	31.8	32.1
Sv, μm	51.8	54.6	51.4	51	52.6	53.1
Sz, μm	85.3	88.8	81.7	79.4	84.3	85.2
Sa, μm	5.49	5.52	5.33	5.23	5.45	5.45
Sal, mm	0.0563	0.0563	0.0585	0.0629	0.0576	0.0563
Str	0.661	0.661	0.68	0.733	0.677	0.661
Std, °	90	90	90	90	90	90
Sdq	0.465	0.614	0.39	0.387	0.419	0.427
Sdr, %	9.72	17.2	7	6.64	8	8.29
Spd, 1/mm^2	12.3	16.5	10.3	9.57	11.2	11.6
Spc, 1/mm	0.211	0.306	0.0816	0.0927	0.096	0.0895
Sk, μm	2.88	3.52	2.92	3.06	3.09	3.3
Spk, μm	8.93	8.4	8.38	7.1	8.84	8.99
Svk, μm	23.4	24.6	22.8	22.5	23.5	24.4
Sbi	0.354	0.346	0.384	0.398	0.375	0.37
Sci	0.494	0.495	0.468	0.424	0.484	0.48
Svi	0.304	0.296	0.301	0.306	0.301	0.301

Table 2. Values of surface topography parameters, from ISO 25178 standard, defined for a primary surface data (start data), containing an HFN (noise data) and after reduction in HFN by various methods: GF, RGF, SF and FFTF, cut-off = 0.025 mm.

	60° LST Detail Analysis after an HFN Removal by Various Methods					
	Start Data	Noise Data	GF	RGF	SF	FFTF
Sq, μm	9.55	9.56	9.28	9.09	9.5	9.5
Ssk	−2.6	−2.59	−2.66	−2.95	−2.62	−2.62
Sku	11.1	11	11.2	12.3	11.1	11.1
Sp, μm	31.7	32.3	27.5	24.2	29.7	29.4
Sv, μm	51	52.2	49.9	50	51.3	53.7
Sz, μm	82.8	84.5	77.3	74.2	81	83.1
Sa, μm	5.71	5.71	5.53	5.36	5.66	5.67
Sal, mm	0.0591	0.0591	0.0613	0.0658	0.0591	0.0591
Str	0.691	0.692	0.711	0.78	0.691	0.691
Std, °	30.7	30.7	30.5	176	30.5	30.5
Sdq	0.455	0.508	0.387	0.432	0.421	0.435
Sdr, %	9.44	11.9	6.99	7.96	8.2	8.67
Spd, 1/mm^2	12.1	14.2	10.7	10.4	11.5	12.6
Spc, 1/mm	0.196	0.238	0.0695	0.0915	0.0873	0.09
Sk, μm	2.22	2.62	2.24	2.11	2.35	2.41
Spk, μm	8.46	8.37	8.2	5.86	8.38	8.31
Svk, μm	26.3	26.6	25	25	26	26.4
Sbi	0.394	0.385	0.455	0.512	0.427	0.43
Sci	0.508	0.509	0.486	0.423	0.505	0.499
Svi	0.305	0.302	0.3	0.306	0.302	0.303

Table 3. Values of surface topography parameters, from ISO 25178 standard, defined for a primary surface data (start data), containing an HFN (noise data) and after reduction in HFN by various methods: GF, RGF, SF and FFTF, cut-off = 0.025 mm.

	120° LST Detail Analysis after an HFN Removal by Various Methods					
	Start Data	Noise Data	GF	RGF	SF	FFTF
Sq, μm	9.65	9.67	9.38	9.2	9.6	9.6
Ssk	−2.59	−2.58	−2.65	−2.92	−2.61	−2.6
Sku	10.9	10.9	11	12	11	11
Sp, μm	31.7	31.4	27.6	24.4	29.6	29.5
Sv, μm	51	52.2	49.7	50	51.3	53.2
Sz, μm	82.6	83.7	77.3	74.4	80.9	82.8
Sa, μm	5.79	5.8	5.62	5.46	5.75	5.76
Sal, mm	0.0601	0.0601	0.0623	0.0662	0.0604	0.0601
Str	0.689	0.689	0.71	0.765	0.692	0.689
Std, °	149	149	149	149	149	149
Sdq	0.458	0.511	0.39	0.438	0.424	0.438
Sdr, %	9.61	12	7.11	8.17	8.36	8.82
Spd, 1/mm^2	12	14.1	10.2	9.91	11.5	12
Spc, 1/mm	0.189	0.234	0.0702	0.0776	0.0882	0.0929
Sk, μm	2.23	2.79	2.26	2.2	2.36	2.57
Spk, μm	8.32	8.67	7.88	6.02	8.31	8.63
Svk, μm	25.4	26.2	24.3	24.4	25.2	25.9
Sbi	0.401	0.405	0.46	0.517	0.435	0.435
Sci	0.506	0.506	0.486	0.425	0.503	0.497
Svi	0.307	0.304	0.302	0.308	0.304	0.304

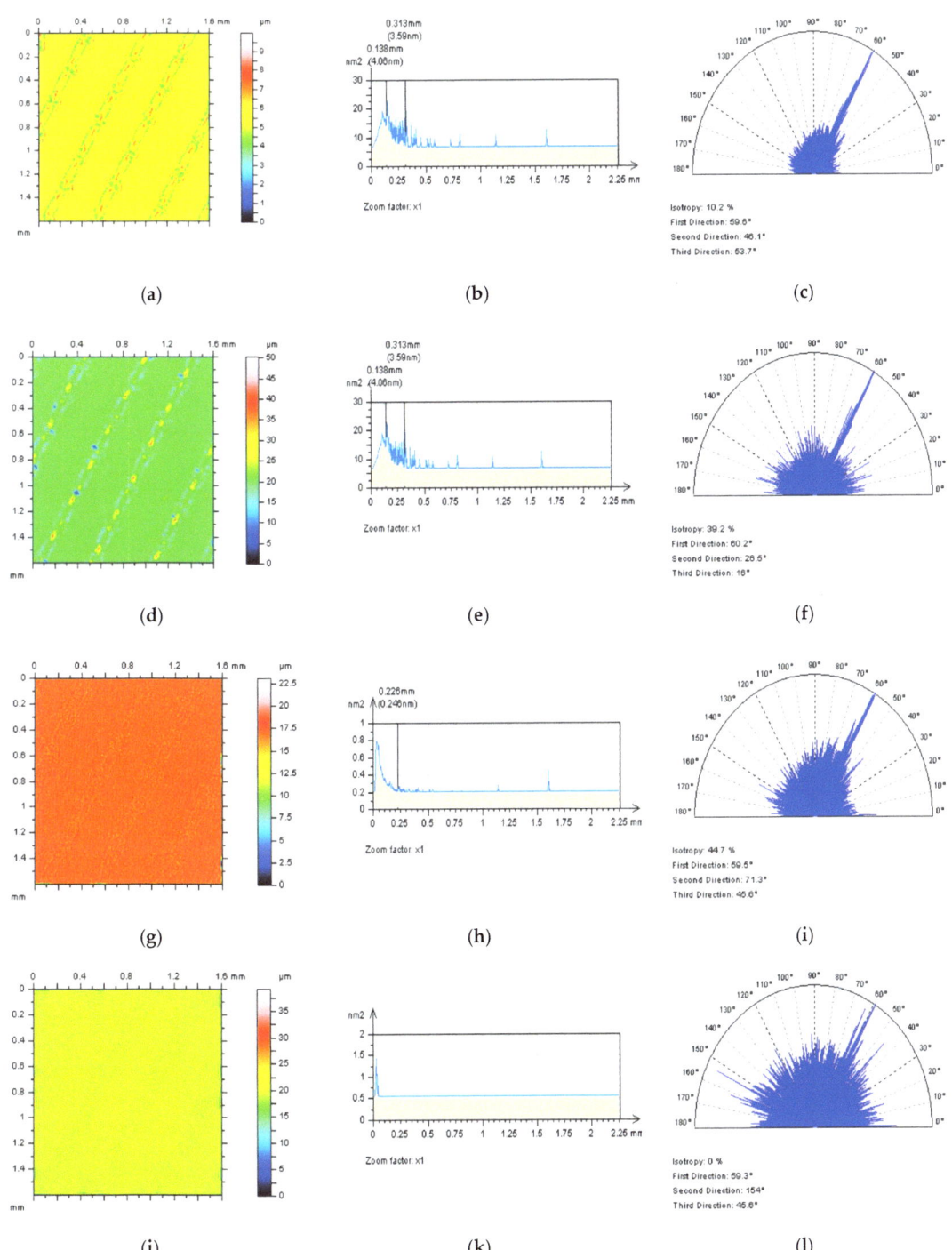

Figure 8. Contour map plots (**left** column), their PSDs (**middle**) and ACFs (**right** column), of a NS received by application of GF (**a**–**c**), RGF (**d**–**f**), SF (**g**–**i**) and FFTF (**j**–**l**) method with the cutoff = 0.025 mm.

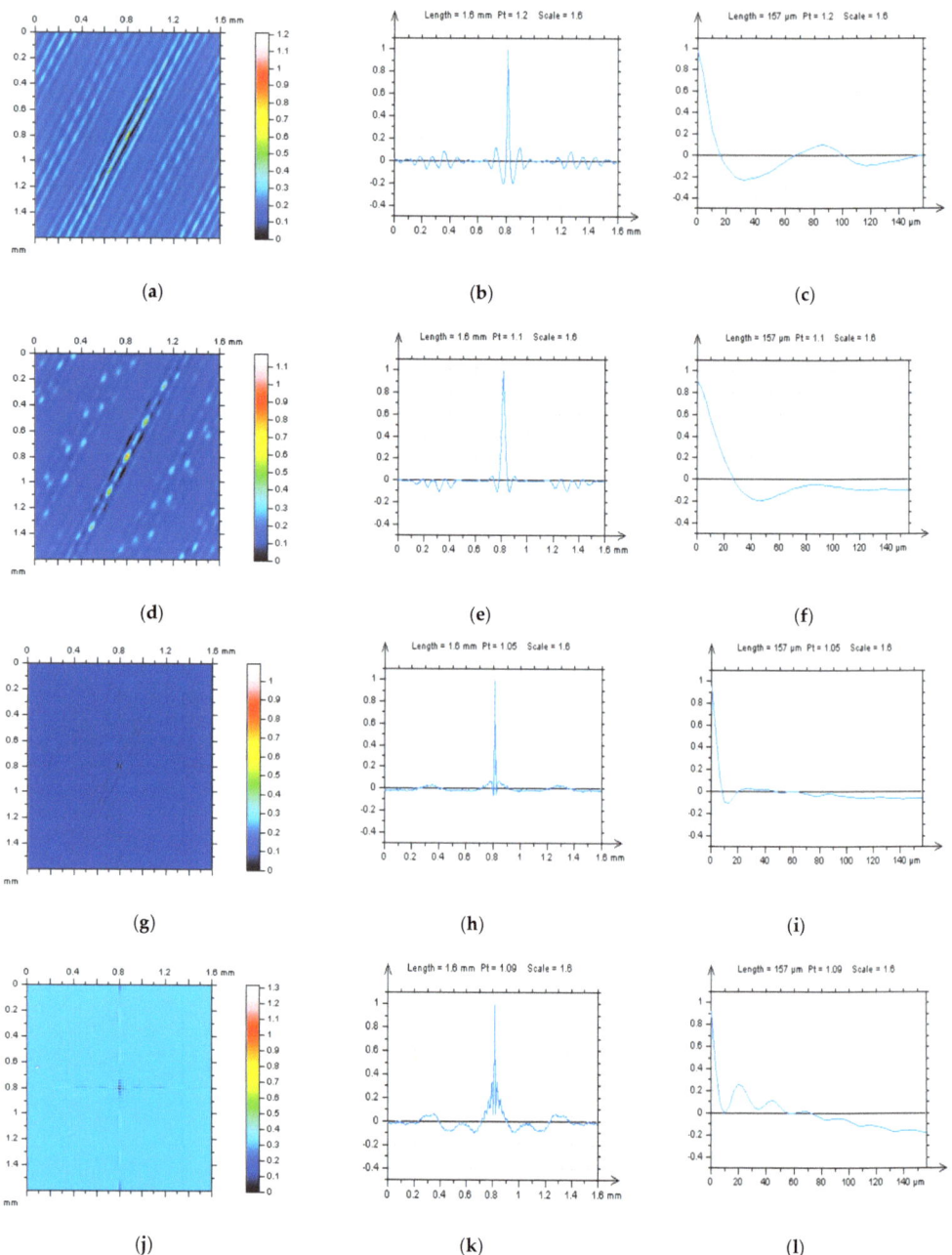

Figure 9. ACFs analysis received for 30° LST detail: ACFs for surface (**left** column), their extracted (**middle**) and enlarged (**right** column) centre parts, defined by application of GF (**a–c**), RGF (**d–f**), SF (**g–i**) and FFTF (**j–l**) method with the cutoff = 0.025 mm.

The profiles (Figure 10) characteristics indicated that both SF and FFTF schemes can be valuable in an HFN extraction from the measured data. All of the applied functions, like PSD and ACF, presented both methods as valuable in an HFN reduction; nevertheless, when SF was used some non-noise features were observed in a profile NS data–it was indicated in Figure 10a,d,g by the highest features B1, B2 and B3, respectively. For that matter, SF could distort the results of surface topography measurement when applied for an HFN removal. From all of the above, FFTF can be significantly improved for a reduction in an HFN from the results of surface topography measurements of LST details.

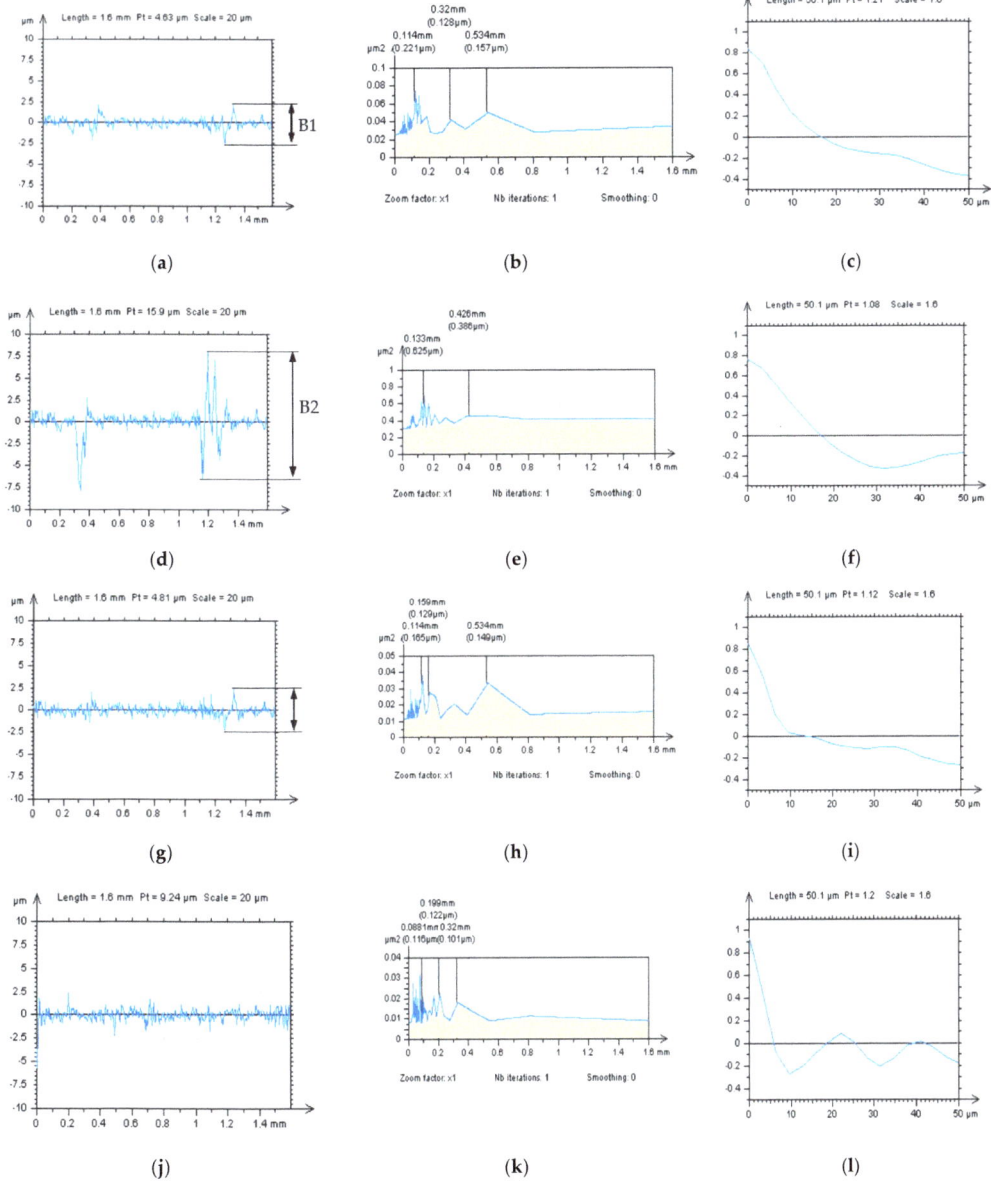

Figure 10. Profiles (**left** column), their PSDs (**middle**) and ACFs (**right** column) received from the NS created from 60° LST by application of GF (**a–c**), RGF (**d–f**), SF (**g–i**) and FFTF (**j–l**) method, cutoff = 0.025 mm.

4. The Outlook

Except for many studies provided for the surface topography measurements analysis, there are still some crucial issues that require more sophisticated studies and must be resolved in future. Of those the most significant are:

1. The effect of size and density of features was not considered for the LST details. In one of the previous studies by the author of this paper, various surface textures were considered with this issue; nevertheless, those with laser texturing were not comprehensively studied. From that issue, the effect of surface topography features sizes and their densities on the process of detection and, respectively, reduction in high-frequency noise should be widely considered;
2. The accuracy of the detection process of high-frequency measurement errors can be strongly affected by the amplitude of the noise. Therefore, some improvements to the proposed approaches must be included with different noise amplitude, which was not analysed in the current paper;
3. The correlation between the amplitude of the high-frequency noise and the height (amplitude) of the analysed detail were also not studied against their influence on the process of noise detection, and correspondingly, reduction. From that point of view, each of the filters, like Gaussian (regular and with robust performance), spline or fast Fourier, can give different results and their validity can be also discussed;
4. Moreover, the influence of amplitude on the high-frequency measurement noise on the results of proposed techniques was not considered in this paper. Furthermore, the effect of the amplitude of high-frequency noise on the results of considered filters application, and also on the results of the calculation of the surface topography parameters (e.g., those from the ISO 25178 standard) must be studied to provide more surface functional advantages.

5. Conclusions

It is extremely difficult to propose appropriate procedures for accurate extraction of the high-frequency measurement errors from raw measured surface topography data; however, the following conclusions may be defined:

1. In the process of detection of the high-frequency measurement noise from the results of laser-textured surface measurements, profile (2D) characteristics may be more convenient than those of areal (3D); however, each of the measured details must be treated individually;
2. The application of PSD characteristics may be valuable in high-frequency noise detection; nevertheless, other methods, like ACF or TD, can be required. The most encouraging technique should be based on a few characteristics, using the PSD, ACF and TD approaches simultaneously;
3. When detection of the high-frequency measurement noise is hampered by the occurrence of the deep/wide features, like treatment traces in the LST details, the application of the thresholding method can provide positive results. When the surface contains deep or wide features or their density is relatively large, the thresholding technique removes those features from analysed detail (profile or areal data). Application of this method with all of the commonly used algorithms, (e.g., PSD and ACF) can give more accurate responses about the presence of high-frequency noise when a thresholding method is applied;
4. Of four general, regular filters available in the commercial software that consider Gaussian methods (regular regression or robust modifications), spline or fast Fourier approaches, this last one can be classified as the most suitable for the reduction in the influence of the high-frequency measurement noise on the results of surface topography measurements. However, suitable application of digital filtering requires careful use so that inappropriately used algorithms can remove necessary data from those raw measurements;

5. Generally, the functions available in the commercial software (like PSD, ACF, TD or GF, RGF, SF and FFTF) can be suitable in the process of detection and, correspondingly, reduction in the high-frequency measurement errors from the laser-textures topographies; nevertheless, the minimising of data processing errors must be classified as a required issue.

Funding: This research received no external funding.

Institutional Review Board Statement: Not applicable.

Informed Consent Statement: Not applicable.

Data Availability Statement: Data sharing is not applicable to this article.

Conflicts of Interest: The author declares no conflict of interest.

Parameters and Abbreviations

The following abbreviations and parameters are used in the manuscript:

ACF	autocorrelation function
FFTF	Fast Fourier Transform Filter
GF	Gaussian filter
HFN	high-frequency noise
L-surface	long-wavelength surface
LST	laser surface texturing
NS	noise surface
PSD	power spectral density
RGF	robust Gaussian filter
S-filter	removes small-scale lateral components
SF	spline filter
Sa	arithmetic mean height Sa, μm
Sal	auto-correlation length, mm
Sbi	surface bearing index
Sci	core fluid retention index
Sdq	root mean square gradient
Sdr	developed interfacial areal ratio, %
Sk	core roughness depth, μm
Sku	kurtosis
Sp	maximum peak height, μm
Spc	arithmetic mean peak curvature, 1/mm
Spd	peak density, $1/\text{mm}^2$
Spk	reduced summit height, μm
Sq	root mean square height, μm
Ssk	skewness
Std	texture direction, °
Str	texture parameter
Sv	maximum valley depth, μm
Svi	valley fluid retention index
Svk	reduced valley depth, μm
Sz	the maximum height of surface, μm

References

1. Segu, D.Z.; Choi, S.G.; Choi, J.H.; Kim, S.S. The effect of multi-scale laser textured surface on lubrication regime. *Appl. Surf. Sci.* **2013**, *270*, 58–63. [CrossRef]
2. Kovalchenko, A.; Ajayi, O.; Erdemir, A.; Fenske, G. Friction and wear behavior of laser textured surface under lubricated initial point contact. *Wear* **2011**, *271*, 1719–1725. [CrossRef]
3. Grabon, W.; Koszela, W.; Pawlus, P.; Slawomir Ochwat, A. Improving tribological behaviour of piston ring–cylinder liner frictional pair by liner surface texturing. *Tribol. Int.* **2017**, *113*, 182–188. [CrossRef]
4. Brinkman, S.; Bodschwinna, H. Characterisation of automotive bore performance using 3D surface metrology. In *Advanced Techniques for Assessment Surface Topography*; Blunt, L., Jiang, X., Eds.; KoganPage Science: London, UK; Sterling, VA, USA, 2003.

5. Ronen, A.; Etsion, I.; Kligerman, Y. Friction-reducing surface texturing in reciprocating automotive components. *Tribol. T.* **2001**, *44*, 359–366. [CrossRef]
6. Ryk, G.; Kligerman, Y.; Etsion, I. Experimental investigation of laser surface texturing for reciprocating automotive components. *Tribol. T.* **2002**, *45*, 444–449. [CrossRef]
7. Faria, D.; Madeira, S.; Buciumeanu, M.; Silva, F.S.; Carvalho, O. Novel laser textured surface designs for improved zirconia implants performance. *Mater. Sci. Eng. C* **2020**, *108*, 110390. [CrossRef]
8. Samanta, A.; Wang, Q.; Shaw, S.K.; Ding, H. Nanostructuring of laser textured surface to achieve superhydrophobicity on engineering metal surface. *J. Laser Appl.* **2019**, *31*, 022515. [CrossRef]
9. Samanta, A.; Wang, Q.; Shaw, S.K.; Ding, H. Roles of chemistry modification for laser textured metal alloys to achieve extreme surface wetting behaviors. *Mater. Des.* **2020**, *192*, 108744. [CrossRef]
10. Hu, T.; Zhang, Y.; Hu, L. Tribological investigation of MoS_2 coatings deposited on the laser textured surface. *Wear* **2021**, *278–279*, 77–82. [CrossRef]
11. Li, X.; Li, Y.; Tong, Z.; Ma, Q.; Ni, Y.; Dong, G. Enhanced lubrication effect of gallium-based liquid metal with laser textured surface. *Tribol. Int.* **2019**, *129*, 407–415. [CrossRef]
12. Ma, C.P.; Guan, Y.C.; Zhou, W. Laser polishing of additive manufactured Ti alloys. *Opt. Laser. Eng.* **2017**, *93*, 171–177. [CrossRef]
13. Hu, T.; Hu, L. The study of tribological properties of laser-textured surface of 2024 aluminium alloy under boundary lubrication. *Lubr. Sci.* **2012**, *24*, 84–93. [CrossRef]
14. Segu, D.Z.; Kim, J.H.; Choi, S.G.; Jung, Y.S.; Kim, S.S. Application of Taguchi techniques to study friction and wear properties of MoS2 coatings deposited on laser textured surface. *Surf. Coat. Tech.* **2013**, *232*, 504–514. [CrossRef]
15. Grzesik, W. Prediction of the Functional Performance of Machined Components Based on Surface Topography: State of the Art. *J. Mater. Eng. Perform.* **2016**, *25*, 4460–4468. [CrossRef]
16. Shao, Y.; Yin, Y.; Du, S.; Xia, T.; Xi, L. Leakage Monitoring in Static Sealing Interface Based on Three Dimensional Surface Topography Indicator. *ASME J. Manuf. Sci. Eng.* **2018**, *140*, 101003. [CrossRef]
17. Morehead, J.; Zou, M. Superhydrophilic surface on Cu substrate to enhance lubricant retention. *J. Adhes. Sci. Technol.* **2014**, *28*, 833–842. [CrossRef]
18. Podulka, P. Improved Procedures for Feature-Based Suppression of Surface Texture High-Frequency Measurement Errors in the Wear Analysis of Cylinder Liner Topographies. *Metals* **2021**, *11*, 143. [CrossRef]
19. Zheng, M.; Wang, B.; Zhang, W.; Cui, Y.; Zhang, L.; Zhao, S. Analysis and prediction of surface wear resistance of ball-end milling topography. *Surf. Topogr. Metrol. Prop.* **2020**, *8*, 025032. [CrossRef]
20. Szala, M.; Świetlicki, A.; Sofińska-Chmiel, W. Cavitation erosion of electrostatic spray polyester coatings with different surface finish. *Bull. Pol. Acad. Sci. Tech. Sci.* **2021**, *69*, e137519. [CrossRef]
21. Macek, W. Correlation between Fractal Dimension and Areal Surface Parameters for Fracture Analysis after Bend-ing-Torsion Fatigue. *Metals* **2021**, *11*, 1790. [CrossRef]
22. Macek, W.; Branco, R.; Szala, M.; Marciniak, Z.; Ulewicz, R.; Sczygiol, N.; Kardasz, P. Profile and Areal Surface Parameters for Fatigue Fracture Characterisation. *Materials* **2020**, *13*, 3691. [CrossRef] [PubMed]
23. Podulka, P. Selection of Methods of Surface Texture Characterisation for Reduction of the Frequency-Based Errors in the Measurement and Data Analysis Processes. *Sensors* **2022**, *22*, 791. [CrossRef] [PubMed]
24. Podulka, P. Reduction of Influence of the High-Frequency Noise on the Results of Surface Topography Measurements. *Materials* **2021**, *14*, 333. [CrossRef] [PubMed]
25. Pawlus, P.; Wieczorowski, M.; Mathia, T. *The Errors of Stylus Methods in Surface Topography Measurements*; Zapol: Szczecin, Poland, 2014.
26. Pawlus, P. Digitisation of surface topography measurement results. *Measurement* **2007**, *40*, 672–686. [CrossRef]
27. Podulka, P. Bisquare robust polynomial fitting method for dimple distortion minimisation in surface quality analysis. *Surf. Interface Anal.* **2020**, *52*, 875–881. [CrossRef]
28. Podulka, P. The effect of valley depth on areal form removal in surface topography measurements. *Bull. Pol. Acad. Sci. Tech. Sci.* **2019**, *67*, 391–400. [CrossRef]
29. Magdziak, M. Selection of the Best Model of Distribution of Measurement Points in Contact Coordinate Measurements of Free-Form Surfaces of Products. *Sensors* **2019**, *19*, 5346. [CrossRef]
30. Podulka, P. Proposal of frequency-based decomposition approach for minimization of errors in surface texture parameter calculation. *Surf. Interface Anal.* **2020**, *52*, 882–889. [CrossRef]
31. De Groot, P.; DiSciacca, J. Definition and evaluation of topography measurement noise in optical instruments. *Opt. Eng.* **2020**, *59*, 064110. [CrossRef]
32. Gomez, C.; Su, R.; de Groot, P.; Leach, R.K. Noise Reduction in Coherence Scanning Interferometry for Surface Topography Measurement. *Nanomanuf. Metrol.* **2020**, *3*, 68–76. [CrossRef]
33. *ISO 2016 25178-600*; Geometrical Product Specification (GPS)—Surface Texture: Areal Part 600: Metrological Characteristics for Areal-Topography Measuring Methods. International Organization for Standardization: Geneva, Switzerland, 2016.
34. Servin, M.; Estrada, J.C.; Quiroga, J.A.; Mosiño, J.F.; Cywiak, M. Noise in phase shifting interferometry. *Opt. Express* **2009**, *17*, 8789–8794. [CrossRef] [PubMed]
35. De Groot, P.J. The Meaning and Measure of Vertical Resolution in Optical Surface Topography Measurement. *Appl. Sci.* **2017**, *7*, 54. [CrossRef]

36. Podulka, P. Suppression of the High-Frequency Errors in Surface Topography Measurements Based on Comparison of Various Spline Filtering Methods. *Materials* **2021**, *14*, 5096. [CrossRef] [PubMed]
37. Podulka, P. Proposals of frequency-based and direction methods to reduce the influence of surface topography measurement and data analysis errors. *Coatings* **2022**, *12*, 726. [CrossRef]
38. Pawlus, P. An analysis of slope of surface topography. *Metrol. Meas. Syst.* **2005**, *12*, 295–313.
39. Santoso, T.; Syam, W.P.; Darukumalli, S.; Leach, R. Development of a compact focus variation microscopy sensor for on-machine surface topography measurement. *Measurement* **2022**, *187*, 110311. [CrossRef]
40. Syam, W.P.; Jianwei, W.; Zhao, B.; Maskery, I.; Elmadih, W.; Leach, R. Design and analysis of strut-based lattice structures for vibration isolation. *Precis. Eng.* **2018**, *52*, 494–506. [CrossRef]
41. Muhamedsalih, H.; Jiang, X.; Gao, F. Vibration compensation of wavelength scanning interferometer for in-process surface inspection. In *Future Technologies in Computing and Engineering: Proceedings of Computing and Engineering Annual Researchers' Conference 2010: CEARC'10*; University of Huddersfield: Huddersfield, UK, 2010; pp. 148–153.
42. Syam, W.P. In-process surface topography measurements. In *Advances in Optical Surface Texture Metrology*; Leach, R.K., Ed.; IOP Publishing: Bristol, UK, 2020.
43. Podulka, P. Comparisons of envelope morphological filtering methods and various regular algorithms for surface texture analysis. *Metrol. Meas. Syst.* **2020**, *27*, 243–263. [CrossRef]
44. Krolczyk, G.M.; Maruda, R.W.; Krolczyk, J.B.; Nieslony, P.; Wojciechowski, S.; Legutko, S. Parametric and nonparametric description of the surface topography in the dry and MQCL cutting conditions. *Measurement* **2018**, *121*, 225–239. [CrossRef]
45. Krolczyk, G.M.; Maruda, R.W.; Nieslony, P.; Wieczorowski, M. Surface morphology analysis of Duplex Stainless Steel (DSS) in Clean Production using the Power Spectral Density. *Measurement* **2016**, *94*, 464–470. [CrossRef]
46. Pawlus, P.; Reizer, R.; Wieczorowski, M.; Krolczyk, G.M. Material ratio curve as information on the state of surface topography—A review. *Precis. Eng.* **2020**, *65*, 240–258. [CrossRef]
47. Pawlus, P.; Reizer, R. Functional importance of honed cylinder liner surface texture: A review. *Tribol. Int.* **2022**, *167*, 107409. [CrossRef]
48. Podulka, P. The Effect of Surface Topography Feature Size Density and Distribution on the Results of a Data Processing and Parameters Calculation with a Comparison of Regular Methods. *Materials* **2021**, *14*, 4077. [CrossRef] [PubMed]
49. Pawlus, P.; Reizer, R.; Wieczorowski, M. Problem on non-measured points in surface texture measurements. *Metrol. Meas. Syst.* **2017**, *24*, 525–536. [CrossRef]
50. Podulka, P.; Pawlus, P.; Dobrzanski, P.; Lenart, A. Spikes removal in surface measurement. *J. Phys. Conf. Ser.* **2014**, *483*, 012025. [CrossRef]
51. Elson, J.; Bennett, J. Calculation of the power spectral density from surface profile data. *Appl. Opt.* **1995**, *34*, 201–208. [CrossRef] [PubMed]
52. Podulka, P. Fast Fourier Transform detection and reduction of high-frequency errors from the results of surface topography profile measurements of honed textures. *Eksploat. Niezawodn.* **2021**, *23*, 84–89. [CrossRef]
53. Tian, H.; Ribeill, G.; Xu, C.; Reece, C.E.; Kelley, M.J. A novel approach to characterizing the surface topography of niobium superconducting radio frequency (SRF) accelerator cavities. *Appl. Surf. Sci.* **2011**, *257*, 4781–4786. [CrossRef]
54. Whitehouse, D.J. Surface metrology. *Meas. Sci. Technol.* **1997**, *8*, 955–972. [CrossRef]
55. Jiang, X.; Senin, N.; Scott, P.J.; Blateyron, F. Feature-based characterisation of surface topography and its application. *CIRP Ann-Manuf. Tech.* **2021**, *70*, 681–702. [CrossRef]
56. Senin, N.; Leach, R.K.; Pini, S.; Blunt, L. Texture-based segmentation with Gabor filters, wavelet and pyramid decompositions for extracting individual surface features from areal surface topography maps. *Meas. Sci. Technol.* **2015**, *26*, 095405. [CrossRef]
57. Newton, L.; Senin, N.; Chatzivagiannis, E.; Smith, B.; Leach, R.K. Feature-based characterisation of Ti6Al4V electron beam powder bed fusion surfaces fabricated at different surface orientations. *Addit. Manuf.* **2020**, *35*, 101273. [CrossRef]
58. Alqahtani, M.A.; Jeong, M.K.; Elsayed, E.A. Multilevel spatial randomness approach for monitoring changes in 3D topographic surfaces. *Int. J Prod. Res.* **2020**, *58*, 5545–5558. [CrossRef]

Article

Investigation of the Surface Roughness and Surface Uniformity of a Hybrid Sandwich Structure after Machining

Elżbieta Doluk *, Anna Rudawska and Izabela Miturska-Barańska

Faculty of Mechanical Engineering, Lublin University of Technology, 20-388 Lublin, Poland
* Correspondence: e.doluk@pollub.pl; Tel.: +48-0507-666-485

Abstract: The parameters of surface roughness Ra, Rz and Rmax as well as surface topography Sa, Sz, Sp and Sv of the two-layer sandwich structure composed of an AW-2024 T3 aluminum alloy (Al) and a carbon-fiber-reinforced polymer (CFRP) were measured to determine an impact of the machining configuration (arrangement of the materials forming a sandwich structure) and the type of tool (presence of the tool coating) on the quality of the surface obtained through circumferential milling. The measurements revealed that milling produced different values of surface roughness for the aluminum alloy and the CFRP composite with values of 2D and 3D surface roughness being higher for the composite layer. The highest value of Ra of 1.10 μm was obtained for the surface of the CFRP composite using the CFRP/Al configuration and a TiAlN-coated tool. The highest values of the Rz (6.51 μm) and Rmax (8.85 μm) surface roughness parameters were also obtained for the composite layer using the same machining configuration and type of tool.

Keywords: circumferential milling; sandwich structure; aluminum alloy; CFRP composite

Citation: Doluk, E.; Rudawska, A.; Miturska-Barańska, I. Investigation of the Surface Roughness and Surface Uniformity of a Hybrid Sandwich Structure after Machining. *Materials* 2022, 15, 7299. https://doi.org/10.3390/ma15207299

Academic Editors: Claude Estournes and Frank Czerwinski

Received: 12 September 2022
Accepted: 18 October 2022
Published: 19 October 2022

Publisher's Note: MDPI stays neutral with regard to jurisdictional claims in published maps and institutional affiliations.

Copyright: © 2022 by the authors. Licensee MDPI, Basel, Switzerland. This article is an open access article distributed under the terms and conditions of the Creative Commons Attribution (CC BY) license (https://creativecommons.org/licenses/by/4.0/).

1. Introduction

A sandwich structure is made up of two basic components: a thinner external layer (facesheet), made of a stiffer material with better resistance properties, and a thicker but light internal layer (core), made of a different material. When joined together, the layers form a light, stiff, and durable structure which is stronger than similar monolithic structures [1–3]. Along with laminates, sandwich structures are classified as layered composites. Depending on their intended use, they can be made of different materials and can have different geometries and methods of stiffening and layer bonding. Sandwich structures can be divided into fully metal, hybrid, and composite structures, depending on the type of materials that make them up. Cavity milling is more difficult with hybrid layered composites than in the case of metal structures due to the heterogeneity and anisotropy of the structures resulting from materials with different mechanical and physical properties being bound together. In addition, bonding such materials requires considerable effort [4]. Moreover, sandwich structures are believed to be difficult to machine. During the machining of a sandwich structure composed of aluminum alloy and a carbon-fiber reinforced polymer (CFRP), it is additionally necessary to overcome difficulties related to the deformation of the upper and lower layers of the material, the interference of metal chips with the composite and the possible accumulation of aluminum shavings or carbon dust between the facings of the sandwich structure if it delaminates [5]. The heavy-duty carbon fibers used in composites are difficult to break and tend to drag, which very often causes them to develop micro-cracks and to delaminate along the operating direction of the tool [6]. Milling of CFRP composite requires frequent replacement of the milling tool, as the abrasive properties of carbon fiber cause premature wear. Furthermore, when machining CFRP composites, the milling tool alternately cuts hard fibers and the softer material of the skin, also accelerating the wear of the milling cutter [7,8].

Although hybrid layered structures are frequently used in many industries, the optimal conditions for machining this type of materials have not yet been adequately determined.

Most studies focus on the damage to sandwich structures and their properties [2], compare their properties with those of similar monolithic structures [9], or review current trends in the use of such materials [10,11]. Additionally, recommendations on the machining of hybrid layered materials are based primarily on studies of CFRP composite processing.

Surface roughness is one of the most important and most commonly used properties that characterize the quality of a milled surface. It affects dimensional accuracy, alignment and the correct operation of machine and device parts [3,12,13]. Layered materials have been widely used in aerospace as structural components and parts of aircraft equipment. Components used in the aerospace industry must meet strict quality criteria due to its narrow tolerance ranges. Protective and decorative coatings are often applied to aviation components. High values of surface roughness parameters may result in insufficient adhesion of the coating and thus reduce its durability [14]. The heterogeneity and anisotropy of the machined material are also extremely important properties. The different surface roughness on the sandwich structure layers can significantly reduce the visual effect of the components. Therefore, it is important to obtain a uniform surface of the sandwich structure with the desired roughness and surface topography. Surface roughness is affected by multiple parameters, such as the level of wear of the cutting tool, the levels of vibration, the stability of the machining process, and the cutting parameters [15–17].

Many researchers have analyzed the roughness and geometric structure of a fiber composite surface after machining. However, no studies have been carried out to examine the roughness of milled hybrid sandwich structures. Some have noted [18,19] that milling strength and surface roughness during the machining of CFRP composites depend on the milling parameters, with feed rate being the main factor that affects the quality of the surface after machining. Suresh et al. [20] made similar conclusions. Ramulu et al. [21] conducted research on the machinability of CFRP composites. They noted that a higher cutting speed results in less surface roughness, which is supported by results published by Çolak et al. [22]. Channdrasekaran and Devarasiddappa [23] noted that the surface roughness of machined fiber composites is linearly dependent on the feed rate and inversely dependent on the cutting speed. Bayraktar et al. [24] determined the impact of the cutting tool (material, geometry, and tool coating) and the process parameters (cutting speed and feed rate) on the cutting speed and surface roughness of CFRP composites after milling. It was noted that feed rate had the greatest impact on the quality of the machined surface among the examined factors. A non-coated cutter produced the lowest cutting strength and surface roughness values, which increased with the number of flutes and the angle of inclination of the cutter's cutting edge. Ramirez et al. [25] examined cutter wear and surface roughness after drilling CFRP composite. They concluded that new criteria must be formulated to assess surface topography after machining due to the anisotropy and heterogeneity of the machined material. Zarrouk et al. [26] examined the impact of machining conditions (rotational speed and machining depth) on the machining strength and surface quality of a milled nomex honeycomb. The results of the experiment were compared to a theoretical model. Eskandari et al. [27] studied the cutting speed, feed rate, and drill bit diameter in terms of their impact on the quality of openings drilled into sandwich structures with a foam core. Grilo et al. [28] investigated the delamination of drilled CFRP composites associated with drill bit diameter and machining parameters (feed rate and rotational speed). They showed that the feed rate was related directly by a proportional relationship to the delamination factor and the adjusted delamination factor. Shunmugesh et al. [29] used Grey Relational Analysis and the Taguchi technique to optimize machining conditions (machining speed, feed rate and drill bit diameter) in the context of delamination and surface quality (surface roughness and the dimensional accuracy of openings) after CFRP composite drilling. Khoran et al. [30] also studied the impact of machining parameters on the quality of openings produced by drilling. The experiment was performed on sandwich structures made of various materials (balsa wood, foamed materials) and used digital technology to measure the delamination and uncut fiber coefficients. Xu et al. [31] studied the effects of cutting parameters and tool geometry on surface roughness and cutting forces after milling of CFRP composites with a diamond tool (PCD). They showed that a higher value of tool rake angel did not affect surface roughness but resulted in lower cutting forces. Oláh et al. [32] reviewed solutions used in the cavity milling of sandwich structures and discussed the results of a simulation for cutting a honeycomb structure using a dedicated tool. The use of numerical methods such as the finite element method (FEM), the boundary element method (BEM), mathematical modeling, and the application of machine learning methods can help reduce the time and cost of testing, which is extremely important when testing relatively expensive composite materials [33,34]. Onyibo and Safaei [35] used the finite element method (FEM) to design sandwich structures with a honeycomb core. The study compared the mechanical properties obtained by modeling structures with different core thicknesses and made of different materials. Lavaggi et al. [36] studied the bonding of a honeycomb core with facesheets made of thermoset materials. They used three TGLM (Theory guided machine learning) models to optimize bonding properties such as the level of facesheets consolidation and bond-line porosity.

Milling an Al/CFRP structure causes heavy wear on the cutting tools, resulting in a poorer-quality machined surface. One way to obtain high-quality surfaces after milling is to use coated tools. Wang et al. [37] studied the tool wear during the drilling of CFRP/Ti structures. The wear of the cutter after machining a sandwich structure was compared against reference samples (used to machine separate composite and metal layers). The results indicated that the wear of the tool on the side of the metal layer was nine times lower than that on the side of the composite. The tests also found that each type of material resulted in a different type of wear. Hosokawa et al. [38] presented the results of CFRP composites milling using diamond-coated tools with a variable inclination angle of the cutting edge. They were able to demonstrate that the occurrence of defects on the surface of the machined material depended on machining forces, whereas tool wear depended on the orientation of the fibers.

The complexity of layered composites can cause many difficulties during the machining process. There are no comprehensive studies on the surface quality of sandwich structures after milling. There are also no standards or recommendations indicating an acceptable value or level of defect uniformity after milling of this type of constructions. Recommendations for cutting conditions (for example, recommended tool geometry, cutting parameters, machining configuration) have also not been developed. It causes machining conditions for this type of materials to be currently chosen based on recommendations for the composite layer, and cutting tools are dedicated for use with specific core types. This approach leads to a non-uniform surface quality of the sandwich structure, and this affects its further operation and aesthetic qualities. This study investigates the effect of the machining configuration (arrangement of layers forming the sandwich structure) and the type of tool (presence of a TiAlN coating) on the surface quality of the Al/CFRP sandwich structure after milling. The surface quality was defined by surface roughness and surface topography. A new and innovative approach is to evaluate the surface quality of the tested structure on the basis of the uniformity of defect distribution on the surface of individual layers of the structure. In addition, the I_R coefficient has been proposed as a tool for assessing the surface uniformity of such materials. The obtained research results are targeted for use in industrial practice. Prior to the study, it was assumed that circumferential milling of a two-layer metal–polymer composite sandwich structure results in differences in surface quality of the layers making up the sandwich structure. Based on the results obtained, the research hypothesis was verified, and the influence of the studied variables on the research object was determined.

2. Materials and Methods

Tests were performed on a two-layer sandwich structure made up of an EN AW-2024 T3 aluminum alloy (Al) and a carbon-fiber-reinforced polymer (CFRP) with an alternating arrangement of fibers. In the experiment, the cutting speed v_c [m/min], feed rate f_z [mm/blade], axial milling depth a_p [mm], radial milling depth a_e [mm], materials forming the structure, shape, and dimensions of the samples were used as constant factors. The output factors were surface roughness, surface topography, and surface uniformity. A two-level total plan was used to plan the experiment. The influence of two independent variables was studied: machining configuration and tool type. The design of the considered sandwich structure limited the machining configuration to the occurrence of two states. In order to limit the research time, two states were also adopted for the second independent variable (tool type). The applied method of planning the experiment made it possible to quickly generate all possible combinations of states of the independent variables.

The used aluminum alloy is characterized by low density and increased yield strength [39]. The alloy has low oxidation resistance, so it is not suitable for solutions where there is a risk of corrosion. The material is machinable but not suitable for anodizing and welding. One area of its application is the aerospace industry. The CFRP composite was created through a vacuum pressure impregnation process, using an autoclave manufactured by Scholz (Coesfeld, Germany). The proportion of carbon fibers in the hardened composite was around 60%. Table 1 shows selected properties of the CFRP composite used in the test.

Table 1. Selected properties of the CFRP composite used in the test [40].

Tensile strength R_m	1900
Bending strength R_{eg} [MPa]	2050
Young's modulus (E) [GPa]	135
Apparent interlaminar shear strength (ILSF) [MPa]	85

Scotch-Weld EC-9323 B/A (3M, Saint Paul, MN, USA) epoxy glue was used to bond the aluminum alloy and the CFRP composite. The polymerization process took place in a vacuum bag under a pressure of 0.1 MPa for a period of 24 h. The plates were then seasoned under ambient conditions for 14 days.

The machined samples had dimensions of 60 × 120 × 12 mm (Figure 1). The thickness of a single layer was 6 mm. Due to low value of the adhesive layer (0.1 ± 0.02 mm), it was disregarded when determining the total thickness of the structure. The experiment analyzed the impact of two machining configurations:

- Al/CFRP—milling starting from the metal layer (Figure 1a)
- CFRP/Al—milling starting from the composite layer (Figure 1b)

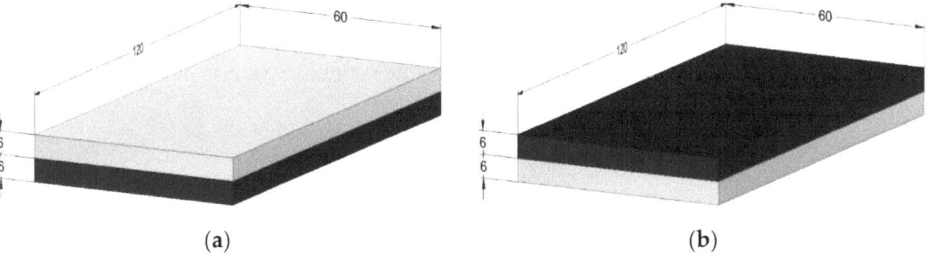

Figure 1. Tested sample: (**a**) Al/CFRP configuration, (**b**) CFRP/Al configuration (Unit: mm).

Simultaneous circumferential milling using a VMC 800 HS machining centre (AVIA, Warsaw, Poland) was employed during the tests. The spindle performed right-hand rotation. The shorter edge of the sample (60 mm) was machined during the tests. Each sample was machined three times and the arithmetic mean of the three measurements was used in the analysis. Figure 2 presents the scheme of the milling process.

Figure 2. Scheme of cutting process.

Manufacturers of machining tools currently tend to make dedicated tools for machining specific types of materials. This makes it particularly difficult to determine the tools and process parameters suitable for machining both the facesheet and core materials. Careful selection of the tool, its geometry and the technological parameters of the process is an important part of the experiment. The experiment used a Garant double-edged, solid carbide endmill by Hoffman Group (Munich, Germany). The shape and dimensions of the tool are presented in Figure 3 and Table 2.

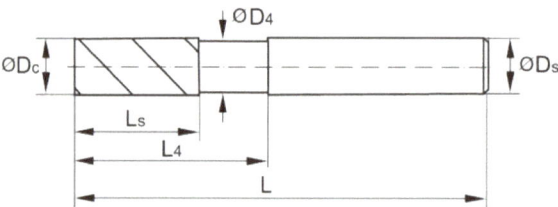

Figure 3. Geometry of endmill used [41].

Table 2. Technical data of tool used [41].

No. of teeth	2
Through-coolant	No
Tool material	Solid carbon (90% WC, 10% Co)
Cutting edge $\varnothing D_c$	12 mm
Shank $\varnothing D_s$.	12 mm
Recess $\varnothing D_4$	11.8 mm
Flute length L_s	26 mm
Overhang length L_4 incl. recess	38 mm
Overall length L	83 mm
Helix angle λ_s	45°
Rake angle γ	16°
Corner chamfer angle	45°
Corner chamfer width at 45°	0.10 mm

To examine the impact of the tool coating on the quality of the surface after milling, an identical endmill was used, this one coated with a 5-μm thick TiAlN coating, applied using the PVD method. The tools and coating type were selected due to their versatility (suitability for machining aluminum alloys and plastics). The machining was carried out using constant milling parameters: a milling speed v_c of 300 m/min, a feed rate f_z of 0.08 mm/blade, an axial milling depth a_p of 12 mm, and a radial milling depth a_e of 4 mm. The parameters were chosen based on values recommended by the manufacturer of the tool for machining aluminum alloys and polymers. The value of a_p was chosen so as to enable the simultaneous milling of both layers of the structure.

The roughness and topography of the surface were recorded using a 3D T8000 RC120-400 (Hommel-Etamic, Jena, Germany) surface roughness, waviness, and topography measurement device. The scheme of the experimental stand is shown in Figure 4.

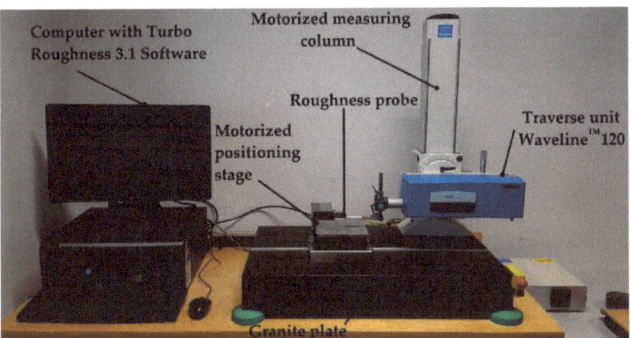

Figure 4. Scheme of experimental stand.

Ra (arithmetic mean deviation of the assessed profile), Rz (maximum height of the profile), and Rmax (maximum roughness depth) surface roughness parameters [42], as well as the Sa (arithmetic mean height), Sz (maximum height), Sv (maximum pit height) and Sp (maximum peak height) surface topography parameters, were analyzed [43]. Measurements were taken at a distance equal to half of the thickness of the sample from the sample's edge so that the measured section was positioned centrally to the layer being examined. The roughness and topography of the surface were measured separately for each layer in the structure, longitudinally to the feed direction. The test involved taking 320 measurements on the surface of each layer. The total length of the measurement l_t, the length of the measurement section l_n and the length of the elementary section l_r were, respectively: l_t = 4.8 mm, l_n = 4.0 mm, l_r = 0.8 mm. The measuring interval was 15 μm. The measurement speed was 0.50 mm/s. The measurement time for one layer (320 measurement points) was about 2 h and 40 min.

The surface quality of the hybrid sandwich structures after milling may also be defined by the uniform or regularity of defect distribution on the machined surface. This is caused by the fact that the metal layer may have a different surface quality than the composite layer. Therefore, the surface uniformity was determined based on the roughness parameters of the surface, depending on the machining configuration and tool used (Figure 5).

It was assumed that the most uniform surface was represented by the lowest difference between the values of a given surface roughness parameter of each layer (h_{min}), whereas the least uniform surface had the highest difference (h_{max}).

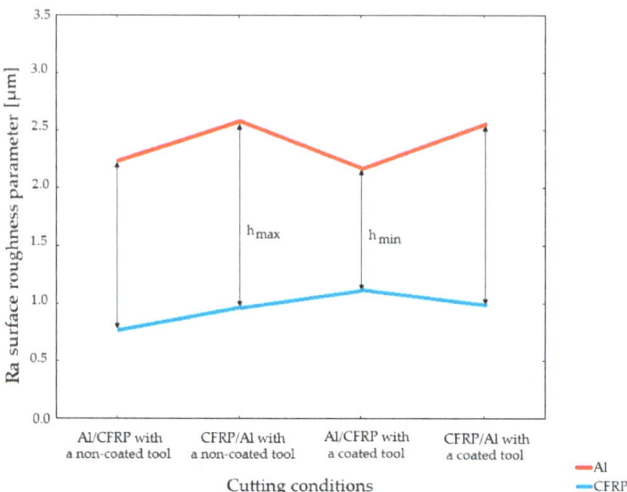

Figure 5. Determining the surface uniformity after milling based on selected surface roughness parameter.

3. Results

3.1. 2D Surface Roughness Parameters

Figures 6–8 illustrate the results of the roughness measurements of the 2D surface.

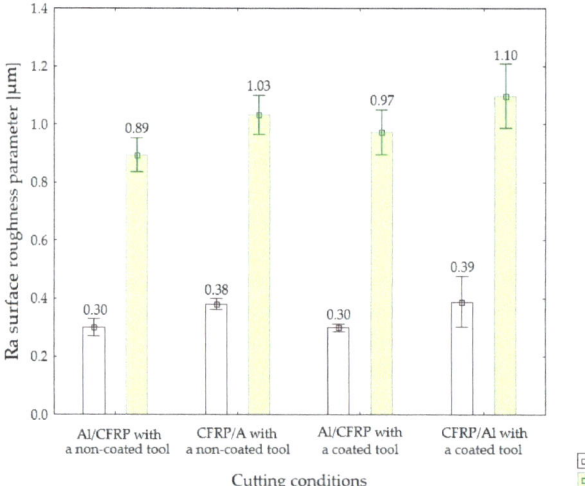

Figure 6. Values of Ra surface roughness parameter depending on the machining configuration and the type of tool.

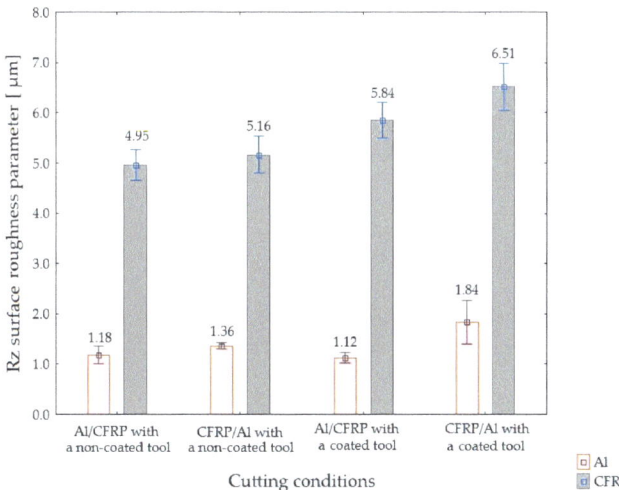

Figure 7. Values of Rz surface roughness parameter depending on the machining configuration and the type of tool.

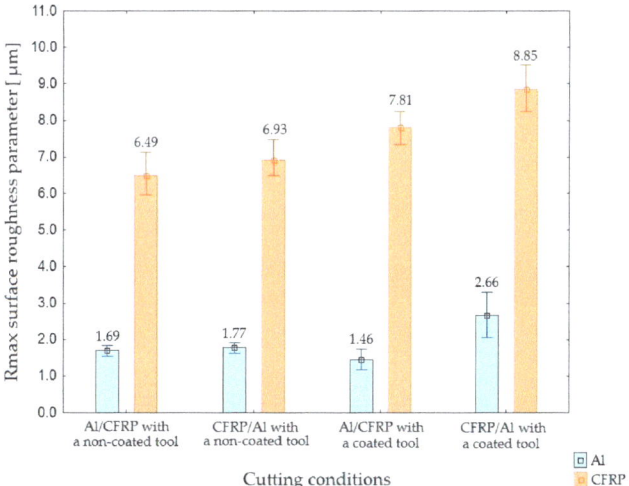

Figure 8. Values of Rmax surface roughness parameter depending on the machining configuration and the type of tool.

The minimum value of Ra parameter was obtained on the surface of the aluminum alloy, while the maximum value was observed on the surface of the CFRP composite. For all the cutting conditions, the values of Ra surface roughness parameter obtained on the surface of the CFRP composite were higher than the values of this parameter obtained for the aluminum alloy. The lowest values for the metal layer (0.30 μm) were measured in the following two machining configurations: Al/CFRP using a non-coated tool and Al/CFRP using a TiAlN-coated tool. The lowest value of this parameter for the composite layer was obtained when using the Al/CFRP machining configuration and a non-coated tool (0.89 μm). The highest values of Ra surface roughness parameter for the two materials (aluminum alloy—0.39 μm; CFRP composite—1.10 μm) were obtained in the CFRP/Al configuration, using a TiAlN-coated tool

3.2. Statistical Analysis of 2D Surface Roughness Parameters

To determine the impact of the machining configuration and the type of tool on the values of Ra, Rz, and Rmax surface roughness parameters, a two-factor analysis of variance (ANOVA) was carried out, the results of which are shown in Tables 3–5.

The data in Table 3 indicate that a statistically significant result for Ra surface roughness parameter measured at the surface of the aluminum alloy was recorded only for the machining configuration ($F_{1,\,1276} = 83.37$; $p < 0.01$). Tool type and A × B interaction had similar values of the test statistic and had no significant impact on Ra surface roughness parameter. In the case of the Ra surface roughness parameter measured at the surface of the CFRP composite, statistically significant results were recorded for machining configuration ($F_{1,\,1276} = 190.52$; $p < 0.01$) and tool type ($F_{1,\,1276} = 59.51$; $p < 0.01$). The impact of the interaction between the two factors was statistically insignificant in this case as well.

Table 3. Two-factor analysis of variance for Ra surface roughness parameter depending on the machining configuration and the type of tool.

Impact	Aluminum Alloy				
	SS	Df	MS	F	p-Value
A: Machining configuration	2.18	1	2.18	83.37	<0.01
B: Tool type	<0.01	1	<0.01	0.13	0.72
A × B interaction	<0.01	1	<0.01	0.12	0.73
Error	33.42	1276	0.03		
Total	35.61	1279			
Impact	CFRP				
	SS	Df	MS	F	p-Value
A: Machining configuration	5.51	1	5.51	190.52	<0.01
B: Tool type	1.72	1	1.72	59.51	<0.01
A × B interaction	0.02	1	0.02	0.70	0.4
Error	36.89	1276	0.03		
Total	44.14	1279			

Table 4. Two-factor analysis of variance for Rz surface roughness parameter depending on the machining configuration and the type of tool.

Impact	Aluminum Alloy				
	SS	Df	MS	F	p-Value
A: Machining configuration	65.12	1	65.12	66.42	<0.01
B: Tool type	14.04	1	14.04	14.32	<0.01
A × B interaction	23.05	1	23.05	23.51	<0.01
Error	1250.55	1276	0.98		
Total	1352.76	1279			
Impact	CFRP				
	SS	Df	MS	F	p-Value
A: Machining configuration	61.03	1	61.03	64.28	<0.01
B: Tool type	404.22	1	404.22	425.74	<0.01
A × B interaction	17.20	1	17.20	18.12	<0.01
Error	1211.51	1276	0.95		
Total	1693.96	1279			

Table 4 lists the results of ANOVA for the Rz surface roughness parameter. The analysis of these results indicates that the machining configuration, type of tool used, and the interaction between these variables had a statistically significant impact on the highest value of the roughness profile of the aluminum alloy and CFRP composite. In the case of the aluminum alloy, the machining configuration had the most significant impact on Rz surface roughness parameter ($F_{1, 1276} = 66.42$; $p < 0.01$), whereas the tool type had the least significant impact ($F_{1, 1276} = 14.32$; $p < 0.01$). ANOVA indicated that the tool type had the most significant impact on Rz surface roughness parameter for the CFRP composite ($F_{1, 1276} = 404.2274$; $p < 0.01$), while the A × B interaction had the least impact ($F_{1, 1276} = 18.12$; $p < 0.01$).

Table 5. Two-factor analysis of variance for Rmax surface roughness parameter depending on the machining configuration and the type of tool.

Impact	Aluminum Alloy				
	SS	Df	MS	F	p-Value
A: Machining configuration	132.08	1	132.08	47.22	<0.01
B: Tool type	34.01	1	34.01	12.16	<0.01
A × B interaction	101.44	1	101.44	36.27	<0.01
Error	3569.08	1276	2.80		
Total	3836.61	1279			
Impact	CFRP				
	SS	Df	MS	F	p-Value
A: Machining configuration	172.20	1	172.20	50.08	<0.01
B: Tool type	836.96	1	836.96	243.40	<0.01
A × B interaction	29.24	1	29.24	8.50	<0.01
Error	4387.60	1276	3.44		
Total	5426	1279			

3.3. Surface Topography

Figures 9 and 10 show 3D maps illustrating the surface topography of the samples after machining. The surface topography of each layer making up the sandwich structure is shown, depending on the machining configuration and tool used.

The surface topography of the metal layer of a sample milled in the Al/CFRP configuration using a non-coated tool (Figure 9a) was characterized by a directed arrangement of irregularities. Projections and indentations were both regular and random. Numerous grooves and ridges were also visible, diversifying the roughness of the surface. In the case of the CFRP composite (Figure 9b), a unidirectional surface with a random arrangement of irregularities was obtained. The values of the Sa, Sz, Sp, and Sv parameters at the surface of the aluminum alloy were 46%, 71%, 78%, and 76% lower, respectively, than at the surface of the composite layer.

After machining the aluminum alloy in the CFRP/Al configuration with a non-coated tool, a directed surface with a mixed structure was produced (Figure 9c). The surface topography was characterized by a diverse arrangement of irregularities. The surface of the CFRP composite was unidirectional, with a random, regular distribution of irregularities (Figure 9d). The values of Sa, Sz, Sp, and Sv parameters measured at the surface of the metal layer were nearly 41, 81%, 80%, and 82% lower, respectively, than the parameters measured at the surface of the composite layer.

Milling using the Al/CFRP configuration and a TiAlN-coated tool produced a directed structure with a determined distribution of irregularities on the surface of the aluminum alloy (Figure 10a). The irregularities were evenly distributed and spaced. In the case of CFRP composite (Figure 10b), a unidirectional surface with a random arrangement of irregularities was obtained. The microgeometry of the surface demonstrated a minor difference in the elevation of the elements of topography. The values of Sa, Sz, Sp, and Sv parameters measured at the surface of the aluminum alloy were 53%, 66%, 45%, and 80% lower than those for the composite layer, respectively.

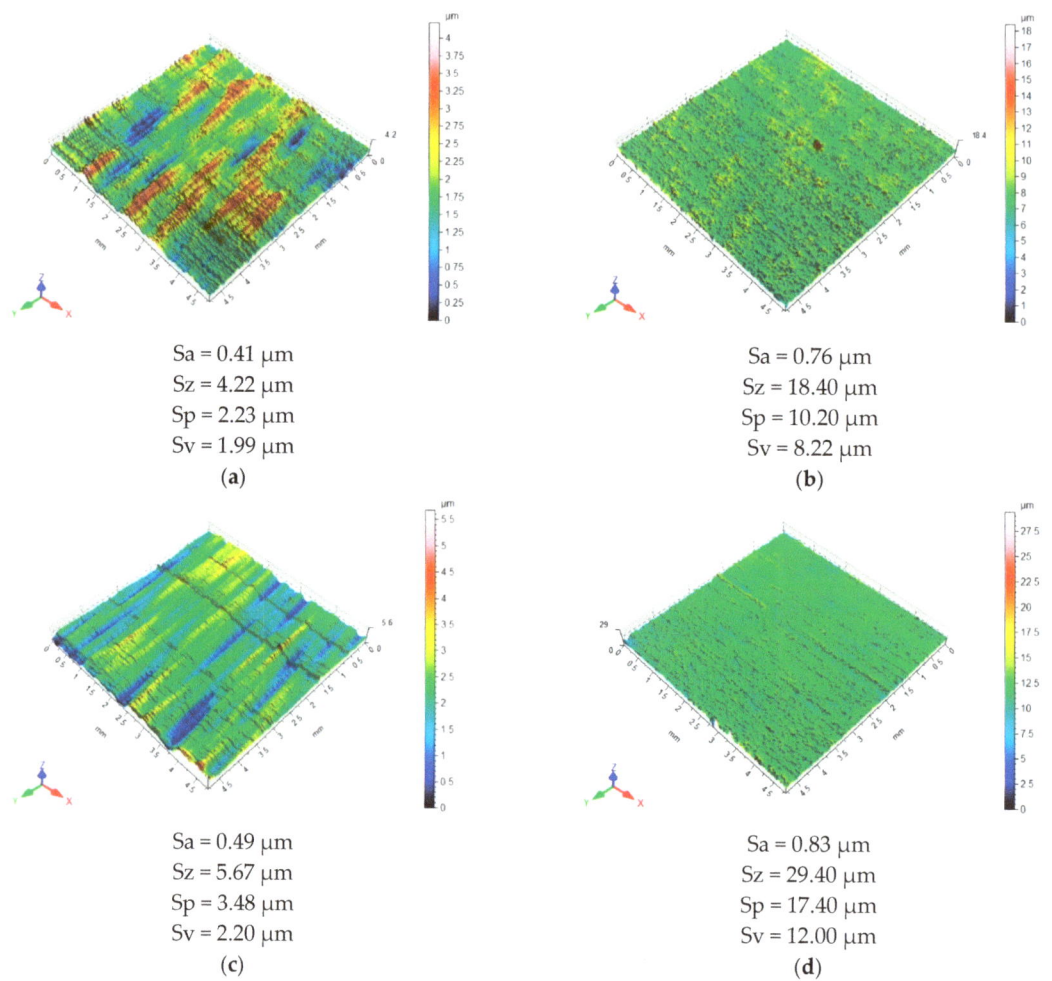

Sa = 0.41 μm
Sz = 4.22 μm
Sp = 2.23 μm
Sv = 1.99 μm
(a)

Sa = 0.76 μm
Sz = 18.40 μm
Sp = 10.20 μm
Sv = 8.22 μm
(b)

Sa = 0.49 μm
Sz = 5.67 μm
Sp = 3.48 μm
Sv = 2.20 μm
(c)

Sa = 0.83 μm
Sz = 29.40 μm
Sp = 17.40 μm
Sv = 12.00 μm
(d)

Figure 9. Surface topography after milling: (**a**) aluminum alloy in Al/CFRP configuration using a non-coated tool, (**b**) CFRP composite in Al/CFRP configuration using a non-coated tool, (**c**) aluminum alloy in CFRP/Al configuration using a non-coated tool, and (**d**) CFRP composite in CFRP/Al composite configuration using a non-coated tool.

Milling the metal layer with the CFRP/Al configuration and a coated tool produced a unidirectional topography with a regular arrangement of irregularities. The surface featured regular deformations: straight-line tracks spaced at similar intervals (Figure 10c). The topography of the composite surface was characterized by a unidirectional arrangement of irregularities. Numerous randomly arranged projections and indentations were visible (Figure 10d). In this case, the Sa and Sz values were 95% and 52% higher than in the case of the surface of the aluminum alloy, respectively. As regards the Sp value for the composite layer, the peaks of micro-irregularities were flattened: The Sp parameter was 32% lower, from 9.55 μm to 6.53 μm. The deepest indentation of the 3D profile (Sv) increased nearly fourfold.

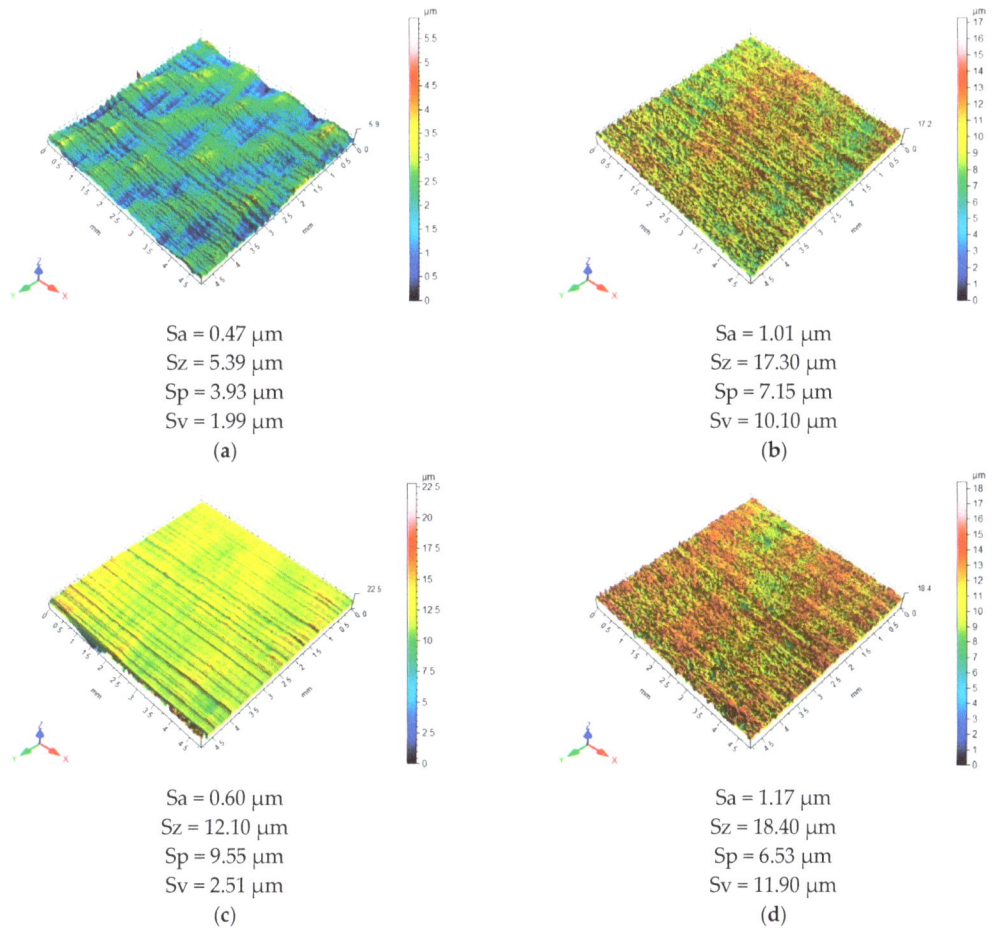

Sa = 0.47 μm
Sz = 5.39 μm
Sp = 3.93 μm
Sv = 1.99 μm
(a)

Sa = 1.01 μm
Sz = 17.30 μm
Sp = 7.15 μm
Sv = 10.10 μm
(b)

Sa = 0.60 μm
Sz = 12.10 μm
Sp = 9.55 μm
Sv = 2.51 μm
(c)

Sa = 1.17 μm
Sz = 18.40 μm
Sp = 6.53 μm
Sv = 11.90 μm
(d)

Figure 10. Surface topography after milling: (**a**) aluminum alloy in Al/CFRP configuration using a coated tool, (**b**) CFRP composite in Al/CFRP configuration using a coated tool, (**c**) aluminum alloy in CFRP/Al configuration using a coated tool, and (**d**) CFRP composite in CFRP/Al composite configuration using a coated tool.

3.4. Surface Uniformity

The surface uniformity of the two-layer structures after machining was determined based on the values of Ra, Rz, and Rmax surface roughness parameters.

In the case of Ra surface roughness parameter, the lowest difference between the layers was observed when using the Al/CFRP configuration and a non-coated endmill, whereas the highest difference was observed when using the CFRP/Al configuration and a TiAlN-coated tool (Figure 11).

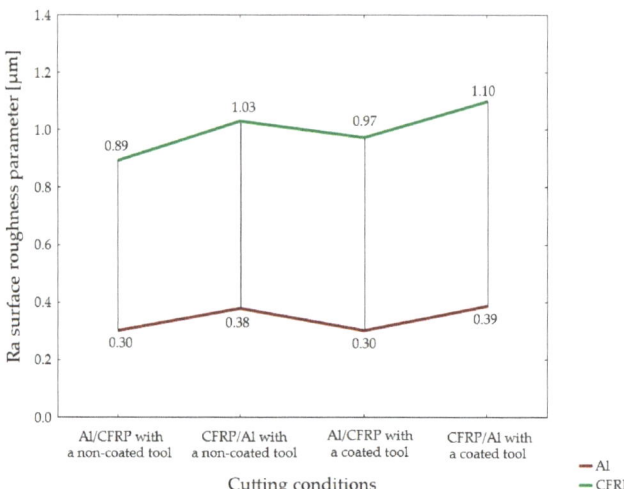

Figure 11. Surface uniformity based on the value of Ra surface roughness parameter obtained depending on the machining configuration and the type of tool.

Based on the values of Rz and Rmax surface roughness parameter, it can be ascertain that the Al/CFRP configuration with a non-coated tool produced the highest surface uniformity (lowest differences between the values obtained for both layers of the sandwich structure), whereas the Al/CFRP configuration plus a coated tool were characterized by the lowest surface uniformity (Figures 12 and 13).

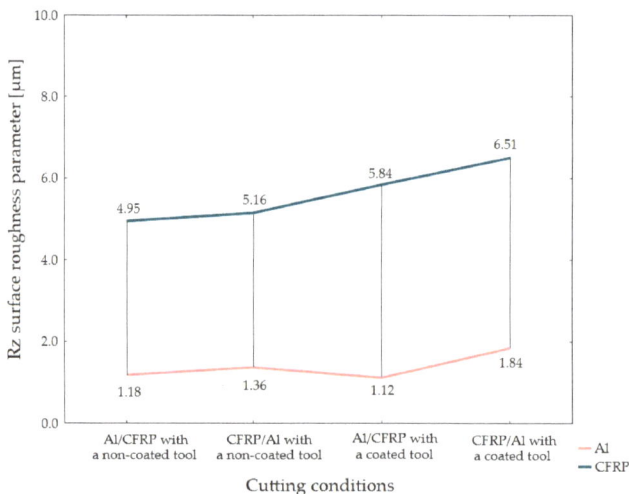

Figure 12. Surface uniformity based on the value of Rz surface roughness parameter obtained depending on the machining configuration and the type of tool.

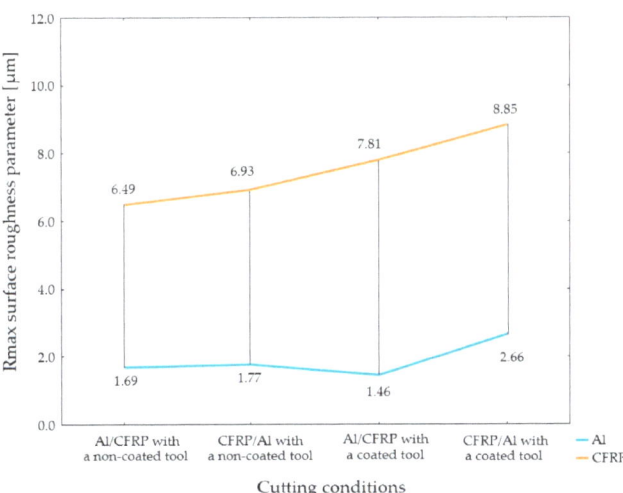

Figure 13. Surface uniformity based on the value of Rmax surface roughness parameter obtained depending on the machining configuration and the type of tool.

Currently, no standards or laws exist that would regulate the acceptable surface quality of sandwich structures after machining. Many researchers have attempted to forecast the roughness of the fiber composite surfaces by creating theoretical models and comparing them to the results of experiments [44,45]. However, there are no standards or papers dealing with hybrid layered structures. Based on the authors' experience in studying the surface quality of sandwich structures after machining, a coefficient that specifies the level of uniform of surface roughness after machining such materials was created. The coefficient was designated I_R and is calculated as follows:

$$I_R = \frac{|R_{Al} - R_{CFRP}|}{|R_{Al} + R_{CFRP}|} \tag{1}$$

where

I_R is the surface uniformity coefficient;
R_{Al} is the aluminum alloy surface roughness parameter;
R_{CFRP} is the CFRP composite surface roughness parameter.

The difference between the values of selected surface roughness parameters of the layers' surfaces was referenced against the total surface roughness of the sandwich structure. This enables the coefficient I_R to assess the surface uniformity of sandwich structure. By using any value of the given surface roughness parameter in formula (1), it can be observed that a higher difference between the values obtained for the aluminum alloy and the CFRP composite results in a lower uniform of the surface of the sandwich structure (h_{max}) and a higher value of I_R. Therefore, for the purposes of predicting the surface uniformity of sandwich structures after machining, it should be assumed that the lower the value of I_R ($I_R \rightarrow min$), the higher the surface uniformity of the sandwich structure (h_{min}).

Table 6 shows the calculated values of the coefficient I_R for the analyzed surface roughness parameters obtained for the adopted cutting conditions.

Table 6. Values of the coefficient I_R for the considered surface roughness parameters depending on the machining configuration and the type of tool.

Cutting Conditions	Surface Roughness Parameter		
	Ra	Rz	Rmax
Al/CFRP with a non-coated tool	0.50	0.61	0.58
CFRP/Al with a non-coated tool	0.46	0.53	0.59
Al/CFRP with a coated tool	0.53	0.68	0.69
CFRP/Al with a coated tool	0.48	0.58	0.54

Analyzing the results in Table 6, it can be seen that the highest I_R values for the considered surface roughness parameters were obtained after milling in the Al/CFRP configuration using a coated tool. The lowest surface uniformity obtained for Rz and Rmax parameters was obtained for the same machining configuration and tool used. However, the minimum I_R values and the highest surface uniformity were not obtained for the same cutting conditions.

4. Discussion

The objective of this study was to determine the impact of machining configuration and tool type on the surface quality of the sandwich structure after milling. The results of the study indicate that similar values were obtained for the Rz and Rmax surface roughness parameters. The minimum and maximum values of the Ra surface roughness parameter were not obtained for the same cutting conditions, as was the case for the other parameters. This may have resulted from the fact that the Ra surface roughness parameter does not account for the presence of defects typical of fiber-reinforced polymers [46].

Using the CFRP/Al machining configuration and a coated tool resulted in a higher value of Ra surface roughness parameter on the aluminum alloy and the CFRP surfaces. However, the statistical analysis indicated that the machining configuration was the only variable that had a statistically significant impact on the value of Ra surface roughness parameter obtained for the metal layer. The statistically significant variables for the composite layer were the machining configuration and the tool type. The interaction of these factors was statistically insignificant for both materials.

The results indicate that a composite material has a lower machinability—for all considered cutting conditions, lower values of the tested parameters were obtained for the surface of the metal layer. This was due to the heterogeneity and abrasive properties of the composite material and the presence of typical defects on its surface (including fiber pull-out and matrix cracking) [47]. The values of Rz and Rmax surface roughness parameters were higher for the CFRP/Al configuration than the Al/CFRP configuration, for both the aluminum alloy and the CFRP composite [48]. Obtaining higher values of the surface roughness parameters in the CFRP/Al configuration is due to the different properties of the materials forming the sandwich structure. The aluminum layer above the composite material stiffened the workpiece, making the machining more stable. Using a coated tool reduced the surface roughness of the aluminum alloy in the Al/CFRP configuration, while it increased the value in the CFRP/Al configuration. Using a coated tool with the composite layer resulted in higher values of Rz and Rmax surface roughness parameters in both configurations. This was the result of the thicker endmill material and the rounding of the milling edge caused by the tool coating [49,50]. The machining configuration had the most significant impact on the values of Rz and Rmax surface roughness parameter for the aluminum alloy, whereas the tool type was the major factor in the case of the CFRP composite. Different values of the tested surface roughness parameters for aluminum alloy and CFRP composite was also due to the anisotropy of the sandwich structure. The tool encountered different cutting resistances during machining—when cutting the CFRP composite, which has a lower density compared to aluminum alloy, a sudden change in cutting resistance occurred, and the tool was pulled deep into the workpiece. The result was the occurrence of a non-uniform quality on the surface of the sandwich structure.

The analysis of surface topography indicated that in most cases higher 3D surface roughness parameters were obtained for the composite layer. In addition, the topographies of the aluminum alloy and CFRP composite surfaces had different micro-irregularity arrangements.

The study went on to create a coefficient for assessing the surface uniformity of sandwich structures. Based on the values of the surface roughness parameters obtained through experimentation, differences in the quality of the layers that made up the sandwich structure were calculated for each variant. The results were compared against the calculated values of the newly-created coefficient I_R, leading to the assumption that the minimal value of I_R determines the most uniform quality of a sandwich structure. Analysis of the results showed that the cutting conditions (machining configuration and type of tool) allowing the lowest values of surface roughness parameters for the materials forming the structure did not guarantee the highest surface uniformity. This shows that surface uniformity is not the same as surface roughness and should still be considered separately. Therefore, it is necessary to continue research aimed at finding a tool and machining conditions to achieve surface roughness and surface uniformity at similar acceptable levels.

5. Conclusions

The following conclusions have been formulated based on the tests and analysis:

1. The composite layer had poorer surface quality than the aluminum layer.
2. The Ra surface roughness parameter is the least suitable of all analyzed surface roughness parameters for assessing the surface quality of sandwich structures after machining.
3. The CFRP/Al configuration increased the values of Ra, Rz and Rmax surface roughness parameters on the surfaces of both materials.
4. The tool coating did not affect the values of Ra parameter obtained on the surface of the aluminum alloy. For the CFRP composite, the presence of the TiAlN coating led to higher values of this parameter.
5. In most cases, a coated tool increased the Rz and Rmax surface roughness parameters.
6. The CFRP/Al configuration and a coated tool increased the values of Sa, Sz, Sp, and Sv 3D surface roughness parameters in the majority of cases.
7. Using the Al/CFRP configuration and a non-coated tool are recommended to receive the most uniform surface of the sandwich structure.
8. Milling with the Al/CFRP configuration and a TiAlN-coated tool resulted in the lowest surface uniformity.

Author Contributions: Conceptualization, E.D. and A.R.; methodology, E.D., A.R. and I.M.-B.; software, E.D. and I.M.-B.; validation, E.D.; formal analysis, A.R.; investigation, E.D.; resources, E.D., A.R. and I.M.-B.; data curation, E.D. and I.M.-B.; writing—original draft preparation, E.D.; writing—review and editing, A.R.; visualization, E.D.; supervision, A.R.; project administration, E.D.; funding acquisition, E.D. All authors have read and agreed to the published version of the manuscript.

Funding: The research was financed in the framework of the project Lublin University of Technology—Regional Excellence Initiative, funded by the Polish Ministry of Science and Higher Education (contract no. 030/RID/2018/19).

Institutional Review Board Statement: Not applicable.

Informed Consent Statement: Not applicable.

Data Availability Statement: The raw/processed data required to reproduce these findings cannot be shared at this time due to technical or time limitations. Data can be made available on individual request.

Acknowledgments: Faculty of Mechanical Engineering, Lublin University of Technology, Lublin, Poland.

Conflicts of Interest: The authors declare no conflict of interest.

References

1. Alkhoder, M.; Iyer, S.; Shi, W.; Venkalesh, T.A. Low frequency acoustic characteristics of periodic honeycomb cellular cores: The effect of relative density and strain fields. *Compos. Struct.* **2015**, *133*, 77–84. [CrossRef]
2. Arbaoui, J.; Schmitt, Y.; Pierrot, J.-L.; Royer, F.-X. Numerical simulation and experimental bending behaviour of multi–layer sandwich structures. *J. Theor. Appl. Mech.* **2014**, *52*, 431–442.
3. Królczyk, G.; Legutko, S.; Niesłony, P.; Gajek, M. Study of the surface integrity microhardness of austenitic stainless steel after turning. *Teh. Vjesn.* **2014**, *21*, 1307–1311.
4. *Sandcore: Best Practice Guide for Sandwich Structures in Marine Applications*; New Rail, University of Newcastle upon Tyne: Tyne, UK, 2005.
5. Hintze, W.; Hartmann, D.; Schütte, C. Occurrence and propagation of delamination during the machining of carbon fibre reinforced plastics (CFRPs)-An experimental study. *Compos. Sci. Technol.* **2011**, *71*, 1719–1726. [CrossRef]
6. Ciecielag, K. Effect of Composite Material Fixing on Hole Accuracy and Defects During Drilling. *Adv. Sci. Technol. Res. J.* **2021**, *15*, 54–65. [CrossRef]
7. Miller, J.; Eneyew, E.D.; Ramulu, M. Machining and Drilling of Carbon Fiber Reinforced Plastic (CFRP) Composites. *SAMPE J.* **2013**, *49*, 36–47.
8. Ciecielag, K. Study on the Machinability of Glass, Carbon and Aramid Fiber Reinforced Plastics in Drilling and Secondary Drilling Operations. *Adv. Sci. Technol. Res. J.* **2022**, *16*, 57–66. [CrossRef]
9. Muc, A.; Nogowczyk, R. Failure modes of sandwich structures with composite faces. *Composites* **2005**, *5*, 31–35.
10. Mousa, M.A.; Uddin, N. Debonding of composites structural insulated sandwich panels. *J. Reinf. Plast. Compos.* **2010**, *28*, 3380–3391. [CrossRef]
11. Yalkin, H.E.; Icten, B.M.; Alpyildiz, T. Enhanced mechanical performance of foam core sandwich composites with through the thickness reinforced core. *Compos. Part B Eng.* **2015**, *79*, 383–391. [CrossRef]
12. Maruda, R.W.; Królczyk, G.M.; Niesłony, P.; Królczyk, J.B.; Legutko, S. Chip formation zone analysis during the turning of austenitic stainless steel 316 L under MQCL cooling condition. *Procedia Eng.* **2016**, *149*, 297–304. [CrossRef]
13. Čep, R.; Janásek, A.; Petrů, J.; Sadílek, M.; Mohyla, P.; Valíček, J.; Harničarová, M.; Czán, A. Surface roughness after machining and influence of feed rate on process. *Key Eng. Mater.* **2014**, *581*, 341–347. [CrossRef]
14. Berardo, A.; Pugno, N.M. A model for hierarchical anisotropic friction, adhesion and wear. *Tribol. Int.* **2020**, *152*, 106549. [CrossRef]
15. Maruda, R.W.; Wojciechowski, S.; Szczotkarz, N.; Legutko, S.; Mia, M.; Gupta, M.K.; Niesłony, P.; Królczyk, G.M. Metrological analysis of surface quality aspects in minimum quantity cooling lubrication. *Measurement* **2021**, *171*, 108847. [CrossRef]
16. Wang, X.L.; Qu, Z.G.; Lai, T.; .Ren, G.F.; Wang, K.W. Enhancing water transport performance of gas diffusion layers through coupling manipulation of pore structure and hydrophobicity. *J. Power Sources* **2022**, *525*, 231121. [CrossRef]
17. Zhang, L.; Han, E.; Wu, Y.; Wang, X.; Wu, D. Surface decoration of short-cut polyimide fibers with multi-walled carbon nanotubes and their application for reinforcement of lightweight PC/ABS composites. *Appl. Surf. Sci.* **2018**, *442*, 124–137. [CrossRef]
18. Davim, J.P.; Reis, P. Damage and dimensional precision on milling carbon fiber-reinforced plastics using design experiments. *J. Mater. Process. Technol.* **2005**, *160*, 160–167. [CrossRef]
19. Sorrentino, L.; Turchetta, S. Cutting forces in milling of carbon fibre reinforced plastics. *Int. J. Manuf. Eng.* **2014**, *2014*, 1–8. [CrossRef]
20. Suresh, P.V.S.; Venkateswara, R.P.; Deshmukh, S.G. A genetic algorithmic approach for optimization of surface roughness prediction model. *Int. J. Mach. Tools Manuf.* **2002**, *42*, 675–680. [CrossRef]
21. Ramulu, M.; Arola, D.; Colligan, K. Preliminary investigation of effects on the surface integrity of fibre reinforced plastics. *ESDA* **1994**, *64*, 93–101.
22. Çolak, O.; Sunar, T. Cutting forces and 3D surface analysis of CFRP milling with PCD cutting tools. *Procedia CIRP* **2016**, *45*, 75–78. [CrossRef]
23. Chandrasekaran, M.D.; Devarasiddappa, D. Development of Predictive Model for Surface Roughness in End Milling of Al-sicp Metal Matrix Composites using Fuzzy Logic. *World Acad. Sci. Eng. Technol.* **2012**, *6*, 7–25.
24. Bayraktar, S.; Turgut, Y. Investigation of the Cutting Forces and Surface Roughness in Milling Carbon Fiber Reinforced Polymer Composite Material. *Mater. Technol.* **2016**, *50*, 591–600. [CrossRef]
25. Ramirez, C.; Poulachon, G.; Rossi, F.; M'Saoubi, R. Tool Wear Monitoring and Hole Surface Quality During CFRP Drilling. *Procedia CIRP* **2014**, *13*, 163–168. [CrossRef]
26. Zarrouk, T.; Salhi, J.E.; Atlati, S.; Salhi, M.; Nouari, M.; Salhi, M. The Influence of Machining Conditions on the Milling Operations of Nomex Honeycomb Structure. *PalArch's J. Archaeol. Egypt/Egyptol.* **2021**, *17*, 11008.
27. Eskandari, H.; Danaee, I.; Noori, S. Investigation the effective parameters on damages induced in composite sandwich structures through drilling. *Iran. J. Manuf. Eng.* **2018**, *4*, 51–60.
28. Grilo, T.J.; Paulo, R.M.F.; Silva, C.R.M.; Davim, J.P. Experimental delamination analyses of CFRPs using different drill geometries. *Compos. Part B Eng.* **2013**, *45*, 1344–1350. [CrossRef]
29. Shunmugesh, K.; Kavan, P. Investigation and optimization of machining parameters in drilling of carbon fiber reinforced polymer (CFRP) composites. *Pigment Resin Technol.* **2017**, *46*, 21–30.

30. Khoran, M.; Ghabezi, P.; Farhani, M.; Besharati, M.K. Investigation of drilling composite sandwich structures. *Int. J. Adv. Manuf. Technol.* **2015**, *76*, 1927–1936. [CrossRef]
31. Xu, Z.; Wang, Y. Study on Milling Force and Surface Quality during Slot Milling of Plain-Woven CFRP with PCD Tools. *Materials* **2022**, *15*, 3862. [CrossRef]
32. Oláh, F.; Andrásfalvy, K.; Lukács, J.; Horváth, R. Manufacturing problems of sandwich composite structures. In Proceedings of the 2021 IEEE 21st International Symposium on Computational Intelligence and Informatics (CINTI), Budapest, Hungary, 18–20 November 2021; pp. 167–172.
33. Szabelski, J.; Karpiński, R.; Machrowska, A. Application of an Artificial Neural Network in the Modelling of Heat Curing Effects on the Strength of Adhesive Joints at Elevated Temperature with Imprecise Adhesive Mix Ratios. *Materials* **2022**, *15*, 721. [CrossRef]
34. Jonak, J.; Karpiński, R.; Wójcik, A.; Siegmund, M. The Influence of the Physical-Mechanical Parameters of Rock on the Extent of the Initial Failure Zone under the Action of an Undercut Anchor. *Materials* **2021**, *14*, 1841. [CrossRef] [PubMed]
35. Onyibo, E.C.; Safaei, B. Application of finite element analysis to honeycomb sandwich structures: A review. *Rep. Mech. Eng.* **2022**, *3*, 283–300. [CrossRef]
36. Lavaggi, T.; Samizadeh, M.; Niknafs Kermani, N.; Khalili, M.M.; Advani, S.G. Theory-guided machine learning for optimal autoclave co-curing of sandwich composite structures. *Polym. Compos.* **2022**, *43*, 5319. [CrossRef]
37. Wang, X.; Kwon, P.Y.; Sturtevant, C.; Kim, D.; Lantrip, J. Comparative tool wear study based on drilling experiments on CFRP/Ti stack and its individual layers. *Wear* **2014**, *317*, 265–276. [CrossRef]
38. Hosokawa, A.; Hirose, N.; Ueda, T.; Furumoto, T. High–quality machining of CFRP with high helix end mill. *CIRP Ann.* **2014**, *63*, 89–92. [CrossRef]
39. *EN 515:2017*; Standard: Aluminium and Aluminium Alloys—Wrought Products—Temper Designations. European Committee for Standardization: Brussels, Belgium, 2017.
40. *DIN EN ISO 14125:2011-05 1.5.2011*; Standard—Fibre-Reinforced Plastic Composites—Determination of Flexural Properties. DIN ISO: Berlin, Germany, 2011.
41. Hoffmann Group. *Catalogue 1 Machining/Clamping Technology*; Hoffmann Group: Munich, Germany, 2019.
42. *ISO 4287:2021*; Standard—Geometrical Product Specifications (GPS)—Surface Texture: Profile Method—Terms, Definitions and Surface Tex-ture Parameters. ISO: Geneva, Switzerland, 2021.
43. *ISO 25178-2:2021*; Standard—Geometrical Product Specifications (GPS)—Surface Texture: Areal—Part 2: Terms, Definitions and Surface Texture Parameters. ISO: Geneva, Switzerland, 2021.
44. Che-Haron, C.H.; Jawaid, A. The effect of machining on surface integrity of titanium alloy Ti-6% Al-4% V. *J. Mater. Process. Technol.* **2005**, *166*, 188–192. [CrossRef]
45. Ramesh, S.; Karunamoorthy, L.; Palanikumar, K. Measurement and analysis of surface roughness in turning of aerospace titanium alloy (gr5). *Measurement* **2012**, *45*, 1266–1276. [CrossRef]
46. Teicher, U.; Müller, S.; Münzner, J.; Nestler, A. Micro-EDM of Carbon Fibre-Reinforced Plastics. *Procedia CIRP* **2013**, *6*, 320–325. [CrossRef]
47. M'Saoubi, R.; Axinte, D.; Soo, S.L.; Nobel, C.; Attia, H.; Kappmeyer, G.; Engin, S.; Sim, W.-M. High performance cutting of advanced aerospace alloys and composite materials. *CIRP Annals* **2015**, *64*, 557–580. [CrossRef]
48. Doluk, E.; Rudawska, A.; Kuczmaszewski, J.; Miturska-Barańska, I. Surface Roughness after Milling of the Al/CFRP Stacks with a Diamond Tool. *Materials* **2021**, *14*, 6835. [CrossRef] [PubMed]
49. Teti, R. Machining of Composite Materials. *CIRP Ann.* **2002**, *51*, 611–634. [CrossRef]
50. Janardhan, P.; Sheikh-Ahmad, J.; Cheraghi, H. Edge Trimming of CFRP with Diamond Interlocking Tools. *SAE Int.* **2006**, *2*, 3173.

Article

Friction Behaviour of 6082-T6 Aluminium Alloy Sheets in a Strip Draw Tribological Test

Tomasz Trzepieciński [1,*], Ján Slota [2], Ľuboš Kaščák [2], Ivan Gajdoš [2] and Marek Vojtko [3]

[1] Department of Manufacturing Processes and Production Engineering, Rzeszow University of Technology, al. Powst. Warszawy 8, 35-959 Rzeszów, Poland
[2] Institute of Technology and Material Engineering, Faculty of Mechanical Engineering, Technical University of Košice, Mäsiarska 74, 040 01 Košice, Slovakia
[3] The Institute of Materials Research, Slovak Academy of Sciences, Watsonova 47, 040 01 Košice, Slovakia
* Correspondence: tomtrz@prz.edu.pl

Citation: Trzepieciński, T.; Slota, J.; Kaščák, Ľ.; Gajdoš, I.; Vojtko, M. Friction Behaviour of 6082-T6 Aluminium Alloy Sheets in a Strip Draw Tribological Test. *Materials* **2023**, *16*, 2338. https://doi.org/10.3390/ma16062338

Academic Editor: Robert Pederson

Received: 15 February 2023
Revised: 8 March 2023
Accepted: 13 March 2023
Published: 14 March 2023

Copyright: © 2023 by the authors. Licensee MDPI, Basel, Switzerland. This article is an open access article distributed under the terms and conditions of the Creative Commons Attribution (CC BY) license (https://creativecommons.org/licenses/by/4.0/).

Abstract: Aluminium alloy sheets cause many problems in sheet metal forming processes owing to their tendency to gall the surface of the tool. The paper presents a method for the determination of the kinematic friction coefficient of friction pairs. The determination of coefficient of friction (COF) in sheet metal forming requires specialised devices that 'simulate' friction conditions in specific areas of the formed sheet. In this article, the friction behaviour of aluminium alloy sheets was determined using the strip drawing test. The 1-mm-thick 6082 aluminium alloy sheets in T6 temper were used as test material. Different values for nominal pressures (4.38, 6.53, 8.13, 9.47, 10.63, and 11.69 MPa) and different sliding speeds (10 and 20 mm/min.) were considered. The change of friction conditions was also realised with several typical oils (hydraulic oil LHL 32, machine oil LAN 46 and engine oil SAE 5W-40 C3) commonly used in sheet metal forming operations. Friction tests were conducted at room temperature (24 °C). The main tribological mechanisms accompanying friction (adhesion, flattening, ploughing) were identified using a scanning electron microscope (SEM). The influence of the parameters of the friction process on the value of the COF was determined using artificial neural networks. The lowest value of the COF was recorded when lubricating the sheet metal surface with SAE 5W40 C3 engine oil, which is characterised as the most viscous of all tested lubricants. In dry friction conditions, a decreasing trend of the COF with increasing contact pressure was observed. In the whole range of applied contact pressures (4.38–11.69 MPa), the value of the COF during lubrication with SAE 5W40 C3 engine oil was between 0.14 and 0.17 for a sliding speed of 10 mm/min and between 0.13 and 0.16 for a sliding speed of 20 mm/min. The value of the COF during dry friction was between 0.23 and 0.28 for a sliding speed of 10 mm/min and between 0.22 and 0.26 for a sliding speed of 20 mm/min. SEM micrographs revealed that the main friction mechanism of 6082-T6 aluminium alloys sheet in contact with cold-work tool steel flattens surface asperities. The sensitivity analysis of the input parameters on the value of COF revealed that oil viscosity has the greatest impact on the value of the COF, followed by contact pressure and sliding speed.

Keywords: 6082-T6; aluminium alloy; coefficient of friction; sheet metal forming; surface topography; ANN

1. Introduction

Sheets made of aluminium and aluminium alloys, due to their favourable ratio of strength to weight, are being used increasingly often in the automotive industry. Adding alloying elements to aluminium can increase its strength properties several times [1]. Alloys obtained in this way are characterised by low density and high impact strength. Nickel and cobalt, as well as magnesium and manganese, increase their 'strength' properties, and titanium and chromium affect grain size reduction [2]. Wrought alloys typically contain up to 5% alloying elements and are used in a hardened and heat-treated condition. In special conditions, cast aluminium alloys can also be processed by plastic working [3]. However, aluminium-based alloys generally have relatively low fatigue strength [4]. The fatigue resistance of aluminium alloys can be improved by adding elements belonging to the group of transition metals, including titanium, vanadium, and zirconium [5]. In this paper, the frictional properties of AW-6082-T6 aluminium alloy sheets are presented. The 6xxx series alloys contain magnesium (0.2–3%) and silicon (0.2–1.8%) as the main alloying additions. Some of the 6xxx series alloys contain manganese (up to 1.4%) and copper (up to 1.2%). Alloys in this group show good formability and are susceptible to machining. Those that do not contain copper have good corrosion resistance and can be anodized. Typical applications of aluminium alloys containing magnesium and silicon include structural elements of motor vehicles, interior fittings, and profiles for the construction industry.

A phenomenon accompanying friction in the formation of aluminium and aluminium alloy sheets is the galling of the surface of tools with the sheet material. In this way, the changing topography of the tool surface causes an unfavourable change in the topography of the formed drawpiece and more resistance to the movement of the sheet metal on the surface of the tool [6]. Surface topography has a decisive influence on friction, wear, and lubrication under mixed lubrication and dry friction conditions [7,8]. In the case of forming sheets made of aluminium, adhesive wear is also of great importance, consisting of metallic local adhesion of the roughness asperities in the micro-areas of plastic deformation. The mechanism of coexistence of abrasive and adhesive wear may result in the extension of frictional contact time, manifesting itself in the intensification of the surface galling [9].

A common way to reduce the coefficient of friction in sheet metal forming is to use lubricants. The purpose of using lubricant is also to reduce the wear intensity of the friction pair elements [6,7]. The lubricant should meet a number of the following requirements, including being easy to apply to the metal being processed and the tool, having high resistance to normal loads, and being easily remove from the surface of the product [8,9]. The grease should be characterized by appropriate viscosity and chemical activity, which is the ability to form a protective layer on the friction surface [10]. Appropriate chemical activity of lubricants is provided by surface-active compounds, such as fatty acids (oleic, stearic, palmitic) and their salts. Different conditions for the implementation of sheet metal forming processes and the variety of plastically processed materials mean that the following parameters should be taken into account when selecting a lubricant for a specific application [11]: processing temperature, value of maximum unit pressures, grade of processed material and grade of tool material, sliding speed, type of protective coating on the tool surface, and tool design. The surface topography of sheet metal has a decisive influence on friction, wear, and lubrication under mixed lubrication and dry friction conditions [12,13].

The basic criteria for the division of lubricants are the consistency of the lubricant, origin (mineral or organic), and intended use [14]. Due to the consistency, lubricants are distinguished in the following groups: liquid lubricants (oils), emulsions (oil mists), and solid lubricants. Lubricating oils are obtained by mixing base oils with enriching additives [15,16].

Mineral oils are obtained by processing base oils, and synthetic oils are produced by chemical synthesis or the processing of mineral oils [17]. Mineral oils are complex mixtures of saturated and aromatic hydrocarbons with a ring or chain structure, containing 20 to 40 carbon atoms in a molecule [18]. Oils resulting from distillation from crude oil differ in viscosity, chemical composition, and physical properties. Due to their viscosity, mineral oils are divided into spindle, machine, motor, and gear oils.

In order to determine the friction and wear of materials, the following types of tests and tribometers may be used: (a) strip drawing test with flat [19,20] and (b) rounded [21] (c) countersamples, (d) bending-under-tension test [22,23], (e) draw-bead test [24,25], and (f) ball-on-disc [26], (g) block-on-disc [27], (h) pin-on-ring [28], and (i) pin-on-disc tribometers [29,30]. Most of the studies, the results of which can be found in the literature, indicate the susceptibility of aluminium sheets to galling and intensification of the flattening mechanism. As analysis of the literature showed, each aluminium alloy has a specific friction performance. Moreover, owing to poor formability of aluminium and aluminium alloys at room temperature, friction tests of alloys focus on plastic working in warm and cold forming conditions. Yahaya and Samion [31] analysed the friction condition of AW-6061 aluminium alloy lubricated with bio-lubricant in a cold forging test. It was found that the palm oil-based lubricant has good performance compared to a mineral-based oil in terms of surface roughness. However, the mineral oil had better friction performance than the palm oil-based lubricant. Trzepieciński [32] investigated the frictional performance of AW-2024-T3 Alclad aluminium alloy sheets using strip drawing test. Analysis of the effect of the friction conditions on the effectiveness of lubrication and change in the surface roughness of the metal sheets were analysed using analysis of variance (ANOVA). The other most commonly used friction tester is the pin-on-disc tribometer. Guezmil et al. [33] investigated the tribological behaviour of anodic oxide layer formed on AW-5754 aluminium alloy in a pin-on-disc tribometer. Different sliding speeds, normal loads, and oxide thicknesses were considered to establish the COF. It was found that increased normal load and sliding speed increased the COF. The effects of initial lubricant volume, temperature, sliding speed and contact pressure on the evolutions of the COF of AW-7075 aluminium alloy and the breakdown phenomenon were investigated by Yang et al. [34] in pin-on-disc tests. It was found that the COF rapidly increased as the lubricant film thickness decreased to a critical value. Das [35] studied the tribological properties of three aluminium as-cast alloy samples, i.e., Al-14 wt%, Al-10 wt%Si, and Al-7 wt%Si aluminium alloys, using a pin-on-disc type wear-testing machine. The wear of aluminium specimens was seen to increase at higher sliding speeds and at higher applied loads. Friction and wear on aluminium–silicon alloys have been extensively tested by Shabel et al. [36]. They identified two major types of wear relevant to industrial applications of Al–Si alloys: sliding wear and abrasive wear depending on the silicon particles, intermetallic constituents, and matrix hardness. Luis Pérez et al. [37] studied the friction properties of AW-5083 and AW-5754 aluminium alloys processed by equal channel angular pressing (ECAP). It was found that both nanostructured aluminium alloys show better wear behaviour if they are compared with conventional isothermal forging. Li et al. [38] also confirmed that ECAP processing leads to a decrease in the COF owing to improved mechanical properties.

Many published works are focused on the analysis of the specific parameter of the friction process on the coefficient of friction or the occurrence of a specific friction mechanism. A considerable amount of factors in the friction process exist that affect the COF value and, as a result, building analytical friction model for specified process conditions is practically impossible. The artificial neural networks (ANNs) allow the researchers to overcome the difficulty arising in the assessment of the complex relationships between friction process parameters and COF. ANNs require a set of experimental training data to work properly. Based on the training process, ANNs acquire the ability to predict the value of the output parameter.

To the best knowledge of the authors, sheet metals made of AW-6082-T6 aluminium alloy have not been tested in strip drawing tests for pressures occurring in the blank holder zone during cold sheet metal forming. The selection of the appropriate lubricant is crucial to ensure appropriate conditions for the sheet metal forming of the automotive components.

Therefore, this article contains the results of the friction of AW-6082-T6 sheet metals. The friction behaviour of AW-6082-T6 aluminium alloy sheets was determined using the strip drawing test. Different values for nominal pressures and different sliding speeds were considered. The change of friction conditions was also realised with three typical lubricants (machine oil, engine oil and hydraulic oil) commonly used in sheet metal forming operations. The experimental design consists of 48 trials performed for two speeds, six pressures and four friction conditions. The tests were repeated three times to determine the average value of the coefficient of friction. The main tribological mechanisms accompanying friction were identified using a scanning electron microscope. Owing to the difficulty in determining the impact of the simultaneous interaction of many friction parameters on the value of the COF, artificial neural networks (ANNs) were used to identify the main relationships between the COF and process parameters. Sliding speed, average unit pressure, and lubricant viscosity were selected as input parameters of multilayer neural network. The output parameter was the value of the COF. The training of the network was carried out using the backpropagation algorithm.

2. Materials and Methods

2.1. Materials

The 1-mm-thick AW-6082 aluminium alloy sheets in T6 temper condition were used as test material. T6 indicates that the alloy has been solution heat-treated and, without any significant cold working, artificially aged to achieve precipitation hardening. The results of the friction tests will be used by the authors for the technological design of the forming process of automotive parts made of 6082-T6 aluminium alloy.

6082-T6 aluminium alloy is a wrought aluminium–magnesium–silicon family medium strength, weldable alloy with excellent corrosion resistance. The strength of this kind of this grade of aluminium alloy is the highest among all the alloys of the 6xxx series. 6082-T6 alloy is used for highly stressed applications in transport and marine frames.

The basic mechanical parameters of the tested sheets (Table 1) were determined in a uniaxial tensile test according to the ISO 6892-1:2009 [39] on specimens that were cut transverse to the rolling direction (90°), along a rolling direction (0°), and at an angle of 45° according to the rolling direction of the sheet metal. Three specimens were tested for each direction and average values of mechanical parameters were determined. Engineering stress–strain curves for the characteristic sample directions are shown in Figure 1.

Table 1. Mechanical parameters of the 6082-T6 aluminium alloy sheets.

Sample Direction	$R_{p0.2}$, MPa	R_m, MPa	A_{80}, %	n	r	Δr
0°	314	342	13.7	0.087	0.528	
45°	307	337	14.2	0.086	0.657	−0.139
90°	313	341	12.0	0.086	0.509	

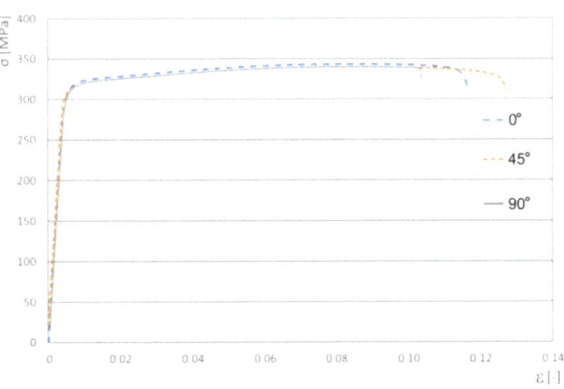

Figure 1. Engineering stress–strain curves for the EN AW-6082-T6 aluminium alloy sheets.

2.2. Friction Testing Procedure

Friction tests were carried out using the device (Figure 2) mounted on a Zwick/Roell Z100 testing machine. The test involves pulling a sample in the form of a strip of sheet metal with a width of w = 18 mm and a length of l = 240 mm clamped between cylindrical counter samples. Sliding speeds were 10 mm/min and 20 mm/min.

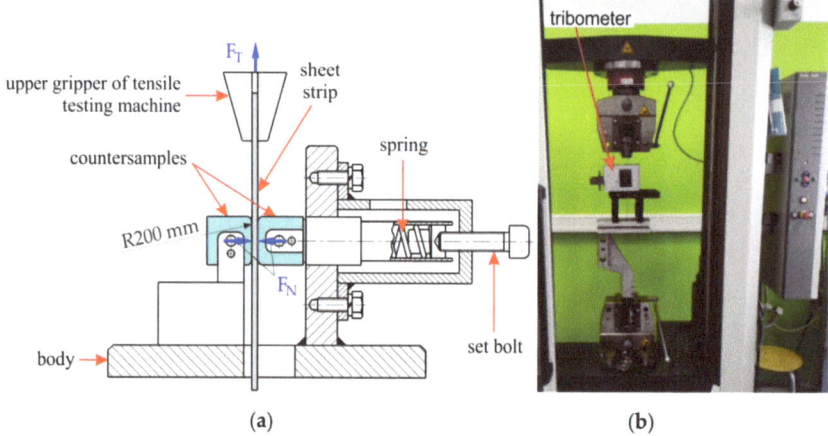

Figure 2. (a) Scheme and (b) view of tribological simulator.

The friction force F_T is measured using the measuring system of the testing machine. During the test, the counterexamples were pressed against the strip sheet with the force F_N, using a spring with a known deflection–axial force characteristic. Based on the values of the forces F_T and F_N, the value of the kinematic coefficient of friction is determined, according to the relationship:

$$\mu = \frac{F_T}{2F_N} \tag{1}$$

Counter samples made of 145Cr6 cold-work steel were used in the tests. To determine the mean contact pressure in the strip drawing test, the formulae proposed by Haar [40] (Equation (2)) was used, which was based on the width of the sample w, the contact force F_N, the radius of the counter samples R = 200 mm, and the elastic properties of the sheet and counter-sample materials allow for determination of mean contact pressure p_{av}.

$$P_{av} = \frac{\pi}{4} \cdot \sqrt{\frac{\frac{F_N}{w} \cdot \frac{2E_1 E_2}{E_2 \cdot (1-v_1^2) + E_1 \cdot (1-v_2^2)}}{2\pi R}} \tag{2}$$

For the material of the steel counter sample, the following values of Young's moduli E_1 and Poisson's ratio v_1 as $E_1 = 2 \cdot 10^5$ GPa [41], $v_1 = 0.3$ [42] were adopted. The values of the same material parameters for the sample material were assumed as follows: $E_2 = 69{,}000$ MPa [42], and $v_2 = 0.33$ [42].

The values of the applied pressure forces were between $p_{av} = 4.38$ MPa and 11.69 MPa, which, according to the literature [43–46], corresponds with the values of pressures occurring in the sheet metal forming operations.

All samples were degreased with acetone before the friction process. The friction tests were carried out in conditions of dry friction and sheet metal surface lubrication, with oils typically used in sheet metal forming. The basic criterion for the selection of lubricants was a wide range of viscosity η variability, which is the basic property of oils used in metal forming. In this way, three oils were selected:

- hydraulic oil LHL 32 (η = 21.9 mm²/s),
- machine oil LAN 46 (η = 43.9 mm²/s),
- and engine oil SAE 5W-40 C3 (η = 81 mm²/s).

The lubricant was applied directly to the surface of the samples.

Experimental design of the friction tests is shown in Table 2. Three specimens were tested for each friction test and average values of the COFs have been determined. Some force vs. time plots are shown in Figure 3.

Table 2. Experiment design.

Test No.	Sliding Speed, mm/min	Pressure, MPa	Friction Conditions
1	10	4.38	dry friction
2	10	6.53	dry friction
3	10	8.13	dry friction
4	10	9.47	dry friction
5	10	10.63	dry friction
6	10	11.69	dry friction
7	20	4.38	dry friction
8	20	6.53	dry friction
9	20	8.13	dry friction
10	20	9.47	dry friction
11	20	10.63	dry friction
12	20	11.69	dry friction
13	10	4.38	hydraulic oil LHL 32
14	10	6.53	hydraulic oil LHL 32
15	10	8.13	hydraulic oil LHL 32
16	10	9.47	hydraulic oil LHL 32
17	10	10.63	hydraulic oil LHL 32
18	10	11.69	hydraulic oil LHL 32
19	20	4.38	hydraulic oil LHL 32
20	20	6.53	hydraulic oil LHL 32
21	20	8.13	hydraulic oil LHL 32
22	20	9.47	hydraulic oil LHL 32
23	20	10.63	hydraulic oil LHL 32
24	20	11.69	hydraulic oil LHL 32
25	10	4.38	machine oil LAN 46
26	10	6.53	machine oil LAN 46
27	10	8.13	machine oil LAN 46
28	10	9.47	machine oil LAN 46
29	10	10.63	machine oil LAN 46
30	10	11.69	machine oil LAN 46
31	20	4.38	machine oil LAN 46
32	20	6.53	machine oil LAN 46
33	20	8.13	machine oil LAN 46
34	20	9.47	machine oil LAN 46
35	20	10.63	machine oil LAN 46
36	20	11.69	machine oil LAN 46
37	10	4.38	engine oil SAE 5W-40 C3
38	10	6.53	engine oil SAE 5W-40 C3
39	10	8.13	engine oil SAE 5W-40 C3
40	10	9.47	engine oil SAE 5W-40 C3
41	10	10.63	engine oil SAE 5W-40 C3
42	10	11.69	engine oil SAE 5W-40 C3
43	20	4.38	engine oil SAE 5W-40 C3
44	20	6.53	engine oil SAE 5W-40 C3
45	20	8.13	engine oil SAE 5W-40 C3
46	20	9.47	engine oil SAE 5W-40 C3
47	20	10.63	engine oil SAE 5W-40 C3
48	20	11.69	engine oil SAE 5W-40 C3

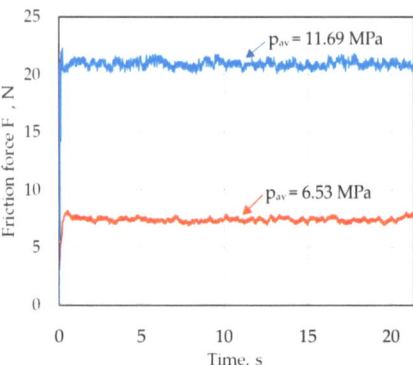

Figure 3. Example of force vs. time plots obtained in a friction test (dry friction conditions).

A T8000-RC surface measuring station was used to characterise the surface roughness of the sheet metals in their as-received state. The surface topography and basic 3D roughness parameters of sheet metal and counterexamples are shown in Figures 3 and 4, respectively.

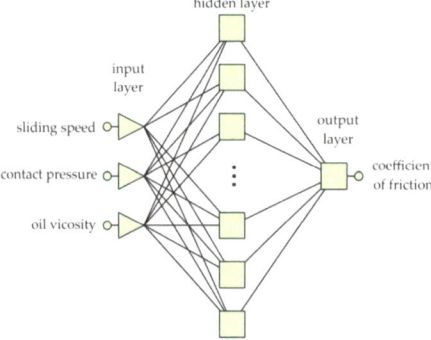

Figure 4. Structure of multilayer perceptron (MLP).

2.3. Artificial Neural Networks

Designing a neural network using Statistica software includes several stages. First, a set of training data consisting of the input parameters and the corresponding value of the output parameter should be selected. Sliding speed, average unit pressure, and lubricant viscosity were selected as input parameters. The output parameter was the value of the coefficient of friction. The values of these parameters have been normalised to the range −1 to 1 [47], using min-max normalisation, which transforms the input data from the range (min, max) to the new range (N_{min}, N_{max}).

Many experiments were then carried out with selected multilayer network structures (Figure 4) in order to obtain the network with the smallest error for the validation set. The training of the network was carried out using the backpropagation algorithm which is commonly used to train multilayer neural networks. From the entire training data set, 15% of the data were selected and assigned to the validation set. The rest of the data comprised the training set. In general, the validation set should consist of between 10 and 20% of the training data. We have assumed that 15% of the data are summed to the validation set. The data of the validation set were used for independent convergence control of the training algorithm.

The quality of the tested neural networks was assessed on the basis of the root mean square (RMS) error.

The network with the smallest RMS error value for the validation set was used for data analysis. The selection of variables affecting the value of the coefficient of friction is difficult due to the synergistic interactions of many parameters often correlated with each other. The sensitivity analysis of the input variables showed a significant impact of all assumed input parameters on the value of the coefficient of friction.

3. Results and Discussion

3.1. Coefficient of Friction

The lowest value of the COF at room temperature (24 °C) was recorded when lubricating the sheet metal surface with SAE 5W40 C3 engine oil. In the whole range of applied contact pressures, the value of the COF during lubrication with this oil was between 0.15 (±0.006) and 0.18 (±0.008) for a sliding speed of 10 mm/min (Figure 5a) and between 0.14 (±0.0078) and 0.17 (±0.0083) for a sliding speed of 20 mm/min (Figure 5b). When lubricated with LAN 46 and SAE oils, an increase in the sliding speed from 10 to 20 mm/min resulted in a reduction of the COF by a maximum of 0.02. Based on the slope of the trend line of changes in the COF with the value of pressure, these lubricants are seen to provide lubrication to a similar degree for each pressure. After exceeding a pressure of 10 MPa, LAN 46 and LHL 32 oils clearly lose their lubricating properties. Under certain conditions, depending on the roughness of the cooperating bodies and the pressure value, the lubricating film breaks. It relates to an intensification of the mechanical cooperation of the roughness summits of the tool and the sheet metal. The results of friction tests of AW-5052 aluminium alloy sheets carried out by Dou and Xia [48] also showed that the values of the COF between the sheet metal and the die generally decrease with increasing sliding speed and normal loads, and the downward trend is slowed down with higher sliding speeds and contact pressure.

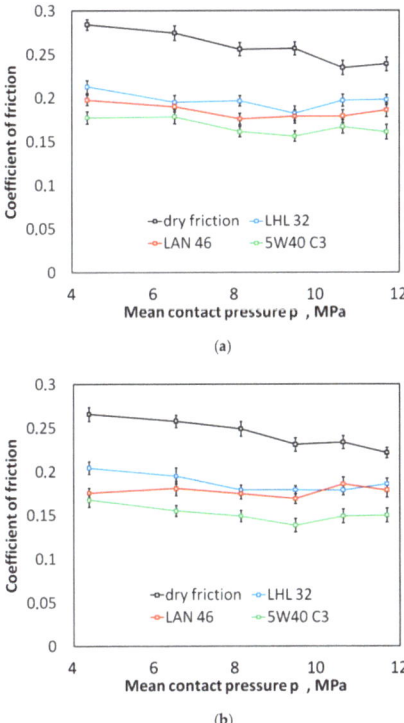

Figure 5. The effect of contact pressure on the coefficient of friction determined at sliding speed (**a**) 10 mm/min and (**b**) 20 mm/min.

To determine the effectiveness of the lubricant in reducing the value of the coefficient of friction, the coefficient of effectiveness of lubrication (EOL) was introduced:

$$\text{EOL} = 100\% - \frac{\mu_{\text{dry friction}} \cdot 100\%}{\mu_{\text{lubrication}}} \qquad (3)$$

Lubrication efficiency is the highest for oil with the highest viscosity 5W40 C3 (Figure 5a,b). For this oil, however, there is a tendency to initial stabilisation of lubrication efficiency at the pressures lower than 10 MPa, and then, after exceeding this pressure, the efficiency decreases. Despite the presence of increased lubricant pressure in the contact zone, the metallic contact is intensified, and the lubricant cannot separate rubbing surfaces very much under high contact pressure. The effectiveness of the other two oils (the LHL 32 oil provides the lowest lubrication efficiency) decreases practically over the entire range of applied contact pressures. In general, the value of the EOL coefficient for a specific pressure value increases with the increase in the sliding speed. The lubrication efficiency of LHL 32, LAN 46, and 5W40 C3 oils at a sliding speed of 10 mm/min (Figure 6a) ranges between 16.0 and 28.9%, 22.1 and 31.2%, and 28.9 and 39.1%, respectively. Meanwhile, at a sliding speed of 20 mm/min. (Figure 6b), the EOL values vary between 16.0 and 28.0%, 19.2 and 33.8%, and 32.3 and 40.2%, respectively. The increase in speed causes an increase in the temperature in the area of the summits of the surface asperities and thus an increase in their plastic properties [49]. The surface asperities are then more susceptible to plastic deformation at a lower value of pressure, causing faster flattening of the sheet surface (Figure 7) and reducing the volume of the lubricant pockets [50]. Liquid lubricants can reduce flattening and friction by filling the surface valleys and carry a substantial amount of the pressure [51].

Aluminium is easily oxidised in the air, so that in the initial period of friction, the oxide film easily separates the two surfaces of the material and there is a slight metallic contact transmitted through the surface asperities [52,53]. The oxide film has low shear strength and breaks quickly. After the load is applied, the top sheet layer cracks and the surfaces come into contact, which increases the bonding strength between the contacting surfaces [49]. Under lubricated conditions, with increasing contact pressure, the synergistic effect of the mechanisms of flattening and ploughing, and the action of the lubricant occurs. After a certain contact pressure is received, the increase in surface roughness and other parameters may reach a steady state value. Therefore, the values of the COF remain constant for the increased contact pressures [49]. This is visible in Figure 5 after exceeding the normal pressure value of 9 MPa.

Figures 7–9 show selected surface morphologies of the surface of sheets tested at a sliding speed of 10 mm/min. The dominant tribological phenomenon occurring during the friction was a flattening of the surface asperities as evidenced by SEM micrographs (Figures 7–9). The phenomenon of flattening the surface asperities occurs both in the conditions of dry friction (Figure 7) and lubrication of sheet surface (Figures 8 and 9). The flattened surfaces are separated by groves, which are remnants of the as-received surface. In the range of the analysed pressures, grooves could be observed, which also occur on the sheet surface in its as-received state. Many cracks on the surface of the sheet in its as-received state (Figure 10) have been seized (Figures 7–9). In general, similar observations can be applied to sheet metals tested at a speed of 20 mm/min (Figure 11).

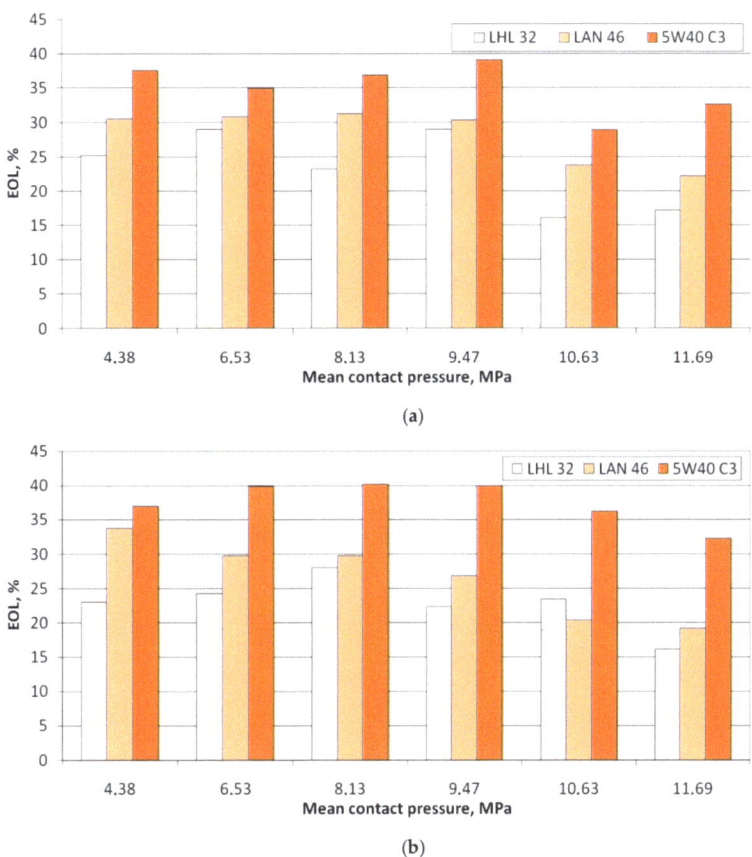

Figure 6. The effect of contact pressure on the EOL for sliding speed. (**a**) 10 mm/min and (**b**) 20 mm/min.

Lubrication is a basic and effective way to reduce the coefficient of friction and wear of mating surfaces. However, it is well-known that iron oxides, due to their low shear stress, act as a solid lubricant and reduce the coefficient of friction [54,55]. The development of smooth oxide 'glazes' consists of fine, crystalline oxide particles on the load-bearing areas of alloys, which can lead to a significant reduction in COF [56]. Writzl et al. [57] detected and characterized the surface oxides using confocal Raman microscopy (power 15 mW, wavelength 633 nm). The outermost layer was composed of iron oxide phases, underlain by a compound layer which consisted of oxides, nitrides, and a high-resistance martensitic layer, with accompanying carbides and nitrides provided high load capacity for the external layers, allowing them to function as solid lubricants [57]. The presence of the oxide phases is fundamental in obtaining the low COF [58]. Wang et al. [59] attributed a lubricated effect during sliding to the formation of surface oxide film containing iron oxides. As also found Brunetti et al. [55], the maintenance of the oxide layer during a sliding wear process reduces the wear rate of the system considerably.

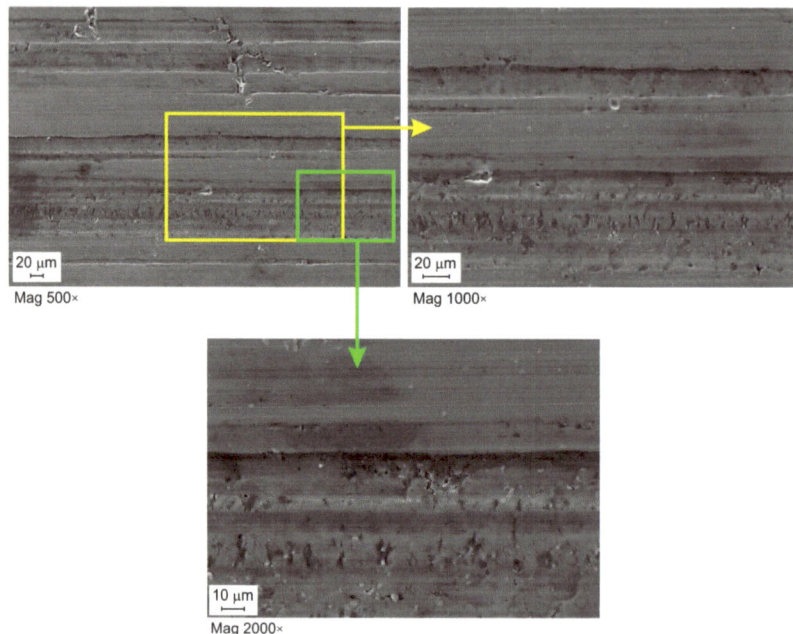

Figure 7. SEM micrographs of the sheet surface tested under dry friction conditions at contact pressure 8.13 MPa and a sliding speed of 10 mm/min.

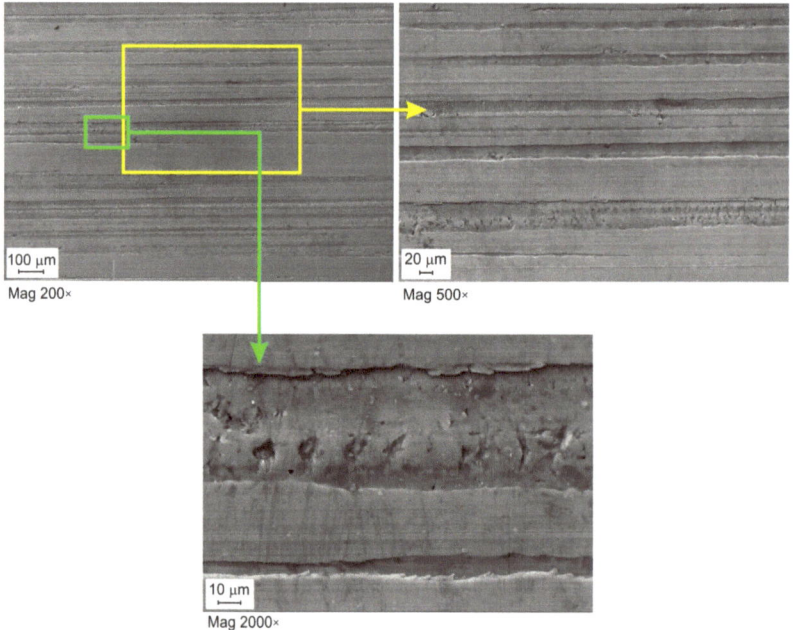

Figure 8. SEM micrographs of the sheet surface tested under lubrication with LHL 32 oil at contact pressure 9.47 MPa and sliding speed 10 mm/min.

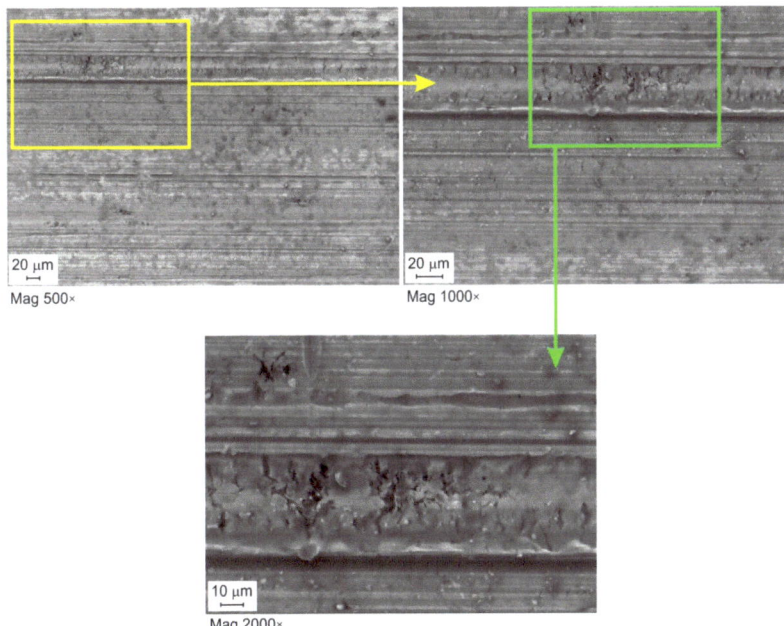

Figure 9. SEM micrographs of the sheet surface tested under lubrication with 5W40 C3 oil at contact pressure 6.53 MPa and sliding speed 10 mm/min.

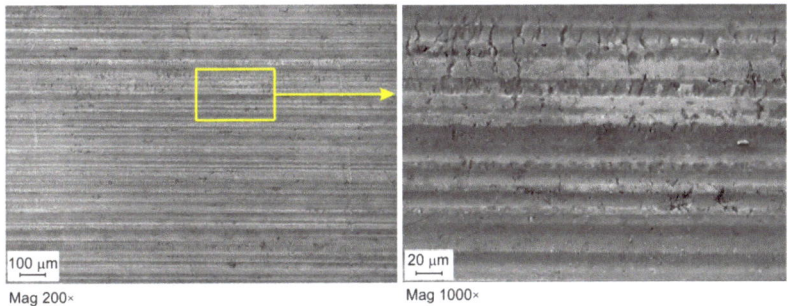

Figure 10. SEM micrograph of the as-received AW-6068-T6 sheet metal.

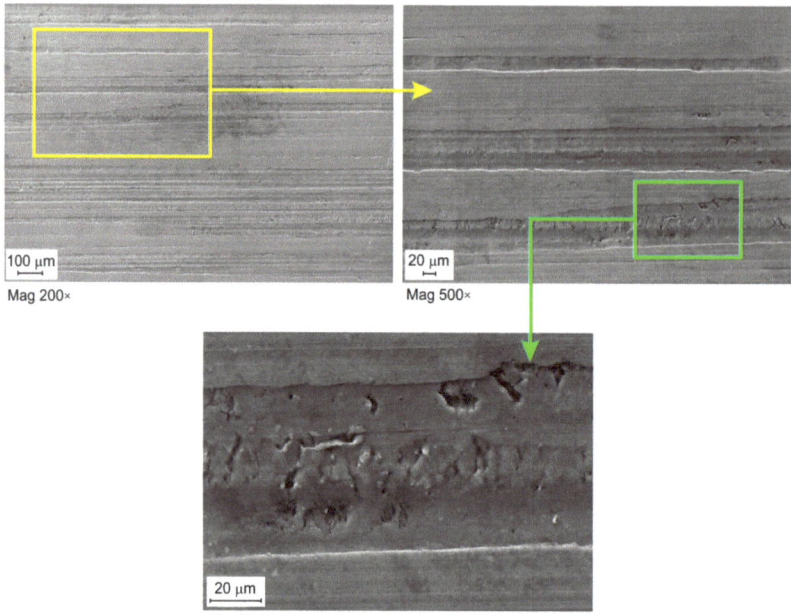

Figure 11. SEM micrographs of the sheet surface tested under lubrication with 5W40 C3 oil at contact pressure 10.63 MPa and sliding speed 20 mm/min.

3.2. ANN Analysis

The smallest value of the RMS error for the validation set and, at the same time, the largest value of Pearson's correlation coefficient R^2 were provided by a network with one hidden layer and nine neurons in the hidden layer (MLP 3-3:9:1-1). The most important regression statistics are presented in Table 3. The correctness of the training process is determined by the similarly high correlation value $R^2 > 0.98$ for both data sets. The prognostic quality of the neural network is determined by the quotient of the standard deviation of errors and the standard deviation of the value of the dependent variable (SD Ratio):

$$\text{SDRatio} = \frac{\text{Error SD}}{\text{Data SD}} \qquad (4)$$

Table 3. Regression statistics for MLP 3-3:9:1-1.

Parameter	Training Set	Validation Set
Data Mean	0.1974275	0.1870625
Data SD	0.03705	0.03334
Error Mean	8.89×10^{-5}	−0.0006155
Error SD	0.005955	0.005251
Abs E. Mean	0.005177	0.003623
SD Ratio	0.1607267	0.157522
Correlation	0.9869992	0.9876078

For networks with very good forecasting capabilities, the SD ratio value should be less than 0.2. An SD ratio value greater than 1 proves that a more accurate estimation of the value of the dependent variable is its arithmetic mean determined based on the training set [60].

The purpose of the sensitivity analysis was to examine the impact of removing individual explanatory variables on the total error of the network. A sensitivity analysis was performed independently for the training set (Table 4) and the validation set (Table 5). The 'Rank' row in these tables lists the variables in order of importance. Oil viscosity has the greatest impact on the value of the coefficient of friction, followed by contact pressure and sliding speed. The error of the network after removing the specified variable from the dataset is given in the 'Error' row. A given parameter is more important the greater the increase in the error value caused by its removal. The 'Ratio' parameter is responsible for the quotient of the error obtained after removing the selected explanatory variable and the error obtained using the network containing all explanatory variables [61].

Table 4. Sensitivity analysis for the training set.

Parameter	Sliding Speed	Oil Viscosity	Contact Pressure
Rank	3	1	2
Error	0.008	0.037	0.011
Ratio	1.372	6.363	1.978

Table 5. Sensitivity analysis for the validation set.

Parameter	Sliding Speed	Oil Viscosity	Contact Pressure
Rank	3	1	2
Error	0.009	0.028	0.010
Ratio	1.902	5.800	2.032

According to the response surfaces, the sliding speed has little effect on the change in the coefficient of friction (Figure 12b,c). As the sliding speed increases, the value of the friction coefficient decreases. Similar results were obtained by Tamai et al. [62] and Wang et al. [63], who studied the influence of the sliding speed on the coefficient of friction of galvanised steel sheets. As the sliding speed increases, the real contact area decreases, limiting the mechanical interaction of the surfaces in contact. The value of the coefficient of friction under lubrication conditions consists of two factors, the coefficient of friction of the solid, and the coefficient of friction of the lubricant [64]. Increasing the sliding speed increases the thermal effect at the summits of the asperities, causing a decrease in the viscosity of the lubricant and thus a decrease in the lubricant's coefficient of friction. As a result, as the sliding speed increases, the coefficient of friction decreases. This effect has also been observed for dry friction conditions. At low sliding speeds, the contact surface is dominated by the elastic contact of the surface asperities, favouring the formation of the phenomenon of fluctuated slip, called the stick-slip phenomenon. A clear tendency of the COF to decrease with increased oil viscosity was observed (Figure 12a,c).

With the increase in the sliding speed, the fluctuations between the slip and friction on the contact surface decrease, with the stabilisation of conditions at higher speeds favouring a reduction in the value of the COF [65]. Increasing the viscosity of the oil reduces the COF (Figure 12a). Viscosity affects the value of the COF in combination with contact pressure. At low pressures, the high-viscosity oil is able to separate or reduce the metallic contact between the tool surface and the sheet metal. In the high-pressure range in the sheet metal formation of the relatively soft workpiece material and the hard tool, metallic contact is unavoidable, but the hydrostatic pressure of the oil is created in the closed lubricant pockets, which acts as a lubricating cushion, separating the rubbing surfaces. Therefore, a loss of lubricating properties of oils occurs faster at higher pressures.

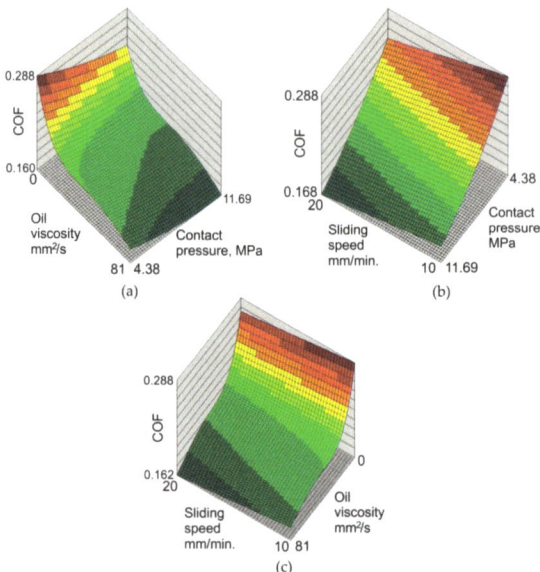

Figure 12. Response surfaces for the effect of (**a**) oil viscosity and contact pressure, (**b**) sliding speed and contact pressure, and (**c**) sliding speed and oil viscosity on the value of the coefficient of friction.

4. Conclusions

This paper presents the friction test results of as-received AW-6082-T6 aluminium alloy sheets in strip draw tribological tests. The tests were carried out for lubricants commonly used in sheet metal forming with viscosity varying between 21.9 and 81 mm^2/s. The synergistic effect of input parameters (oil viscosity, contact pressure, and sliding speed) on the COF was analysed using multilayer perceptron. Based on the results of experimental investigations, including analyses using ANNs and SEM observations, the following conclusions can be drawn:

- The lowest value of the coefficient of friction was recorded when lubricating the sheet metal surface with SAE 5W40 C3 engine oil, which is characterised as the most viscous of all tested lubricants.
- In general, the coefficient of EOL for both analysed sliding speeds was similar and varied between 16.0 and 28.9%, 19.2% and 33.8%, and 28.9 and 40.2% for LHL 32 hydraulic oil, LAN 46 machine oil, and 5W40 C3 engine oil, respectively.
- When considering the effect of contact pressure on the lubrication efficiency, it can be said that at high pressures the lubrication efficiency is lower than at low pressures. This is due to the intensification of surface flattening in high contact pressures and the reduced volume of the valleys (also known as lubricant pockets) on the sheet surface that can hold the lubricant.
- In dry friction conditions, a decreasing trend of the coefficient of friction with increasing contact pressure was observed.
- In lubricated conditions, the value of the coefficient of friction was more stable in the range of contact pressures between 4 and 9 MPa; after exceeding this value, the lubricant film was broken and a slight increase in coefficient of friction was observed.
- SEM micrographs revealed that the main friction mechanism of tested sheets in contact with the surface of the cold-work tool steel is the flattening of surface asperities.
- The trained model of MLP was characterised by high Pearson correlation ($R^2 > 0.98$) for both training and validation sets. The sensitivity analysis of the input parameters on the value of coefficient of friction revealed that oil viscosity has the greatest impact

on the value of the coefficient of friction, followed by contact pressure and sliding speed.

Author Contributions: Conceptualization, T.T.; methodology, T.T.; investigation, T.T., J.S. and M.V.; validation, T.T., J.S., Ľ.K. and I.G.; formal analysis, T.T., J.S., Ľ.K., I.G. and M.V.; resources, J.S., Ľ.K., I.G. and M.V.; data curation, T.T., J.S., Ľ.K. and I.G.; writing—original draft preparation, T.T. and J.S.; writing—review and editing, T.T. and J.S.; project administration, J.S.; funding acquisition, J.S. All authors have read and agreed to the published version of the manuscript.

Funding: This work was supported by Slovak Research and Development Agency and Polish National Agency for Academic Exchange, project title: Research into innovative forming and joining methods of thin-walled components", project numbers: SK-PL-21-0033 and BPN/BSK/2021/1/00067/U/00001. The authors are also grateful for the support in the experimental work to the Grant Agency of the Ministry of Education, Science, Research, and Sport of the Slovak Republic (grant number VEGA 1/0539/23).

Institutional Review Board Statement: Not applicable.

Informed Consent Statement: Not applicable.

Data Availability Statement: Data is contained within the article.

Conflicts of Interest: The authors declare no conflict of interest.

References

1. Huber, G.; Djurdjevic, M.B.; Manasijevic, S. Determination some thermo-physical and metallurgical properties of aluminum alloys using their known chemical composition. *Int. J. Heat Mass Transf.* **2019**, *139*, 548–553. [CrossRef]
2. Hattori, C.S.; Almeida, G.F.C.; Gonçalves, R.L.C.; Santos, R.G.; Souza, R.C.; da Silva, W.C.; Cunali, J.R.C.; Couto, A.A. Microstructure and fatigue properties of extruded aluminum alloys 7046 and 7108 for automotive applications. *J. Mater. Res. Technol.* **2021**, *14*, 2970–2981. [CrossRef]
3. Puga, H. Casting and forming of advanced aluminum alloys. *Metals* **2020**, *10*, 494. [CrossRef]
4. Xiang, P.; Jia, L.J.; Shi, M.; Wu, M. Ultra-low cycle fatigue life of aluminum alloy and its prediction using monotonic tension test results. *Eng. Fract. Mech.* **2017**, *186*, 449–465. [CrossRef]
5. Shaha, S.K.; Czerwinski, F.; Kasprzak, W.; Friedman, J.; Chen, D.L. Improving high-temperature tensile and low-cycle fatigue behavior of Al-Si-Cu-Mg alloys through micro-additions of Ti, V, and Zr. *Metall. Mater. Trans. A* **2015**, *46*, 3063–3078. [CrossRef]
6. Rao, K.P.; Wei, J.J. Performance of a new dry lubricant in the forming of aluminum alloy sheets. *Wear* **2001**, *249*, 85–92. [CrossRef]
7. Yang, T.S. Prediction of surface topography in lubricated sheet metal forming. *Int. J. Mach. Tools Manuf.* **2008**, *48*, 768–777. [CrossRef]
8. Bay, N.; Olsson, D.D.; Andreasen, J.L. Lubricant test methods for sheet metal forming. *Tribol. Int.* **2008**, *41*, 844–853. [CrossRef]
9. Meiler, M.; Pfestorf, M.; Geiger, M.; Merklein, M. The use of dry film lubricants in aluminum sheet metal forming. *Wear* **2003**, *25*, 1455–1462. [CrossRef]
10. Skoblik, R.; Wilczewski, L. *Technologia Metali. Laboratorium*; Wydawnictwo Politechniki Gdańskiej: Gdańsk, Poland, 2006.
11. Gierzyńska, M. *Tarcie, Zużycie i Smarowanie w Obróbce Plastycznej Metali*; Wydawnictwa Naukowo-Techniczne: Warsaw, Poland, 1983.
12. Podulka, P. Resolving Selected Problems in Surface Topography Analysis by Application of the Autocorrelation Function. *Coatings* **2023**, *13*, 74. [CrossRef]
13. Podulka, P. Selection of Methods of Surface Texture Characterisation for Reduction of the Frequency-Based Errors in the Measurement and Data Analysis Processes. *Sensors* **2022**, *22*, 791. [CrossRef]
14. Hol, J.; Meinders, V.T.; de Rooij, M.B.; van den Boogaard, A.H. Multi-scale friction modeling for sheet metal forming: The boundary lubrication regime. *Tibology Int.* **2015**, *81*, 112–128. [CrossRef]
15. Nyholm, N.; Espallargas, N. Functionalized carbon nanostructures as lubricant additives–A review. *Carbon* **2023**, *201*, 1200–1228. [CrossRef]
16. Tonk, R. The science and technology of using nano-materials in engine oil as a lubricant additives. *Mater. Today Proc.* **2021**, *37*, 3475–3479. [CrossRef]
17. Podstawy Techniki Smarowniczej. Available online: https://totalenergies.pl/system/files/atoms/files/rozdzial_02_podstawy_techniki_smarowniczej.pdf (accessed on 20 December 2022).
18. Płaza, S.; Margilewski, L.; Celichowski, G. *Wstęp do Tribologii i Tribochemii*; Wydawnictwo Uniwersytetu Łódzkiego: Łódź, Poland, 2005.
19. Szewczyk, M.; Szwajka, K. Assessment of the Tribological Performance of Bio-Based Lubricants Using Analysis of Variance. *Adv. Mech. Mater. Eng.* **2023**, *40*, 31–38. [CrossRef]

20. Szewczyk, M.; Szwajka, K.; Trzepieciński, T. Frictional Characteristics of Deep-Drawing Quality Steel Sheets in the Flat Die Strip Drawing Test. *Materials* **2022**, *15*, 5236. [CrossRef]
21. Fejkiel, R.; Goleń, P. Application of the Finite Element Method to Simulate the Friction Phenomenon in a Strip Drawing Test. *Adv. Mech. Mater. Eng.* **2023**, *40*, 39–46. [CrossRef]
22. Nielsen, C.V.; Legarth, B.N.; Niordson, C.F.; Bay, N. A correction to the analysis of bending under tension tests. *Tribol. Int.* **2022**, *173*, 107625. [CrossRef]
23. Folle, L.F.; Schaeffer, L. Evaluation of Contact Pressure in Bending under Tension Test by a Pressure Sensitive Film. *J. Surf. Eng. Mater. Adv. Technol.* **2016**, *6*, 201–214. [CrossRef]
24. Trzepiecinski, T.; Kubit, A.; Slota, J.; Fejkiel, R. An Experimental Study of the Frictional Properties of Steel Sheets Using the Drawbead Simulator Test. *Materials* **2019**, *12*, 4037. [CrossRef]
25. Gil, I.; Mendiguren, J.; Galdos, L.; Mugarra, E.; de Andragoña, E.S. New drawbead tester and numerical analysis of drawbead closure force. *Int. J. Adv. Manuf. Technol.* **2021**, *116*, 1855–1869. [CrossRef]
26. Xi, Y.; Björling, M.; Shi, Y.; Mao, J.; Larsson, R. Application of an inclined, spinning ball-on-rotating disc apparatus to simulate railway wheel and rail contact problems. *Wear* **2017**, *374*, 46–53. [CrossRef]
27. Pürçek, G.; Savaşkan, T.; Küçükomeroğlu, T.; Murphy, S. Dry sliding friction and wear properties of zinc-based alloys. *Wear* **2002**, *252*, 894–901. [CrossRef]
28. Venci, A.; Mrdak, M.; Cvijović, I. Microstructures and tribological properties of ferrous coatings deposited by APS (Atmospheric Plasma Spraying) on Al-alloy substrate. *FME Trans.* **2006**, *34*, 151–157.
29. Salguero, J.; Vazquez-Martinez, J.M.; Sol, I.D.; Batista, M. Application of Pin-On-Disc Techniques for the Study of Tribological Interferences in the Dry Machining of A92024-T3 (Al–Cu) Alloys. *Materials* **2018**, *11*, 1236. [CrossRef]
30. Meyveci, A.; Karacan, I.; Çalıgülü, U.; Durmuş, H. Pin-on-disc characterization of 2xxx and 6xxx aluminium alloys aged by precipitation age hardening. *J. Alloys Compd.* **2010**, *491*, 278–283. [CrossRef]
31. Yahaya, A.; Samion, S. Friction condition of aluminum alloy AA6061 lubricated with bio-lubricant in cold forging test. *Ind. Lubr. Tribol.* **2022**, *74*, 378–384. [CrossRef]
32. Trzepieciński, T. Experimental Analysis of Frictional Performance of EN AW-2024-T3 Alclad Aluminium Alloy Sheet Metals in Sheet Metal Forming. *Lubricants* **2023**, *11*, 28. [CrossRef]
33. Guezmil, M.; Bensalah, W.; Khalladi, A.; Elleuch, K.; de-Petris Wery, M.; Ayedi, H.F. Effect of Test Parameters on the Friction Behaviour of Anodized Aluminium Alloy. *Int. Sch. Res. Not.* **2014**, *204*, 795745. [CrossRef]
34. Yang, X.; Zhang, Q.; Zheng, Y.; Liu, X.; Politis, D.; El Fakir, O.; Wang, L. Investigation of the friction coefficient evolution and lubricant breakdown behaviour of AA7075 aluminium alloy forming processes at elevated temperatures. *Int. J. Extrem. Manuf.* **2021**, *3*, 25002. [CrossRef]
35. Das, V.V. Tribological Studies on Aluminium Alloys. Bachelor's Thesis, National Institute of Technology, Rourkela, India, 5 May 2011.
36. Shabel, B.S.; Granger, D.A.; Truckner, W.G. Friction and Wear of Aluminum-Silicon Alloys. In *ASM Handbook: Friction, Lubrication, and Wear Technology*; Blau, P.J., Ed.; ASM International: Materials Park, OH, USA, 1992; pp. 785–794.
37. Luis Pérez, C.J.; Luri Irigoyen, R.; Puertas Arbizu, I.; Salcedo Pérez, D.; León Iriarte, J.; Fuertes Bonel, J.P. Analysis of Tribological Properties in Disks of AA-5754 and AA-5083 Aluminium Alloys Previously Processed by Equal Channel Angular Pressing and Isothermally Forged. *Metals* **2020**, *10*, 938. [CrossRef]
38. Li, J.; Wongsa-Ngam, J.; Xu, J.; Shan, D.; Guo, B.; Langdon, T.G. Wear resistance of an ultrafine-grained Cu-Zr alloy processed by equal-channel angular pressing. *Wear* **2015**, *326*, 10–19. [CrossRef]
39. ISO 6892-1; Metallic Materials—Tensile Testing—Part 1: Method of Test at Room Temperature. International Organization for Standardization: Geneva, Switzerland, 2009.
40. Ter Haar, R. Friction in Sheet Metal Forming, the Influence of (Local) Contact Conditions and Deformation. Ph.D. Thesis, Universiteit Twente, Enschede, The Netherlands, 17 May 1996.
41. Graba, M. Characteristics of selected measures of stress triaxiality near the crack tip for 145Cr6 steel-3D issues for stationary cracks. *Open Eng.* **2020**, *10*, 571–585. [CrossRef]
42. 6082-T6 Aluminum. Available online: https://www.makeitfrom.com/material-properties/6082-T6-Aluminum (accessed on 10 January 2023).
43. Kirkhorn, L.; Frogner, K.; Andersson, M.; Ståhl, J.E. Improved tribotesting for sheet metal forming. *Procedia CIRP* **2012**, *3*, 507–512. [CrossRef]
44. Cillaurren, J.; Galdos, L.; Sanchez, M.; Zabala, A.; de Argandoña, S.; Mendiguren, J. Contact pressure and sliding velocity ranges in sheet metal forming simulations. In Proceedings of the 24th International Conference on Material Forming ESAFORM 2021, Liège, Belgium, 14–16 April 2021.
45. Recklin, V.; Dietrich, F.; Groche, P. Influence of Test Stand and Contact Size Sensitivity on the Friction Coefficient in Sheet Metal Forming. *Lubricants* **2018**, *6*, 41. [CrossRef]
46. Vollertsen, F.; Hu, Z. Tribological Size Effects in Sheet Metal Forming Measured by a Strip Drawing Test. *Ann. CIRP* **2006**, *55*, 291–294. [CrossRef]
47. Kołodziejczyk, J. Podstawy sztucznej inteligencji-Sztuczne sieci neuronowe (SSN). Available online: http://wikizmsi.zut.edu.pl/uploads/d/d7/PSI_ZIP_S1_W6.pdf (accessed on 16 December 2022).

48. Dou, S.; Xia, J. Analysis of Sheet Metal Forming (Stamping Process): A Study of the Variable Friction Coefficient on 5052 Aluminum Alloy. *Metals* **2019**, *9*, 853. [CrossRef]
49. Nuruzzaman, D.M.; Chowdhury, M.A. Effect of Load and Sliding Velocity on Friction Coefficient of Aluminum Sliding Against Different Pin Materials. *Am. J. Mater. Sci.* **2012**, *2*, 26–31. [CrossRef]
50. Bech, J.; Bay, N.; Eriksen, M. Astudy of mechanisms of liquid lubrication in metal forming. *CIRP Ann.* **1998**, *47*, 221–226. [CrossRef]
51. Zwicker, M.; Spangenberg, J.; Martins, P.; Nielsen, C.V. Investigation of material strength and oil compressibility on the hydrostatic pressure build-up in metal forming lubricants. *Procedia CIRP* **2022**, *115*, 78–82. [CrossRef]
52. ASM International. *ASM Handbook, Volume 5, Surface Engineering*; ASM International: Materials Park, OH, USA, 1994.
53. Goryacheva, I.G. *Contact Mechanics in Tribology*; Kluwer Academic Publishers: Dordrecht, The Netherlands, 1998.
54. Brunetti, C.; Leite, M.V.; Pintaude, G. Effect of specimen preparation on contact fatigue wear resistance of austempered ductile cast iron. *Wear* **2007**, *263*, 663–668. [CrossRef]
55. Brunetti, C.; Belotti, L.P.; Miyoshi, M.H.; Pintaúde, G.; D'Oliveira, A.S.C.M. Influence of Fe on the room and high-temperature sliding wear of NiAl coatings. *Surf. Coat. Technol.* **2014**, *258*, 160–167. [CrossRef]
56. Stott, F.H.; Wood, G.C. The influence of oxides on the friction and wear of alloys. *Tribol. Int.* **1978**, *11*, 211–218. [CrossRef]
57. Writzl, V.; Rovani, A.C.; Pintaude, G.; Lima, M.S.F.; Guesser, W.L.; Borges, P.C. Scratch resistances of compacted graphite iron with plasma nitriding, laser hardening, and duplex surface treatments. *Tribol. Int.* **2020**, *143*, 106081. [CrossRef]
58. Saeidi, F.; Taylor, A.A.; Meylan, B.; Hoffmann, P.; Wasmer, K. Origin of scuffing in grey cast iron-steel tribo-system. *Mater. Des.* **2017**, *116*, 622–630. [CrossRef]
59. Wang, Y.X.; Yan, M.F.; Li, B.; Guo, L.X.; Zhang, C.S.; Zhang, Y.X.; Bai, B.; Chen, L.; Long, Z.; Li, R.W. Surface properties of low alloy steel treated by plasma nitrocarburizing prior to laser quenching process. *Opt. Laser Technol.* **2015**, *67*, 57–64. [CrossRef]
60. Lula, P.; Tadeusiewicz, R. Statistica Neural Networks. In *User's Course in Examples*; Statsoft: Cracow, Poland, 2001.
61. Statsoft Polska. *Introduction to Neural Networks*; Statsoft Polska Sp. z o.o.: Cracow, Poland, 2001.
62. Tamai, Y.; Inazumi, T.; Manabe, K. FE forming analysis with nonlinear friction coefficient model considering contact pressure, sliding velocity and sliding length. *J. Mater. Process. Technol.* **2016**, *227*, 161–168. [CrossRef]
63. Wang, L.G.; Zhou, J.F.; Xu, Y.T. Study on friction characteristics of galvanized sheet by strip drawing test. *J. Plast. Eng.* **2016**, *2*, 87–91.
64. Wen, S.Z.; Huang, P. *Tribology Principle*; Tsinghua University Press: Beijing, China, 2002; pp. 35–67.
65. Ludema, K.C. *Friction, Wear, Lubrication. A textbook in Tribology*; CRC Press LLC: Boca Raton, FL, USA, 2000.

Disclaimer/Publisher's Note: The statements, opinions and data contained in all publications are solely those of the individual author(s) and contributor(s) and not of MDPI and/or the editor(s). MDPI and/or the editor(s) disclaim responsibility for any injury to people or property resulting from any ideas, methods, instructions or products referred to in the content.

Article

Cold Drawing of AISI 321 Stainless Steel Thin-Walled Seamless Tubes on a Floating Plug

Krzysztof Żaba [1,*] and Tomasz Trzepieciński [2,*]

[1] Department of Metal Working and Physical Metallurgy of Non-Ferrous Metals, Faculty of Non-Ferrous Metals, AGH—University of Science and Technology, al. Adama Mickiewicza 30, 30-059 Cracow, Poland
[2] Department of Manufacturing Processes and Production Engineering, Faculty of Mechanical Engineering and Aeronautics, Rzeszow University of Technology, al. Powst. Warszawy 8, 35-959 Rzeszów, Poland
* Correspondence: krzyzaba@agh.edu.pl (K.Ż.); tomtrz@prz.edu.pl (T.T.)

Citation: Żaba, K.; Trzepieciński, T. Cold Drawing of AISI 321 Stainless Steel Thin-Walled Seamless Tubes on a Floating Plug. *Materials* **2023**, *16*, 5684. https://doi.org/10.3390/ma16165684

Academic Editor: Francesco Iacoviello

Received: 2 July 2023
Revised: 14 August 2023
Accepted: 16 August 2023
Published: 18 August 2023

Copyright: © 2023 by the authors. Licensee MDPI, Basel, Switzerland. This article is an open access article distributed under the terms and conditions of the Creative Commons Attribution (CC BY) license (https:// creativecommons.org/licenses/by/ 4.0/).

Abstract: The paper presents the results of an analysis of the process of drawing AISI 321 stainless steel thin-walled seamless tubes on a floating plug. The influence of the geometry of dies and plugs, drawing velocity, and lubricants on the possibility of carrying out the pipe drawing process without a loss of strength of the lubricating film and, consequently, disturbance of the forming process and tube cracking, and also on the temperature in the drawing process, the mechanical properties of the tubes drawn, and the microhardness and roughness of the inner and outer surface of the tubes was investigated. The parameters of the drawing tools used were as follows: angle of drawing dies $\alpha = 16°$ and floating plugs with angles of inclination of the conical part of the plug $\beta = 11.5°$, $13°$, and $14°$. The drawing dies and floating plugs were made of G10 sintered carbide. Drawing speed was varied over the range 1 to 10 m/min. The study used several lubricants. Tubes with dimensions (outer diameter D_0, wall thickness g_0 before drawing process) $D_0 = 19$ mm, $g_0 = 1.2$ mm and $D_0 = 18$ mm, $g_0 = 1.2$ mm were drawn to produce tubes with dimensions (outer diameter D_k, wall thickness g_k after drawing process) $D_k = 16$ mm, $g_k = 1.06$ mm on a drawbench with the same total elongation, while the diameter and wall thickness were changed. During the process, continuous measurements were made of the drawing force and temperature in the deformation zone and on the tube surface. It was found that the drawing process causes a decrease in the roughness parameters Ra and Rz of the inner surface of the tubes. Moreover, after drawing, an increase of 30–70% was observed in the microhardness of the tube material in relation to the microhardness of the charge material. Based on the test results, it can be concluded that the work of frictional forces is the main direction of optimization of tube drawing on a floating plug process of hard-deforming materials.

Keywords: cold tube drawing; floating plug; AISI 321 stainless steel; lubricants; mechanical properties; microhardness

1. Introduction

Tube drawing using a floating plug is an important and widely-used technology. A characteristic feature of the process is high intensity heat emission in the deformation zone, especially during drawing hard-deforming materials with high yield strength and work hardening [1,2]. Therefore, the potential of the process depends on operations to minimise the amount of heat emission while improving the quality of the lubricant, thus ensuring a high resistance to very high unit pressure and high temperature in the deformation zone [3]. There are three designs of plug that could be of interest to a producer of stainless steel tubing: a ceramic plug, an oscillating plug, and an adjustable plug [4]. The advantage of the floating-plug drawing (FPD) process is the possibility of using a very high velocity of drawing and the ability to achieve a very high productivity [5].

The frictional forces in the FPD process occur at the contact of the outer surface of the tube and the die in the zones of free reduction of diameter, the zone of simultaneous reduction of the diameter and the tube wall, and in the calibration cone of the drawing die. The internal friction surface is the zone of simultaneous reduction of the diameter, the wall of the tube, and the surface of the cylindrical part of the plug in the calibration zone (Figure 1).

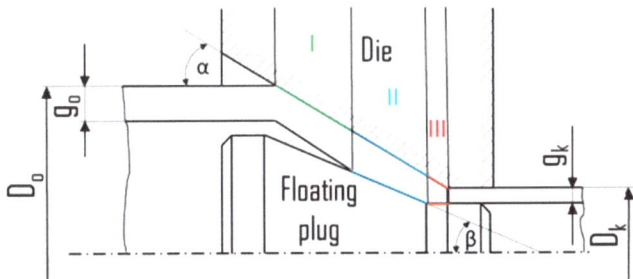

Figure 1. Schematic diagram of the FPD process with marking free reduction of diameter zone (I), drawing zone on the conical (II) and cylindrical (III) plug parts, outer diameter and wall thickness of the tube before (D_0, g_0) and after (D_k, g_k) the drawing process, as well as die (α) and plug (β) angles.

The work of frictional forces on the above-mentioned surfaces is completely transformed into heat, resulting in a local increase in the temperature of the drawing die, tube, and floating plug, as well as the lubricating film on the outer and inner surfaces of the tube [6,7]. The intensity of the heat released is a function of the following factors [8,9]:

- kind of tube material drawn, its yield point, and the intensity of work hardening,
- friction conditions and tribological properties of the lubricant used,
- drawing speed,
- geometric parameters of the deformation zone (the angle of the die cone, the angle of the conical zone of the floating plug, the material strain in the diameter reduction zone and in the zone of simultaneous diameter reduction and wall thinning, and the width of the calibration strip.
- A feature that distinguishes the process of drawing tubes with a floating plug is the high speed of drawing tubes made of materials with a low yield point, low work hardening intensity, and the need to reduce the drawing speed of hard-to-deform materials.

Analysis of the issues shows that previously proposed solutions related to the influence of various factors on the stability of the plug in the deformation zone [10,11], the impact of dependency and the geometrical dimensions of dies and plugs on the intensification of deformation and process velocity [5–7,10,12,13], tool materials [6], minimising stress during drawing [14,15], and the influence of friction and lubricants applied by various methods in carrying out the process [16–19]. After introducing the technology of tube drawing with a floating plug into production, it was often believed that tube drawing was only possible under the condition that $D_T > D_1$ (where D_T is the plug head diameter and D_1 is the diameter of the die) [20]. However, it was found [7,21] that thick-wall tubes can be drawn, but only under the condition that $D_T < D_1$.

Floating plug drawing has been the research topic of many authors in terms of the improvement of the shape and dimensional accuracy of tubes, ensuring adequate lubrication, the improvement of the surface finish, and the minimisation of drawing forces [22]. Danckert and Endelt [23] proposed a new plug with a circular profiled plug instead of a conventional cylindrical plug, which forms a cylindrical bearing channel between the die and the plug. The results of numerical FE-based analysis showed that the drawing force can be reduced and that the drawing force is nearly independent of the length of the die and of small variations in the angle of the die. Damodaran et al. [24] recommended a short bearing channel in order to increase the stability of the drawing process. Yoshida and Furuya [25] determined an optimal shape for the plug using the finite element method (FEM). They investigated the fabrication of fine tubes used in catheters and stents by means of plug drawing and mandrel drawing. Experiments using a ball-type plug, different from the conventional conical plug, allowed the frequency of breaks during drawing to be reduced. Wang et al. [26] proposed a new mathematical model to predict the behaviour of tube drawing with a floating plug in hollow copper tubes. This model incorporates the plug semi-angle, die semi-angle, wall thickness reduction ratio, and friction coefficient. The results confirmed that the model developed is more accurate than the conventional slab model, especially at reduction ratios of up to 40%. Şandru and Camenschi [27] developed a theory of high speed tube drawing with a floating plug. Schrek et al. [28] focused on an analysis of factors negatively affecting the FPD of tubes made from austenitic stainless steel. The results obtained showed very good agreement between the simulations and the experimental results.

Partial or comprehensive research on the process of drawing tubes on a floating plug was carried out for relatively easily deformable materials (aluminium [23], copper [29,30], brass [15,31–33], cobalt–chromium alloys [34], carbon and alloy steel [23,35]) and with the use of tools and lubricants appropriate for these materials and drawing speeds. These studies do not answer a number of questions about the same process using hard-to-deform materials.

In order to reveal the effect of the area of the friction contact surface on the work of external frictional forces and the surface temperature, it was decided to change the difference in the cone angles of the die α and the angle of inclination of the conical part of the plug β within the range $(\alpha - \beta) = 2° - 4.5°$, determined on the basis of previous research. Literature analysis and information collected from lubricant manufacturers made it possible to select the appropriate lubricants. Strain gauge measurement of the drawing force, continuous multi-point measurement of the surface temperature of the die, and contact measurement of the temperature of the tube surface behind the drawing die were the tasks selected to evaluate the influence of the parameters of the drawing process on the value of the work of the friction force. An evaluation of the influence of the parameters of the FPD process on tubes was made on the basis of measurements of surface roughness, microhardness, and the mechanical properties of the tubes.

2. Experimental

Drawing tests were carried out on AISI 321 steel tubes with the following outer diameter D_0 and wall thickness g_0 (D_0 = 19 mm, g_0 = 1.2 mm, D_0 = 18 mm, g_0 = 1.2 mm). The outer surface of the tubes was prepared by grinding after etching (Figure 2a,b). These treatments were carried out to create lubricant pockets on the outer surface of the tube, allowing more lubricant to be delivered to the deformation zone. The purpose of the application of this treatment was to improve the lubrication conditions, which resulted in the reduction of the amount of heat released during the drawing process, as well as the reduction of the irregularity of the deformation and a reduction of the residual stresses.

Tubes were drawn on a floating plug using a drawing machine with a maximum drawing force of 70 kN. The final outer diameter of tubes was D_k = 16 mm, and the wall thickness was g_k = 1.06 mm. The elongation ratios in diameter were λ_D = 1.19 and λ_D = 1.12, and the elongation ratios in wall thickness were λ_g = 1.13 and λ_g = 1.2. Maintaining the same total elongation values at different elongation ratios resulted in the reduction of the diameter and the tube wall thickness, giving the possibility of differentiating the total friction surface and revealing the stress in the tube material due to the reduction in diameter. As a drawing tool, a tungsten carbide G10 die was used with a half angle of α = 16° and a diameter of D = 16 mm.

Drawing dies were fabricated with special holes on the ring to allow the temperature to be measured (Figure 2c,d) at four points in the deformation zone using thermocouples. During the trials, the temperature of the outer surface of the tube was also continuously measured at the outlet of the die using a Flir TG54 pyrometer (Teledyne FLIR Company).

Floating plugs were made of G10 tungsten carbide with a half angle of the angle of inclination of the conical part of the plug β = 11.5°, 13°, and 14°, which enabled various angles between the die and the plug to be obtained in the range (α − β) = 2° − 4.5° (Figure 2e–h).

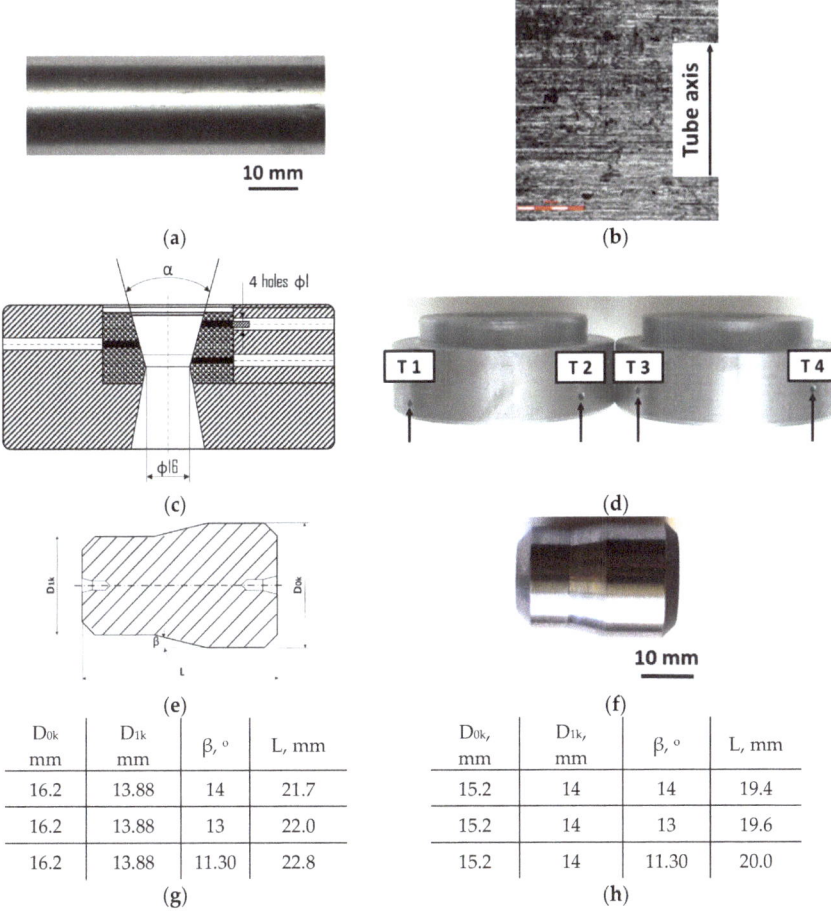

D_{0k} mm	D_{1k} mm	β, °	L, mm
16.2	13.88	14	21.7
16.2	13.88	13	22.0
16.2	13.88	11.30	22.8

(g)

D_{0k}, mm	D_{1k}, mm	β, °	L, mm
15.2	14	14	19.4
15.2	14	13	19.6
15.2	14	11.30	20.0

(h)

Figure 2. AISI 321 stainless steel tube with ground surface: (a) macro and (b) micro observation, (c) drawing die with holes to allow temperature measurement in the deformation zone, (d) view of dies with holes drilled in them, (e) schematic drawing of a floating plug, (f) sample of floating plug, floating plugs dimension according to the scheme of tube drawing (g) D_0 = 19 mm, g_0 = 1.2 mm and D_k = 16 mm, g_k = 1.06 mm, and (h) D_0 = 18 mm, g_0 = 1.2 mm and D_k = 18 mm, g_k = 1.06 mm.

During the tests, the drawing velocity was varied in the range 1–10 m/min. WISURA DSO 7010, Fuchs Oil Corp., further in the text marked as (W), Prolong EP 2, Prolong Super Lubricants (P), Ferokol EPS-220, Naftochem, Corp. (F), Masterdraw 7194AC, ETNA Products, Inc. (M), and Tubol 1962JR, Metalube Limited (T), lubricants were used in the drawing experiments. The lubricants selected for testing are recommended for drawing tubes by world-renowned manufacturers. To measure the force and the temperature, a measuring circuit consisting of a force sensor, a Spider 8 Hottinger Baldwin Messtechnik measuring amplifier, K-type thermocouples, and a personal computer Dell Latitude 5520 were used. The outer diameter of the pipes was measured using an electronic calliper. The wall thickness was measured using an electronic micrometer. In addition, for comparison and verification, the outer diameter of the tubes and the wall thickness were measured using the Atos Core 200 optical 3D scanner from GOM company. The surface topography and roughness of the inner and outer surfaces of the tubes were measured using LEXT OLS4100 confocal laser microscope with 405 nm UV laser light (Leica Microsystems). Surface roughness results are presented by the two most commonly used parameters, i.e., Ra—arithmetic average of the absolute values of the profile heights over the evaluation length and Rz—the average value of the absolute values of the heights of five highest-profile peaks and the depths of five deepest valleys within the evaluation length. The ground surface of the batch material and the surfaces of the tubes after drawing were analysed. Vickers microhardness measurements were made using a Shimadzu HMV-2T tester. The investigations of mechanical properties were carried out using a Zwick Roell Z005 uniaxial tensile test machine. The microstructure studies were carried out on the Olympus GX51 light microscope. The test specimens were cold mounted in a Struers FixiForm container using Struers EpoFix epoxy resin. The samples prepared in this way were then ground on sandpaper with a grit of 240–2000. For polishing, diamond suspensions were used: DP-Suspension P 9 µm, 3 µm, 1 µm. The specimens were made on a RotoPol-11 grinder–polisher with a RotoForce-1 Struers head. Finishing polishing was carried out with the OP-S polishing slurry from Struers. To reveal the AISI 321 stainless steel microstructure, the specimens were electrolytically etched in 10% nitric acid at a voltage of 1.5 V for 20 s at room temperature.

3. Results and Discussion

3.1. Drawing Force and Temperature

The results of the drawing force and the temperature at various angles β between the die and the floating plug (a drawing velocity v_c in the range of 1–10 m/min with (W) lubricant) are presented in Tables 1 and 2. An increase in the drawing velocity caused an increase in the value of the drawing force, reaching a value of F_{max} = 30.15 kN for a drawing velocity of v_c = 6 m/min (Table 1).

Table 1. Drawing forces and maximum temperature of tube and drawing die—D = 19 mm, α = 16°, (W) lubricant.

β, °	v_c, m/min	F_{max}, kN	$F_{śr}$, kN	T_{max} (Drawing Die), °C	T_{max} (Tube), °C	Comment
11.5	1	41.5	34.9	-	-	The phenomenon of periodic pulsation of force (the so-called "bamboo" effect). Tube breakage.
11.5	2	27.2	25.7	78.3	112.3	Partial phenomenon of periodic pulsation of force (the so-called "bamboo" effect).
11.5	3	27.5	25.6	77.8	104.8	
11.5	4	27.7	24.8	81.4	109.4	
11.5	6	30.2	27.6	84.4	133.2	
11.5	10	28.6	26.7	78.3	111.1	
13	1	34.1	29.1	70.2	85.1	
13	2	29.1	25.3	74.4	101	
13	4	31.3	27.3	83.9	123.1	
14	1	28.6	26.4	69.5	77.7	
14	2	27.4	24.6	75.9	96.9	
14	4	25.4	24.9	77.8	111.2	

Table 2. Temperature of die in points T1–T4—D = 19 mm, α = 16°, (W) lubricant.

β, °	v_c, m/min	T_1, °C	T_2, °C	T_3, °C	T_4, °C
11.5	1	-	-	-	-
11.5	2	78.2	77.1	70.4	67.4
11.5	3	77.7	73.8	71.5	68.7
11.5	4	81.4	75.5	71.8	65.3
11.5	6	84.4	78.4	71.9	68.7
11.5	10	76.5	74.5	58.8	54.9
13	1	70.2	70.2	66.7	64.9
13	2	74.4	74.7	71.5	65.6
13	4	83.9	78.1	74.8	69.2
14	1	69.5	68.5	67.1	64.3
14	2	75.9	74.5	67.6	65.2
14	4	77.8	76,1	69.2	66.4

A large friction surface between the plug and the inner surface of the tube was associated with differences in the angles of the die α and the angle of inclination of the conical part of the plug β (α − β) equal to 4.5°. This causes an increase in unit pressure and a deterioration in friction conditions. In addition, for such a difference in the angles, too big a gap is formed between the barrel surface of the plug and the inner surface of the tube, preventing the development of a hydrodynamic effect at low drawing velocity.

Graphs of the course of the temperature change and drawing strength allow one to assess the stability of the drawing process as a function of time, and they also contain information about disturbances, which prove to be linked to the interruption of the lubricant film. Analysing the maximum and average drawing force and the tube surface after drawing using the lubricant W showed that the "bamboo" effect occurred with drawing conditions α = 16°, β = 11.5°, and v_c = 1 m/min. A partial "bamboo" effect occurred with velocity v_c = 2 m/min (Figure 3). On the other hand, there was no such effect for velocity v_c = 3 or 4 m/min (Figure 4) and 6 m/min. This may indicate a major impact of drawing velocity, die geometry, and the floating plug and temperature emissions on the occurrence of instabilities during the process.

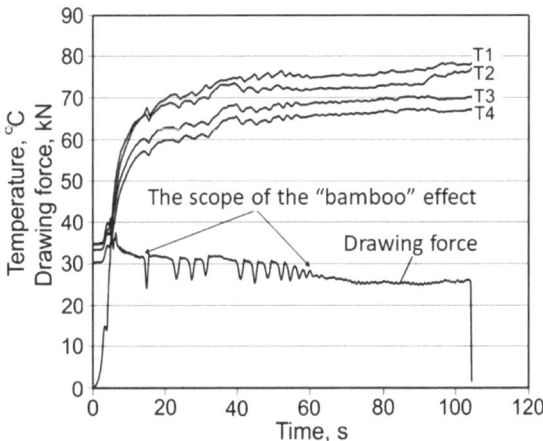

Figure 3. Evolution of both the temperature of die and drawing force F_c for α = 16°, β = 11.5°, lubricant: W, v_c = 2 m/min.

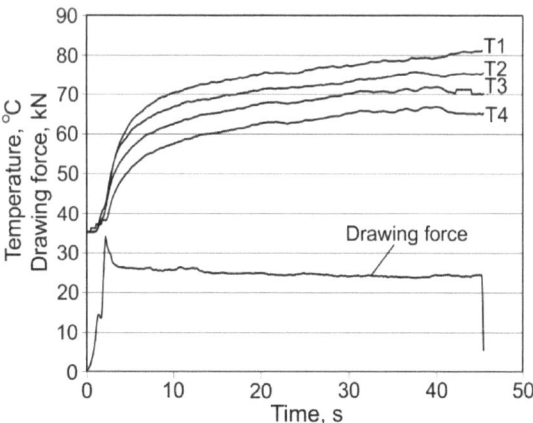

Figure 4. Evolution of both the temperature of the die and drawing force F_c for $\alpha = 16°$, $\beta = 11.5°$, lubricant (W), $v_c = 4$ m/min.

Variation in temperature and drawing force for the die angle $\alpha = 16°$ and plug angles $\beta = 13°$ and $14°$ are similar to the curves shown in Figure 4. When drawing using dies with an angle $\alpha = 16°$ and plug angle $\beta = 13°$, first, a clear decrease and then an increase in drawing force occur with an increase in drawing velocity. The strongest force $F_{max} = 34.1$ kN was for the velocity $v_c = 1$ m/min, while the weakest force $F_{max} = 29.1$ kN was for a velocity $v = 2$ m/min (Table 1). For the velocity $v_c = 4$ m/min, there is an increase in drawing force to the value of $F_{max} = 31.3$ kN. For angles $\alpha = 16°$ and $\beta = 13°$ and velocity $v_c = 2$ m/min, the average drawing force is $F_{av} = 25.2$ kN. For drawing with a die angle $\alpha = 16°$ and plug angle $\beta = 14°$, an increase in drawing velocity causes a decrease in the maximum drawing force, which takes the value $F_{max} = 28.6$ kN for $v_c = 1$ m/min and $F_{max} = 25.3$ kN for velocity $v_c = 4$ m/min (Table 1).

Analysis of the temperature occurring in the die and on the tube surface indicates that, for $\alpha = 16°$ and $\beta = 11.5°$, $13°$, and $14°$, there is an increase in those temperatures with increasing drawing velocity. The lowest maximum temperatures, $T_{max} = 69.5$ °C in the die and $T_{max} = 77.7$ °C on the tube surface, were measured for the angle $\alpha = 16°$ and $\beta = 14°$ with a drawing velocity $v_c = 1$ m/min (Table 2), while the highest maximum temperatures, $T_{max} = 84.4$ °C in the die and $T_{max} = 133.2$ °C on the tube surface, were recorded for $\alpha = 16°$ and $\beta = 11.5°$ at a drawing velocity $v_c = 6$ m/min. This demonstrates that the parameter having the largest impact on the temperature emitted during the process is that of drawing velocity. A small temperature difference between the die and the tube surface (15 °C for $\alpha = 16°$, $\beta = 13°$, and $v_c = 1$ m/min) proved sufficient for the die and lubricant to release heat. With an increase in drawing velocity, a distinct difference was observed between the temperature in the die and on the tube surface (50 °C for $\alpha = 16°$, $\beta = 11.5°$, and $v_c = 6$ m/min). This indicates that it is impossible for both the die and the lubricant to give up heat due to the lack of die cooling and the fact that the heat capacity of the lubricant is too small.

Temperature change tends to increase during drawing. This increase is greater at lower drawing velocity. With an increase in drawing velocity, the curves are flatter. But, they failed to achieve a constant temperature during the drawing of tubes with lengths of 2000 mm, thus providing the conditions for a stationary process.

When the "bamboo" effect occurred, which presented a periodic build-up on the tube surface (Figure 5a), an oscillation of temperature was observed (Figure 3). The temperature curves are composed of characteristic "teeth", corresponding to the oscillations of the drawing force, starting and ending with the "bamboo" effect.

During tube drawing on dies with angles $\alpha = 16°$ and $\beta = 11°30'$, $13°$, and $14°$ over the complete range of drawing velocity using lubricants (P) and (F), unsatisfactory results were obtained. Tube breakages were observed after 50 s from the start of the drawing process, i.e., after the tube was drawn for about 1.5 m (Figure 5b). The occurrence of tube breakages is a problem during drawing, with an approximate 20% rate in production [24].

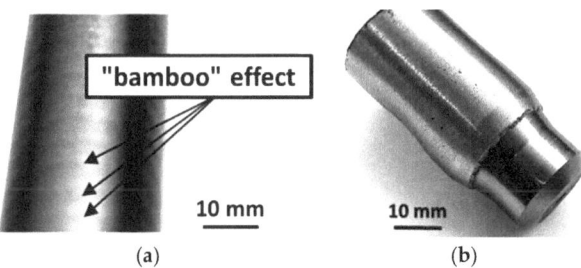

Figure 5. (a) "Bamboo" effect formed on the tube surface as a result of the instability of the process conditions, and (b) view of the broken tube with a floating plug inside.

A characteristic feature of the loss of lubricant properties was a clear increase in the temperature in the die and a corresponding increase in force (Figure 6a). Thus, a clear deterioration in physical properties and the viscosity of the lubricant under the influence of an increase in temperature can be observed in the deformation zone, revealing a rupture of the lubricant film and an increase in the drawing force.

Completely unsatisfactory results were obtained when drawing tubes with a dimension $D_0 = 19$ mm, $g_0 = 1.2$ mm using a die with dimensions $\alpha = 16°$ and $\beta = 11.5°$, $13°$, and $14°$ while using the lubricants (M) and (T) (Figure 6b).

When analysing the influence of the strain values applied, it should be noted that positive test results were obtained for the drawing of tubes with dimensions $D_0 = 19$ mm, $g_0 = 1.2$ mm into tubes with dimensions $D_k = 16$ mm, $g_k = 1.06$ mm, while unsatisfactory results were obtained for the drawing of tubes from dimensions $D_0 = 18$ mm, $g_0 = 1.2$ mm into dimensions $D_k = 16$ mm, $g_k = 1.06$ mm, with the same total strain amounting to $\lambda = 1.34$. The difference was in the partial deformations of the wall thickness and diameter of the tube. Satisfactory results were obtained for the drawing of tubes from dimensions $D_0 = 19$ mm, $g_0 = 1.2$ mm into dimensions $D_k = 16$ mm, $g_k = 1.06$ mm, because it was possible to obtain conditions allowing the production of natural back-tension, thus lowering the pressure of the metallic workpiece on the die and the plug with no increase in the drawing force.

The intended use of the lubricants used is strictly defined, not only in terms of strength properties, viscosity as a function of temperature, and the wettability of the surfaces on the tube, drawing die, and floating plug, but also due to the use of a specific material with specific properties. The (W) lubricant can be successfully used for drawing tubes made of AISI 321 steel on a floating plug. On the other hand, (T) and (M) lubricants did not fulfil their task completely when drawing this type of austenitic steel, despite the fact that they are successfully used for drawing copper and brass.

The parameter determining the feasibility of the FPD process is the strength of the lubricating film on the external and internal friction surfaces. The strength of the lubricant film depends on the type of lubricant and the concentration of surface-active agents in the lubricant, and, most of all, on the temperature in the friction zones. If the critical temperature is exceeded, the lubricating film is ruptured and the tube is almost immediately ruptured. Thus, in the case of drawing tubes made of both soft and hard-to-deform materials, the necessary condition for the drawing process is to ensure that temperatures in the deformation zone are lower than the critical temperature for the lubricant used. Optimisation of the geometry of the tools in order to ensure that the minimum work of frictional forces is fully converted into heat will minimise the drawing stress, permitting the FPD process to be carried out.

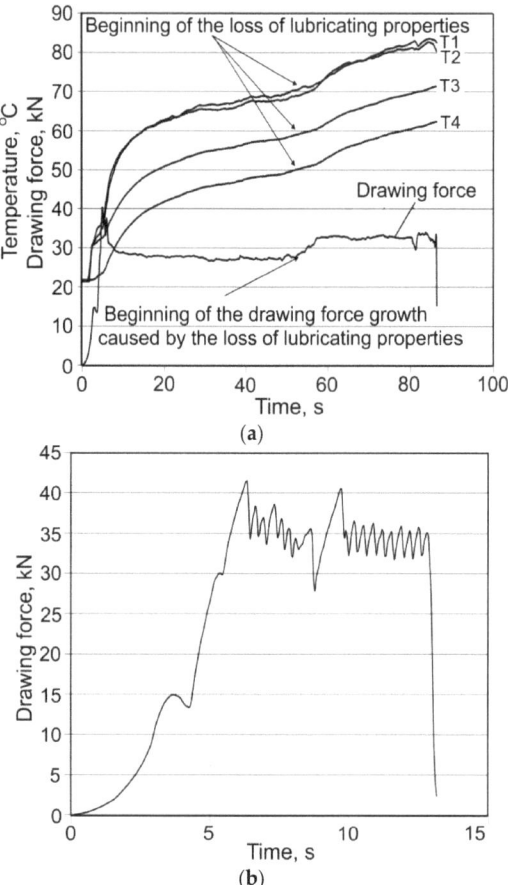

Figure 6. Variation of temperature and of drawing force during drawing at α = 16°, β = 14°, v_c = 2 m/min, and with the use of (P) lubricant (**a**) and (T) lubricant (**b**).

3.2. Microstructure

Table 3 shows examples of the topography of the outer and inner surfaces, as well the microstructure, on the cross section of tubes before and after drawing with β = 11.5°, 13°, 14°, α = 16°, v_c = 4 m/min, and with the use of (W) lubricant.

Microstructural observations of the charging tubes revealed large grains in the central part of the cross-section, with refined grains on the inner and outer surfaces of the tubes. This indicates a poor selection of heat treatment parameters (supersaturation) in the process of preparing the material for drawing. This is also confirmed by the results of the average microhardness measurements on the cross-section of the charge tubes. Microhardness in the middle part of the wall thickness was 160–173 HV, while the microhardness in the vicinity of the inner and outer surfaces of tube was 180–190 HV.

Table 3. Topography of outer (a, d, g, j) and inner (b, e, h, k) surfaces with roughness measurement lines, as well microstructure (c, f, i, l) of tubes before and after drawing with β = 11.5°, 13°, 14°, α = 16°, v_c = 4 m/min, and with the use of (W) lubricant.

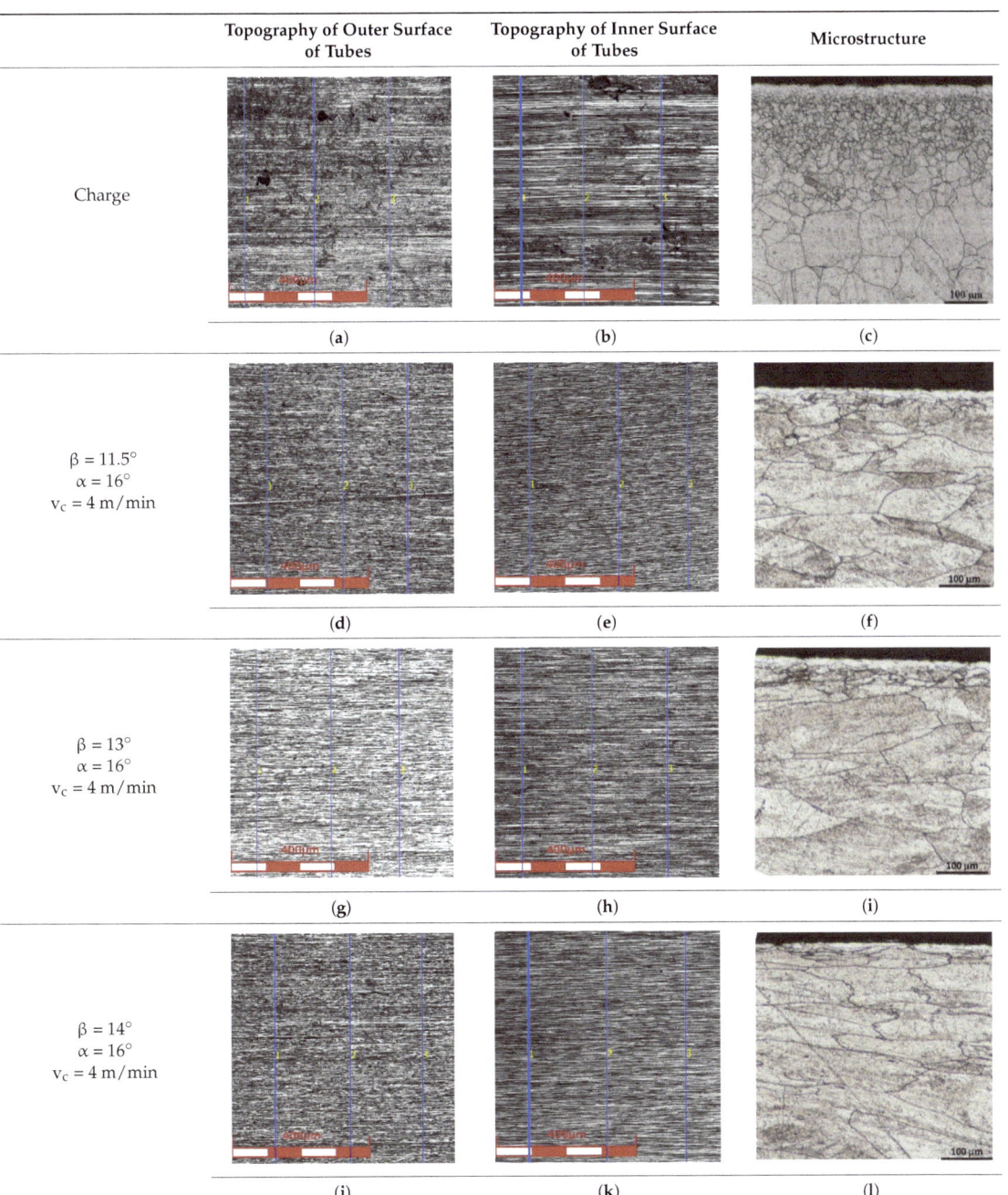

Grinding the outer surface of the charge tubes causes the formation of lubricant pockets that allow more lubricant to be supplied to the deformation zone, improving the frictional conditions of the process (reducing the temperature and the amount of heat generated during drawing). Microscopic observation showed that the scratches forming transverse to the tube axis after grinding remain visible on the surface of the product regardless of the drawing process conditions; however, they are clearly flattened. Longitudinal scratches visible on the inner and outer surfaces of the finished product indicate a lack of the appropriate amount of lubricant or the fact that the fine particles remaining after the grinding process caused a roughening of the tube surface. Varying the geometry of the tools used for drawing and changing the speeds of the process do not significantly change the microstructure of the tubes.

3.3. Surface Roughness

3.3.1. Experimental Results

The results of the mean roughness Ra and Rz measured on the inner and outer surfaces of the tubes, depending on the angle of the floating plug and the drawing speed, are presented in Table 4 and Figures 7 and 8. Surface roughness parameters were measured on the tubular surfaces in the axial direction.

Table 4. Roughness parameters Ra and Rz of inner and outer surfaces of tubes—D = 19 mm, $\alpha = 16°$, (W) lubricant.

Angle of Floating Plug β, °	Drawing Speed v_c, m/min	Outer Surface		Inner Surface	
		Ra, µm	Rz, µm	Ra, µm	Rz, µm
11.5	1	-	-	-	-
	2	0.491	4.794	0.136	1.391
	3	0.475	5.382	0.185	2.067
	4	0.639	6.146	0.233	2.762
	6	0.734	6.471	0.267	2.904
	10	0.762	6.455	0.302	3.348
13	1	0.458	4.699	0.116	1.583
	2	0.461	4.631	0.264	2.677
	4	0.475	4.572	0.353	2.711
14	1	0.432	4.776	0.232	2.104
	2	0.614	5.733	0.247	2.138
	4	0.539	5.272	0.255	2.746

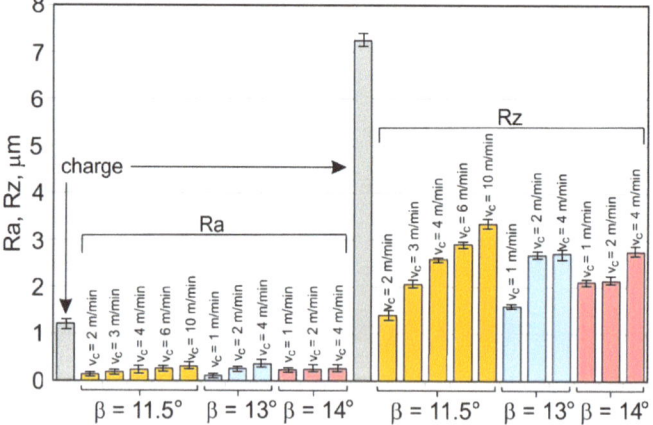

Figure 7. Micro-geometry of the inner surface of the tubes (D = 19 mm, $\alpha = 16°$, (W) lubricant).

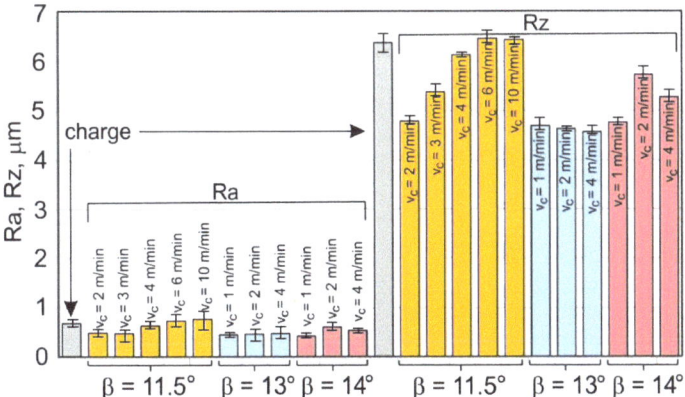

Figure 8. Micro-geometry of the outer surface of the tubes (D = 19 mm, α = 16°, (W) lubricant).

The roughness measurements of the tubes after drawing confirmed the significant influence of the process parameters on the surface roughness of the inner and outer surfaces of the tubes. After the drawing process, there was a significant improvement in the smoothness of the (especially inner) surface of the tubes in relation to the smoothness of the surface of the charging tubes. In the case of the charging tubes, the mean roughness of the internal surface was Ra = 1.25 μm, and, for the external surface, was Ra = 0.7 μm. After drawing, the mean roughness of the tube surface was between Ra = 0.43 μm and Ra = 0.93 μm for the outer surface of the tubes and Ra = 0.09–0.35 μm for the inner surface of the tubes. The mean roughness and 10-point peak–valley surface roughness of the inner surface of the tubes was significantly reduced (Figure 7). The drawing process led to a reduction in the surface roughness parameters Ra and Rz measured on the outer surface of the tubes, but only for the angles of floating plug β = 13° and β = 14° (Figure 8).

With the use of the die angle α = 16° and the angle of floating plug β = 11.5°, 13°, and 14°, along with the increase in the drawing speed, the roughness of the inner surface of the tube increases, which demonstrates the improvement of the lubrication conditions and the increase in the thickness of the lubricating film. FPD of tubes significantly improves the roughness of the inner surface of the tubes in relation to drawing the tubes without a floating plug, in which the circumferential compressive stresses promote an increase in the roughness of the inner surface. The contact of the inner surface of the tube with the floating plug leads to a radical improvement in roughness. It can be concluded that the increase in speed and angle of the die increases the effect of hydrodynamic drawing, which results in a reduction of the friction coefficient that mainly occurs on the inner surface of the tube [12,13].

3.3.2. Analysis of Variance

First, analysis of variance was used to model the relation between the angle of floating plug β, the drawing speed v_c, the measurement side (inner and outer surfaces of tubes), and the value of the roughness parameter Ra. THe measurement side is considered to be a categoric variable.

The model F-value of 66.04 implies the model is significant (Table 5). The model terms are significant when p-values are less than 0.0500. In this case, B and C are significant terms. Values greater than 0.1000 indicate the model terms are not significant. A signal-to-noise ratio (an adequacy precision ratio) of 21.288 indicates an adequate signal for the model (Table 6). If the adequacy precision is greater than 4, the ANOVA model can be used to navigate the design space. The predicted R^2 of 0.8178 is in statistically reasonable agreement with the adjusted R^2 of 0.8610.

Table 5. Results of ANOVA for the roughness parameter Ra.

Source	Sum of Squares	Degrees of Freedom	Mean Square	F-Value	p-Value	Meaning
Model	0.6451	2	0.3225	66.04	<0.0001	significant
A—drawing speed	0.0915	1	0.0915	18.72	0.0004	
B—measurement side	0.5536	1	0.5536	113.35	<0.0001	
Residual	0.0928	19	0.0049			
Total correlation	0.7379	21				

Table 6. Fit statistics of the regression model for the roughness parameter Ra.

Standard deviation	0.0699	R^2	0.8742
Mean	0.3895	Adjusted R^2	0.8610
Coefficient of variation. %	17.94	Predicted R^2	0.8178
		Adequacy precision	21.2885

The relation between the input parameters and the roughness parameter Ra (in terms of the coded factors) is as follows:

$$Ra = 0.4400 + 0.1161A + 0.1586B \qquad (1)$$

The analysis of variance was also used to model the relation between the angle of floating plug β, the drawing speed v_c, the measurement side (inner and outer surfaces of tubes), and the value of the roughness parameter Rz. The measurement side is considered to be a categoric variable.

The model F-value of 69.94 implies the model is significant (Table 7). The model terms are significant when p-values are less than 0.0500. In this case, C, AB, B^2 are significant terms. Values greater than 0.1000 indicate the model terms are not significant. A signal-to-noise ratio (an adequacy precision ratio) of 22.6581 indicates an adequate signal for the model (Table 8). If the adequacy precision is greater than 4, the ANOVA model can be used to navigate the design space. The predicted R^2 of 0.9274 is in statistically reasonable agreement with the adjusted R^2 of 0.9426.

Table 7. Results of ANOVA for the the roughness parameter Rz.

Source	Sum of Squares	Degrees of Freedom	Mean Square	F-Value	p-Value	Meaning
Model	54.52	5	10.90	69.94	<0.0001	significant
A—angle of floating plug	0.5763	1	0.5763	3.70	0.0725	
B—drawing speed	0.0229	1	0.0229	0.1472	0.7063	
C—measurement side	48.04	1	48.04	308.14	<0.0001	
AB	0.9475	1	0.9475	6.08	0.0254	
B^2	1.22	1	1.22	7.82	0.0129	
Residual	2.49	16	0.1559			
Total correlation	57.01	21				

Table 8. Fit statistics of the regression model. for the roughness parameter Rz.

Standard deviation	0.3948	R^2	0.9562
Mean	3.88	Adjusted R^2	0.9426
Coefficient of variation. %	10.19	Predicted R^2	0.9274
		Adequacy precision	22.6581

The relation between the input parameters and the roughness parameter Rz (in terms of the coded factors) is as follows:

$$Rz = 4.18 - 0.5062A + 0.1367B + 1.48C - 1.04AB - 1.01B^2 \qquad (2)$$

3.4. Microhardness

3.4.1. Experimental Results

The results of microhardness measurements of the tubes show a significant increase of 30–70% in the microhardness value on the cross-section of the tubes after drawing, in relation to the microhardness of the charge material. A characteristic feature is the differentiation in the amount of microhardness across the thickness of the wall. In the middle zone of the wall, this increase in value is less than for the values in the outermost zones. The differentiation of angles of the dies and floating plugs, as well as the drawing speed, do not significantly affect the microhardness value (Figure 9).

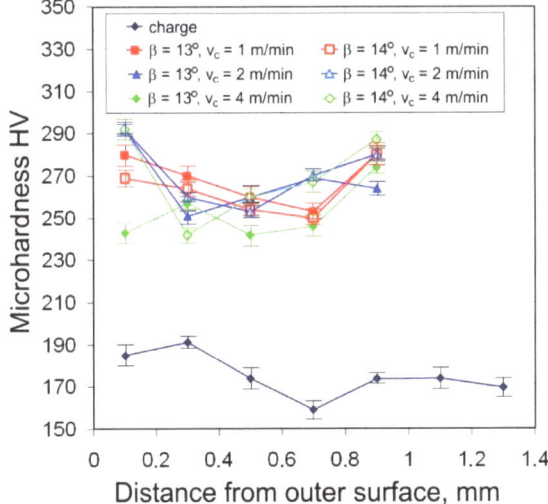

Figure 9. Micro-hardness across the tube cross section (D = 19 mm, α = 16°, (W) lubricant).

The average microhardness of the tubes after drawing is in the range of 240–310 HV. Higher values of microhardness 250–310 HV occur for the die angle α = 16°. With the increase in the difference (α − β), microhardness values increase—the largest occur for α = 16°, β = 11.5° (α − β = 4.5°) (Figure 10). This is due to the greater share of bending forces at tube entry and at the exit of the tube from the die.

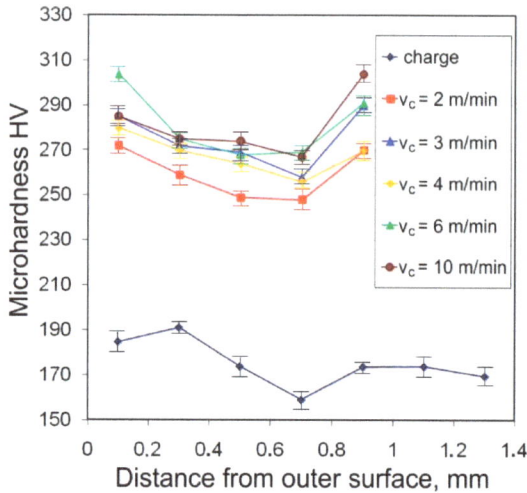

Figure 10. Effect of drawing speed v_c on the microhardness across the tube cross section (D = 19 mm, $\alpha = 16°$, $\beta = 11.5°$, (W) lubricant).

3.4.2. Analysis of Variance

The model F-value of 12.93 implies the model is significant (Table 9). The model terms are significant when p-values are less than 0.0500. In this case B, A^2, C^2 are significant terms. Values greater than 0.1000 indicate the model terms are not significant. A signal-to-noise ratio (an adequacy precision ratio) of 12.8640 indicates an adequate signal for the model (Table 10). If the adequacy precision is greater than 4, the ANOVA model can be used to navigate the design space. The predicted R^2 of 0.4470 is in statistically reasonable agreement with the adjusted R^2 of 0.5248.

Table 9. Results of ANOVA for the hardness.

Source	Sum of Squares	Degrees of Freedom	Mean Square	F-Value	p-Value	Meaning
Model	7122.09	5	1424.42	12.93	<0.0001	significant
A—angle of floating plug	2.72	1	2.72	0.0247	0.8757	
B—drawing speed	495.39	1	495.39	4.50	0.0391	
C—distance from outer surface	19.66	1	19.66	0.1784	0.6746	
A^2	447.17	1	447.17	4.06	0.0495	
C^2	5494.81	1	5494.91	49.87	<0.0001	
Residual	5399.40	49	110.19			
Total correlation	12,521.49	54				

Table 10. Fit statistics of the regression model.

Standard deviation	10.50	R^2	0.5688
Mean	268.51	Adjusted R^2	0.5248
Coefficient of variation. %	3.91	Predicted R^2	0.4470
		Adequacy precision	12.8640

The relation between input parameters and microhardness (in terms of the coded factors) is as follows:

$$HV = 254.11 - 0.3038A + 6.38B - 0.8455C + 7.02A^2 + 23.89C^2 \tag{3}$$

3.5. Mechanical Properties

The ultimate tensile stress of the tubes after drawing is almost doubled as compared to the charge material (Figure 11). On the other hand, plastic properties decrease more than three times (Figure 12), which proves a significant work hardening of the material has taken place during the drawing process.

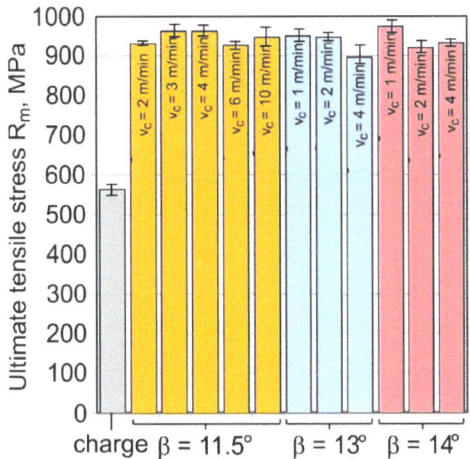

Figure 11. Ultimate tensile strength R_m of the tube material (D = 19 mm, α = 16°, (W) lubricant).

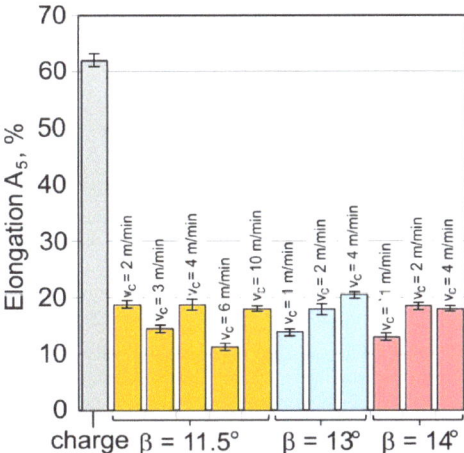

Figure 12. Elongation A_5 of the tube material (D = 19 mm, α = 16°, (W) lubricant).

There is no clear correlation between the parameters of the drawing process and the results obtained for the mechanical properties. When analysing the influence of the floating-plug drawing process on the mechanical properties of the material, it should be noted that, for the total deformation of λ = 1.34, the strength properties increase from the level of $R_m \approx 560$ MPa for the charge material to $R_m \approx 900-960$ MPa for the tubes after drawing, with a simultaneous decrease in elongation from $A_5 = 62\%$ to $A_5 = 12.5-20\%$. This demonstrates a huge increase in the strengthening of cold-processed material with a decrease in plastic properties.

Analysis of Variance

In this section, two models of ANOVA were considered. First, analysis of variance was used to model the relation between the angle of the floating plug β, the drawing speed v_c, and the value of uniaxial tensile strength R_m. In the second model, the elongation A_5 was used as the output. In both cases, the model F-values imply that the models are not significant relative to the noise. Therefore, the influence of the input parameters on the two analysed mechanical parameters can be described by the average value of these parameters.

4. Conclusions

The following conclusions can be drawn on the basis of the analysis of the results obtained:

- The application of an appropriate deformation in the reduction zone of the drawing die results in the creation of back tension, which reduces the pressure of charge metal on the die and plug.
- The reduction of unit pressure, as a consequence, reduces the work of forces of external friction.
- The use of higher values of the angles of the die and floating plug, while maintaining the difference between the angles $(\alpha - \beta) = 4.5°$, reduces the friction surface and the work of friction forces.
- An increase in the drawing speed, while maintaining an appropriate gap between the inner surface of the tube and the cylindrical part of the floating plug, significantly improves the friction conditions due to the increasing effect of the hydrodynamic lubrication phenomenon.
- An increase in the roughness of the inner surface, with an increase in the drawing speed for high values of angles of the die and plug, demonstrates the hydrodynamic effect during the drawing process.
- It is recommended that the materials used for tools should guarantee a small value of coefficient of friction.
- It is recommended that lubricants be used that have tribological properties that create conditions ensuring adequate strength of the lubricating film at a temperature of 90–150 °C.
- The drawing process causes a decrease in the roughness parameters Ra and Rz of the inner surface of the tubes.
- The surface roughness of the outer surface of the tubes does not change much in relation to the surface roughness of the charging tubes. The lowest values of the Ra and Rz surface roughness parameters measured on the outer surface of the tubes occur for the die angle $\alpha = 16°$ and the floating plug angle $\beta = 13°$.
- For the die angle $\alpha = 16°$ and plug angles $\beta = 11.5°$, $13°$, and $14°$, the values of the roughness parameters increase with an increase in drawing speed, although this increase is small. This demonstrates the increase in the thickness of the lubricating film and, thus, the improvement of the lubrication conditions.
- After drawing, an increase of 30–70% was observed in the microhardness of the tube material in relation to the microhardness of the charge material. In the middle zone of the wall thickness, the increase in microhardness is less than in the outer zones. The different angles of the dies and plugs, as well as the drawing speed, do not significantly affect the microhardness value.
- The value of ultimate tensile stress doubled compared to the property in the charge material. However, the plastic properties decreased by more than three times due intensive strengthening of the material. At the same time, no correlation was observed between the drawing process parameters and material properties.
- Limitations of the floating-plug pipe drawing method result from plug geometry and improperly selected lubricants.
- Future works will focus on the possibility of increasing the drawing speed and on the selection of both lubricants and lubrication methods, which determine the possibility of intensifying and increasing the efficiency of the drawing process.

Author Contributions: Conceptualization, K.Ż.; data curation, K.Ż.; formal analysis, T.T.; investigation, K.Ż. and T.T.; methodology, K.Ż. and T.T.; validation, K.Ż.; visualization, K.Ż. and T.T.; writing—original draft, K.Ż.; writing—review and editing, T.T. All authors have read and agreed to the published version of the manuscript.

Funding: This research received no external funding.

Institutional Review Board Statement: Not applicable.

Informed Consent Statement: Not applicable.

Data Availability Statement: The data presented in this study are available on request from the corresponding author.

Conflicts of Interest: The authors declare no conflict of interest.

References

1. Kwan, C.T. A generalized velocity feld for axisymmetric tube drawing through an arbitrarily curved die with an arbitrarily curved plug. *J. Mater. Process. Technol.* **2002**, *122*, 213–219. [CrossRef]
2. Rubio, E.; Camacho, A.; Pérez, R.; Marín, M. Guidelines for Selecting Plugs Used in Thin-Walled Tube Drawing Processes of Metallic Alloys. *Metals* **2017**, *7*, 572. [CrossRef]
3. Orhan, S.; Öztürk, F.; Gattman, J. Effects of the semi die/plug angles on cold tube drawing with a fixed plug by FEM for AISI 1010 steel tube. *Sak. Univ. J. Sci.* **2017**, *21*, 886–892. [CrossRef]
4. Rees, T.W. Multiple component floating plug designs for tube drawing. *J. Appl. Met. Work.* **1981**, *1*, 53–57. [CrossRef]
5. Świątkowski, K.; Hatalak, R. Study of the new floating-plug drawing process of thin-walled tubes. *J. Mater. Process. Technol.* **2004**, *151*, 105–114. [CrossRef]
6. Rubio, E.M. Analytical methods application to the study of tube drawing processes with fixed conical inner plug: Slab and Upper Bound Methods. *J. Achiev. Mater. Manuf. Eng.* **2006**, *14*, 119–130.
7. Farahani, N.D.; Parvizi, A.; Barooni, A.; Naeini, S.A. Optimum curved die profile for tube drawing process with fixed conical plug. *Int. J. Adv. Manuf. Technol.* **2018**, *97*, 1–11. [CrossRef]
8. Yoshida, K.; Yokomizo, D.; Komatsu, T. Production of Special Tubes with a Variety Cross-Sectional Shapes by Bunch Drawing and Fluid-Mandrel Drawing. *Key Eng. Mater.* **2014**, *622*, 731–738. [CrossRef]
9. Hartl, C. Review on Advances in Metal Micro-Tube Forming. *Metals* **2019**, *9*, 542. [CrossRef]
10. Bramley, A.N.; Smith, D.J. Tube drawing with a floating plug. *Met. Technol.* **1976**, *3*, 322–331. [CrossRef]
11. Sadok, L.; Pietrzyk, M. Analysis of the work of a floating plug in the area of deformation. *Hutnik* **1981**, *2*, 62–65. (In Polish)
12. Bisk, M.B.; Szwejkin, W.W. *Wołoczenije Trub na Samoustanawliwajuszczejsja Oprawkie*; Mietałłurgizat: Moskwa, Russia, 1963.
13. Pernis, R. Floating-plug drawing of tubes. *Rudy I Met. Nieżelazne* **2001**, *46*, 305–311.
14. Avitzur, B. *Drawing of Precision Tubes by the Floating Plug Technique*; Lehigh University: Bethlehem, Palestine, 1975.
15. Hatalak, R.; Kruczek, J. Influence of tool geometry on the actual stresses in the process of drawing pipes made of Cu and its alloys on a floating plug. In Proceedings of the Conference Proceeding, Metal Forming'93, Krynica, Poland, 5–7 May 1993. (In Polish)
16. Jones, R.; Harvey, S.J. *An Analysis of Plug Chatter in Fixed and Floating Plug Tube Drawing*; Palgrave Macmillan: London, UK, 1980.
17. Łuksza, J.; Sadok, L. *Selected Drawing Problems*; AGH University Scripts, 1025: Kraków, Poland, 1986. (In Polish)
18. Łuksza, J. *Drawing Elements*; University Scientific and Didactic Publishers: Kraków, Poland, 2001. (In Polish)
19. Shey, J.A. Tribology in metalworking. In *Friction, Lubrication and Wear*; American Society for Metals: Metals Park, OH, USA, 1982.
20. Pernic, R.; Kasala, J. The influence of the die and floating plug geometry on the drawing process of tubing. *Int. J. Adv. Manuf. Technol.* **2013**, *65*, 1081–1089. [CrossRef]
21. Pernis, R. Ťahanie rúr na plávajúcom tŕni so záporným presahom. *Hut. Listy* **1999**, *54*, 18–25. (In Slovak)
22. Palengat, M.; Chagnon, G.; Favier, D.; Louche, H.; Linardon, C.; Plaideau, C. Cold drawing of 316L stainless steel thin-walled tubes: Experiments and finite element analysis. *Int. J. Mech. Sci.* **2013**, *70*, 69–78. [CrossRef]
23. Danckert, J.; Endelt, B. LS_DYNA used to analyze the drawing of precision tubes. In Proceedings of the 7th European LS-DYNA Conference, Salzburg, Austria, 14–15 May 2009; pp. 1–13.
24. Damodaran, D.; Shivputi, R.; Wibowo, F. Invesigation of zipper derects in the floating mandrel drawing of small diameter copper tubes. In Proceedings of the 24th NAMRC Conference, Ann Arbor, MI, USA, 21–23 May 1996; pp. 96–112.
25. Yoshida, K.; Furuya, H. Mandrel drawing and plug drawing of shape-memory-alloy fine tubes used in catheters and stents. *J. Mater. Process. Technol.* **2004**, *153–154*, 145–150. [CrossRef]
26. Wang, C.S.; Wang, Y.C. The theoretical and experimental of tube drawing with floating plug for micro heat-pipes. *J. Mech.* **2008**, *24*, 111–117. [CrossRef]
27. Şandru, N.; Camenschi, G. A mathematical model of the theory of tube drawing with floating plug. *Int. J. Eng. Sci.* **1988**, *26*, 569–585. [CrossRef]
28. Schrek, A.; Brusilová, A.; Švec, P.; Gábrišová, Z.; Moravec, J. Analysis of the Drawing Process of Small-Sized Seam Tubes. *Metals* **2020**, *10*, 709. [CrossRef]

29. Gorecki, W.; Krochmal, W.; Krochmal, K. The influence of sinking thick-walled copper tubes on the behaviour of metal on the zone of deformation. *Rudy I Met. Nieżelazne* **2004**, *49*, 508–513.
30. Wang, S.W.; Chen, Y.; Song, H.W.; El-Aty, A.A.; Liu, J.S.; Zhang, S.H. Investigation of texture transformation paths in copper tube during floating plug drawing process. *Int. J. Mater. Form.* **2021**, *14*, 563–575. [CrossRef]
31. Blazyński, T. *Metal Forming, Tool Profiles and Flow*; Macmillan Press: London, UK, 1976.
32. Świątkowski, K.; Hatalak, R. Application of modified tools in the process of thin-walled tube drawing. *Arch. Metall. Mater.* **2006**, *51*, 193–197.
33. Hatalak, R.; Świątkowski, K. The wall deformation and the diameter reduction in the process of tube drawing with a floating plug. *Arch. Metall.* **2002**, *47*, 275–285.
34. Linardon, C.; Favier, D.; Chagnon, G.; Gruez, B. A conical mandrel tube drawing test designed to assess failure criteria. *J. Mat. Process. Technol.* **2014**, *214*, 347–357. [CrossRef]
35. Gattmah, J.; Ozturk, F.; Orhan, S. Experimental and finite element analysis of residual stresses in cold tube drawing process with a fixed mandrel for AISI 1010 steel tube. *Int. J. Adv. Manuf. Technol.* **2017**, *93*, 1229–1241. [CrossRef]

Disclaimer/Publisher's Note: The statements, opinions and data contained in all publications are solely those of the individual author(s) and contributor(s) and not of MDPI and/or the editor(s). MDPI and/or the editor(s) disclaim responsibility for any injury to people or property resulting from any ideas, methods, instructions or products referred to in the content.

Article

Tribological Performance of Anti-Wear Coatings on Tools for Forming Aluminium Alloy Sheets Used for Producing Pull-Off Caps

Kamil Czapla [1], Krzysztof Żaba [2,*], Marcin Kot [3], Ilona Nejman [4], Marcin Madej [5] and Tomasz Trzepieciński [6]

1. Canpack Metal Closures, ul. Kochanowskiego 28b, 33-100 Tarnów, Poland; kamil.czapla@canpack.com
2. Department of Metal Working and Physical Metallurgy of Non-Ferrous Metals, Faculty of Non-Ferrous Metals, AGH—University of Science and Technology, al. Adama Mickiewicza 30, 30-059 Cracow, Poland
3. Faculty of Mechanical Engineering and Robotics, AGH—University of Science and Technology, al. Adama Mickiewicza 30, 30-059 Cracow, Poland; kotmarc@agh.edu.pl
4. Department of Materials Science and Engineering of Non-Ferrous Metals, Faculty of Non-Ferrous Metals, AGH—University of Science and Technology, al. Adama Mickiewicza 30, 30-059 Cracow, Poland; inejman@agh.edu.pl
5. Faculty of Metals Engineering and Industrial Computer Science, AGH—University of Science and Technology, al. Adama Mickiewicza 30, 30-059 Cracow, Poland; mmadej@agh.edu.pl
6. Department of Manufacturing Processes and Production Engineering, Faculty of Mechanical Engineering and Aeronautics, Rzeszow University of Technology, al. Powst. Warszawy 8, 35-959 Rzeszów, Poland; tomtrz@prz.edu.pl
* Correspondence: krzyzaba@agh.edu.pl

Abstract: Ensuring adequate reliability of the production process of packaging closures has made it necessary to study the effect of annealing and varnishing variants on the strength and structural properties of the stock material. As a test material, EN AW-5052-H28 aluminium alloy sheets with a thickness of 0.21 mm were used. The surface treatment of the test material involved varnishing the sheet metal surface using various varnishes and soaking the sheet metal. The coefficient of friction and the abrasion resistance of the coatings were determined using the T-21 ball-and-disc tribotester. The tested sheets were subjected to tribological analysis by the T-05 roller-block tribotester using countersamples made of Caldie and Sverker 21 tool steels. The results of the tests showed differences in mechanical and structural properties depending on the method of sample preparation. Based on the test results, significant differences in the adhesion of anti-wear coatings were found. The results revealed that the most favourable friction conditions are provided by the CrN coating. The (AlTi)N interlayer in the (AlTi)N/(AlCr)N coating adheres to the substrate over the entire tested area and no detachment from its surface was observed, which proves good bonding at the substrate/coating interface. The tested AlTiN/TiAlSiXN coating is characterised by a more homogeneous, compact microstructure compared to the (AlTi)N/(AlCr)N coating.

Keywords: aluminium alloy; coefficient of friction; surface topography; wear; wear resistance

1. Introduction

The possibility of shaping the properties of materials by heat treatment and the selection of chemical composition is limited for structural steel materials. Improvement to these properties is possible by applying modifications to the surface layer [1]. One possibility is to create protective coatings on the surface of the material. Physical Vapour Deposition (PVD) and Chemical Vapour Deposition (CVD) methods differ primarily in process parameters, temperature and time, as well as their mechanisms of protective layer formation [2,3]. PVD methods are practically used to cover the surfaces of tools with titanium nitride TiN (less often with titanium carbide TiC) in order to achieve an increase in their durability. The increase in the tool's durability can be obtained as a result of complex treatments, e.g., nitriding with additional TiN coating or alternating coating with different layers of TiN + TiC [4]. The layers obtained by the PVD method have a thickness of about 3–5 μm and a hardness of about 2000–3500 HV [4]. Tools coated with PVD methods have a far longer service life than tools manufactured conventionally without protective coatings [5]. Coatings produced by the PVD process can be divided into simple (consisting of a single material, metal or phase) and complex, which consists of more than one material. Composite coatings include multi-component, multi-layer, multi-phase, composite and gradient coatings [6].

The application of protective coatings on tools for plastic working processes is primarily aimed at increasing wear resistance [7]. The wear process of plastic-forming tools carrying high mechanical and friction-wear loads is a complex phenomenon [8]. The process of tool wear consists of mechanical and thermal fatigue. Improper design of tools can be a source of stress accumulation, which initiates cracks under the influence of dynamic loads [9,10]. Another parameter-determining tool wear is the wrong selection of substrate material. Knowledge of the abrasion resistance of thin coatings can help in their correct selection for applications where abrasion plays a major role in their degradation [11]. The basic methods used to determine the durability of coatings are tribological tests using scratch tests [10] or tribotesters with various types: ball-on-disc [12], pin-on-disc [13], block-on-ring [14], etc.

Improving the anti-wear properties of AlTiN coatings has been the subject of many works over recent years. Wang et al. [15] found that the high coefficient of friction of AlTiN coatings was caused by the formation of a tribo-film on the wear track. Meng et al. [16] indicated that the synergistic effect of micro-textures and AlTiN coatings enhanced the anti-wear properties. The hybrid $Al_xTi_{1-x}N$ (x = ~0.65) coatings fabricated using cathodic arc evaporation and magnetron sputtering exhibited enhanced wear resistance and mechanical properties [17]. He et al. [18] investigated the tribological properties of the AlTiN coatings with different Al/Ti atomic ratios (73/27, 70/30, 67/33, 60/40, 50/50) and it was found that all coated inserts possessed an improved wear behaviour under wet cooling conditions. The indentation and wear tests on CrN-coated M50 disks indicated the relationships between applied loads and material removal patterns [19]. Biava et al. [20] experimentally analysed the high temperature corrosion behaviour of CrN, AlCrN and TiAlN coatings deposited by the arc evaporation PVD process onto Waspaloy Ni-based superalloy. The TiAlN and AlCrN and coatings showed a higher elastic modulus and hardness than the CrN coating. The wear investigations conducted by Navinšek and Panjan [21] indicated that a CrN coating (approx. 5 μm thick) with a Cr intermediate layer 0.2 μm thick between the substrate and the coating improved the quality of the surface finish of the products and increased tool life. Polok-Rubiniec et al. [22] found a correlation between hardness and adhesion of the CrN PVD coatings to the plasma-nitrided X37CrMoV5-1 and heat treated steel substrates. The analysis of the structure of CVD and PVD TiAlSiN coatings deposited on cemented carbide tools indicated that both coatings consisted of nanocrystals embedded in amorphous SiN_x [23]. Schulz et al. [24] deposited wear-resistive TiAlCrSiN coatings on a WC/Co metal substrate. The TiAlCrSiN coating showed a significant reduction in coefficient of friction due to the addition of Cr and Si to the coating which reduces adhesion. The wear behaviour of PVD coatings (TiN, TiAlN, TiC, TiAlN and Si_3N_4/AlTiN) in three ball-on-disc tests was analysed by Merklein et al. [25]. The TiC and TiAlCN coatings revealed a higher wear resistance than the coatings of AlCrN and TiN. The tribological performance of the AlTiN-TiSiN deposited by the PVD process was investigated by Claver et al. [26]. The experimental tests using the pin-on-disc test revealed that the combination of the nitriding process with the bonding layer deposited by high-power impulse magnetron sputtering improved the adhesion properties of AlTiN-TiSiN coatings. Das et al. [27] developed a nanocomposite AlTiSiN coating deposited using scalable pulsed-power plasma, which improved the surface finish of AISI D6 steel samples during turning. Liu et al. [28] investigated the microstructure and oxidation resistance of three nanocoatings (TiAlSiN/AlCrN multilayer, AlCrN monolayer and TiAlSiN monolayer). It was found that the AlCrN coating has the highest adhesion strength, while the TiAlSiN coating has the lowest adhesion strength among the coatings tested. Previous studies on AlCrN and TiAlSiN coatings were focused on thermal stability [29], high-temperature friction [30], microstructure [31], mechanical properties [32,33] and oxidation resistance [34,35]. Sousa et al. in a review paper [36] indicated the wear mechanisms of TiAlN-based coatings. Both AlCrN and AlTiN coatings are characterised by high oxidation resistance due to the formation of aluminium oxide surface layers [37].

This article attempts to characterise the coatings used on tools that form the pull-off caps made of EN AW-5052-H28 [38] aluminium alloy sheets. The technological process to produce food closures usually consists of precise cutting and stamping operations. The parameters of the input material, such as the coefficient of friction, are very important to ensure the proper course of the production process. In the food industry, particular emphasis is placed on product quality [39]. In this paper, the effect of the surface treatment of the EN AW-5052-H28 aluminium alloy sheets, consisting of varnishing the sheet metal surface using various varnishes and soaking the sheet metal, on the tribological properties, drawability, mechanical properties and wear resistance of test sheets is investigated. So far, there are no such studies available in the literature regarding EN AW-5052-H28 aluminium alloy sheets used in the food industry. The results of the tests showed differences in the mechanical and microstructural properties of the sheet depending on the method of varnishing. Based on the experimental results, significant differences in the adhesion of anti-wear coatings were found, despite the analogous method of preparing the steel substrate. The tests also showed fundamental differences in the microstructure of the coatings, and for some of them defects in the form of cracks parallel to the substrate and material losses were identified. The results of ball-on-disc tests concluded that the most favourable friction conditions among tested coatings are provided by the CrN coating. The low CoF of the CrN coating is associated with the smallest coefficient of volumetric wear of the sample and ball-shaped countersample.

The obtained results will allow for future research in industrial conditions aimed at determining the impact of the use of selected anti-wear coatings on the geometric quality of products and, consequently, on their application.

2. Materials and Methods

2.1. Test Material

The test material consisted of uncoated EN AW-5052-H28 aluminium alloy sheets with a thickness of 0.21 mm (Table 1), which were subjected to appropriate surface and/or heat treatment (Table 2). Selecting the appropriate sheet preparation technology is crucial in the manufacturing of pull-off caps (Figure 1) from the test material. The surface treatment consisted of varnishing the sheet metal surface using various varnishes and soaking the sheet metal at a temperature between 185 °C and 200 °C for 13 min. The inner surface of the pull-off cap was varnished. However, the outer surface was not varnished. To simplify the identification of lacquered samples, the sides of the sample are marked with colours: 'white' for non-lacquered surfaces and 'yellow' for lacquered surfaces. As a reference, the as-received sheet metal (sample no. 1 in Table 2) was used.

Table 1. Chemical composition [% wt.] of the EN AW-5052-H28 aluminium alloy.

Si	Fe	Cu	Mn	Mg	Cr	Zn	Ti
0.4	0.5	0.1	0.5–1.1	1.6–2.5	0.3	0.2	0.1

Table 2. Parameters of surface treatment of samples.

Number of Sample	Lacquering			Soaking Temperature and Duration			Colour of Surface	
	Adhesive Varnish Salchi VI 1106	Coating Varnish Salchi ANC 6001	Overprint Varnish Salchi VE2028	200 °C, 13 min	190 °C, 13 min	185 °C, 13 min	Inner Side	Outer Side
1	no	no	no	no	no	no	'white'	'white'
2	yes	no	no	yes	no	no	'yellow'	'white'
3	yes	yes	no	yes	yes	no	'yellow'	'white'
4	yes	yes	yes	yes	yes	yes	'yellow'	'white'
5	no	no	no	yes	no	no	'white'	'white'
6	no	no	no	yes	yes	no	'white'	'white'
7	no	no	no	yes	yes	yes	'white'	'white'

Figure 1. Exemplary geometry of lacquered pull-off cap.

2.2. Surface Roughness Measurement

Surface roughness of the samples was measured on a laboratory stand equipped with a T1000 (Hommel-Etamic Jenoptik, Jena, Germany) roughness-measuring instrument. Surface roughness results are presented by the two most commonly used parameters, i.e., Ra—arithmetic average of the absolute values of the profile heights over the evaluation length and Rz—the average value of the absolute values of the heights of the five highest-profile peaks and the depths of the five deepest valleys within the evaluation length. The surface roughness measurement was carried out over a length of 4.8 mm. The test was performed for three orientations of the sample in relation to the sheet rolling direction: 0°, 45° and 90°. Five measurements were made for each direction. The tests were carried out for both sides of the analysed specimens.

2.3. Abrasion Resistance Roller-Block Test

The abrasion resistance test of various grades of tool steels against the EN AW-5052-H28 aluminium alloy countersample was carried out on the T-05 roller-block tester. The abrasion resistance of samples presented in Table 2 was also tested. The measurement was carried out at an ambient temperature with translational motion in dry friction conditions. The principle of operation of the tester is shown schematically in Figure 2. The self-adjusting attachment of block (1), which is the sample holder (4) and the hemispherical insert (3), ensures that the block adheres to the roller (2) and evenly distributes the pressure on the contact surface. The tester used for the tests enables tests to be carried out in accordance with the ASTM D 2714, D 3704, D 2981 and G 77 standards. Countersamples from cold-work tool steels Caldie and Sverker 21 manufactured by Uddeholms AB (Hagfors, Sweden) were used in the tests.

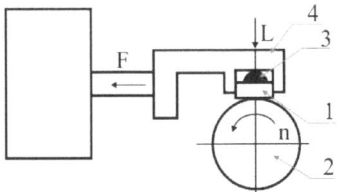

Figure 2. Principle of operation of T-05 roller-block tester: 1—block, 2—roller, 3—insert, 4—sample holder, F-friction force, L-load, n- rotation speed.

The test samples were in the form of sheet metal strips with dimensions 20 × 4 × 0.25 mm. The average value of the average roughness Ra of the test samples, measured using a confocal microscope in accordance with ISO 21920-1:2021 [40], was 0.665 µm. Ring countersamples with a diameter of 49.5 mm made of Caldie and Sverker 21 steels applied with various coatings (Table 3) were used in the test. All measurements were performed at a constant ring rotational speed of 136 rpm. During the tests of the aluminium alloy samples (Table 2), a load of FN = 10 N was used and the friction path was 100 m. While testing the countersamples of Caldie and Sverker 21 steel, the load was 20 N and the friction path was 50 m.

Table 3. Types of coatings used on Caldie and Sverker 21 steel samples.

Sample Designation	Type of Coating
PR1	MTec (AlTi)N
PR2	CrN
PR3	MPower (AlTiN/TiAlSiXN). X = Cr, B, Y
PR4	MForce ((AlTi)N/(AlCr)N)

During the abrasive test, the friction force FT was continuously recorded, which was used to determine the CoF µ according to Equation (1).

$$\mu = \frac{F_T}{F_N} \quad (1)$$

The measure of abrasion resistance is the mass loss of the tested material in relation to the friction path and the applied load. The mass loss expressed in grams was determined in accordance with Equation (2), while the mass percentage of loss was determined in accordance with Equation (3).

$$\Delta m = m_p - m_k \quad (2)$$

$$\Delta m = \frac{m_p - m_k}{m_p} \cdot 100\% \quad (3)$$

where mp is the initial mass of the sample and mk is the final mass of the sample.

Observations of the surface morphology after friction test, measurements of surface roughness and identification of friction mechanisms were carried out using the LEXT OLS 4100 confocal microscope (Olympus Europe SE & Co. KG, Hamburg, Germany). The device is fully automatic in a simple, bottom-table arrangement for observation in reflected light. This microscope uses UV laser light with a wavelength of 405 nm.

2.4. Determination of the Coefficient of Friction (Ball-on-Disc Tribometer)

The CoF and the wear index were determined in accordance with the ISO 20808 standard. For this purpose, a T-21 ball-on-disc tribotester ((Łukasiewicz Research Network—Institute for Sustainable Technologies, Radom, Poland) designed to study the tribological properties of cooperating materials during sliding friction was used. The scheme of the T-21 ball-and-disc tribotester is shown in Figure 3. It consists of a rotating disc (sample) and a statically fixed countersample in the form of a pin with a spherical tip made of Al$_3$O$_4$. The pin is loaded with the force F.

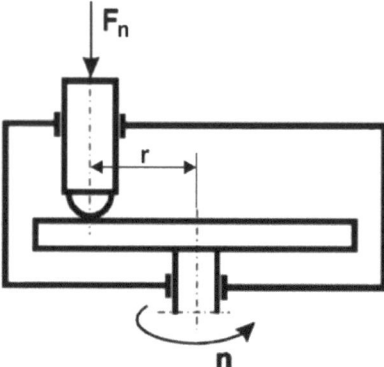

Figure 3. Schematic diagram of the ball-on-disc tribotester, Fn-pressure force, r-distance from the axis, n- rotation speed.

During the test, the values of the friction force, rotational speed of the disc and depth of the wear track were recorded using a computer program. The normal load Fn is applied with weights placed on the countersample mounting lever. After the tests were completed, the samples were cooled to the ambient temperature and cleaned, and then the wear tracks were examined with a profilometer to determine the coefficient of volumetric wear Wv according to the Equation (4):

$$W_v = \frac{V}{F_n \cdot s} \left[\frac{mm^3}{N \cdot m}\right] \qquad (4)$$

where V is volume of the used material [mm^3] and s is friction path [m].

The test was carried out with the following parameters: normal force Fn = 5 N, number of sample revolutions—15,000, radius of friction path r = 4, 5 and 6 mm. The volume of the worn material was determined based on the measurement of the cross-sectional area of the wear track. Measurements of the track profile after wear tests were carried out using an optical interferometric profilometer ProFilm3D (Filmetrics, San Diego, CA, USA).

2.5. Mechanical Properties

The basic mechanical properties of the sheet metals were determined in a static tensile test using the Zwick/Roell Z020 uniaxial tensile testing machine (Zwick/Roell, Ulm, Germany). The tests were carried out in accordance with the EN ISO 6892-1:2022 standard [41]. The tests were carried out for samples cut in the following directions: perpendicular, parallel and at an angle of 45° to the direction of sheet rolling. Three samples of each type were tested and average values of the mechanical parameters were determined.

2.6. Analysis of Microstruture

The analysis of the microstructure of the samples was carried out using a metallographic microscope Axio Vert.A1 Mat equipped with an Axiocam 305 (ZEISS, Jena, Germany) camera, a metallographic microscope GX51 (Olympus Europe SE & Co. KG, Hamburg, Germany) and a scanning electron microscope SU 70 (Hitachi Ltd., Tokyo, Japan). The analyses of the chemical composition in the form of maps of the distribution of elements and of micro-areas of individual materials were performed using the energy dispersion spectroscopy (EDS) method.

Samples for microstructural analysis were cut out and then positioned in epoxy resin by Struers (Copenhagen, Danmark). Samples for optical microscopy were ground with #220,

#500, #800, #1200, #2000 and #4000 grit SiC papers. Then the samples were polished with MD MOL (Struers) woven wool metal-backed polishing cloths using diamond powders with diameters of 6 μm and 3 μm. Finally, the specimens were polished for 2 min using metal-backed coarse- and fine-ground MD cloths in the presence of a suspension of silicon dioxide, organic solvents and water OP-S (Struers).

2.7. Hardness Testing

Nanohardness tests were carried out using the Step 500 device (Anton Paar Gmbh, Ostfildern, Germany), which is equipped with the nanoindenter NHT3 module (nanohardness tester) meeting the requirements of the ASTM-E2546-15 standard [42]. Hardness was determined by the Berkovich method at a load of 20 and 50 mN. The standard Berkovich indenter geometry with a centerline-to-face angle of 65.3° was used.

3. Results

3.1. Surface Roughness of Samples

The tests showed significant differences in the surface roughness of the sample materials. It was found that the Ra and Rz roughness parameters of the analysed samples depend on both the material and the sample orientation in relation to the sheet rolling direction (Figures 4–6). In the case of samples no. 2–4, the side ('white' or 'yellow') of the sheet that was tested was also important. In general, it was found that, in the case of non-lacquered materials (samples no. 1, 5–7), the average values of the roughness parameters on both sides of the sheet are similar (Figure 5). For samples no. 2 and 4, Ra and Rz were higher for the 'yellow' side, while the 'white' side of sample 3 was characterised by much higher roughness.

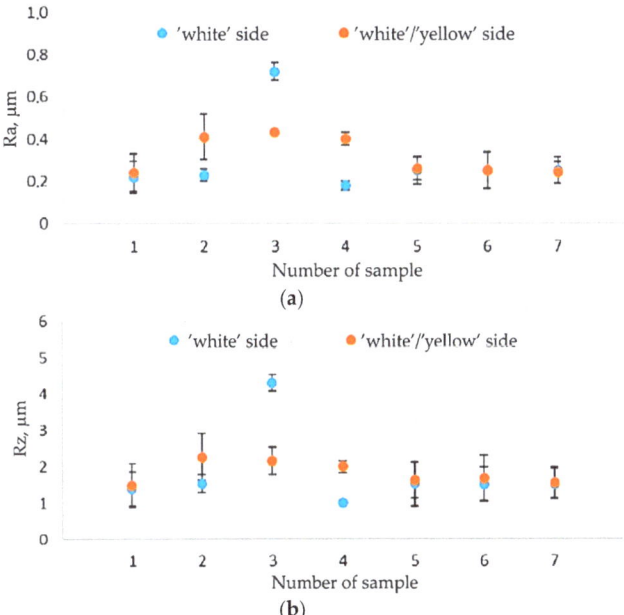

Figure 4. Average values of roughness parameters for the tested samples: (**a**) Ra, (**b**) Rz.

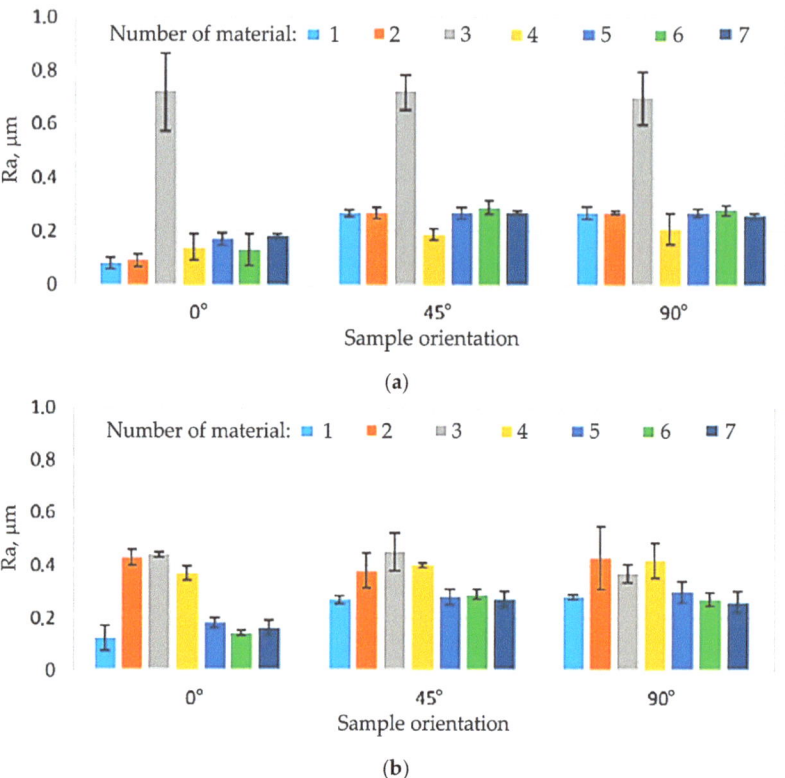

Figure 5. Values of the roughness parameter Ra depending on the orientation angle in relation to the sheet RD on (**a**) 'white' and (**b**) 'white'/'yellow' sides.

For all the analysed sheets, a significant influence of the measurement direction on the value of surface roughness parameters was found, especially for the side without varnish. For samples no. 1 and 5–7, this relationship was observed for both non-lacquered sides of the samples. The lowest values of roughness parameters were recorded for the orientation 0° in relation to the sheet rolling direction (RD). For the remaining orientations (45°, 90°), the values of Ra and Rz were similar (Figures 6 and 7). The values of surface roughness parameters of the lacquered samples no. 2–4 was different from the other samples and differed depending on the surface of the samples tested. In the case of sample 3, it can be concluded that there are no significant differences in the values of the Ra and Rz parameters for the tested directions in relation to the RD. Only slightly lower values of roughness parameters were observed for the 90° orientation and the 'yellow' side of the sample. In the case of samples no. 2 and 4, lower values of roughness parameters were found for the RD and the 'white' side of the sheet. On the other hand, for the 'yellow' side, the values of Ra and Rz were close to each other.

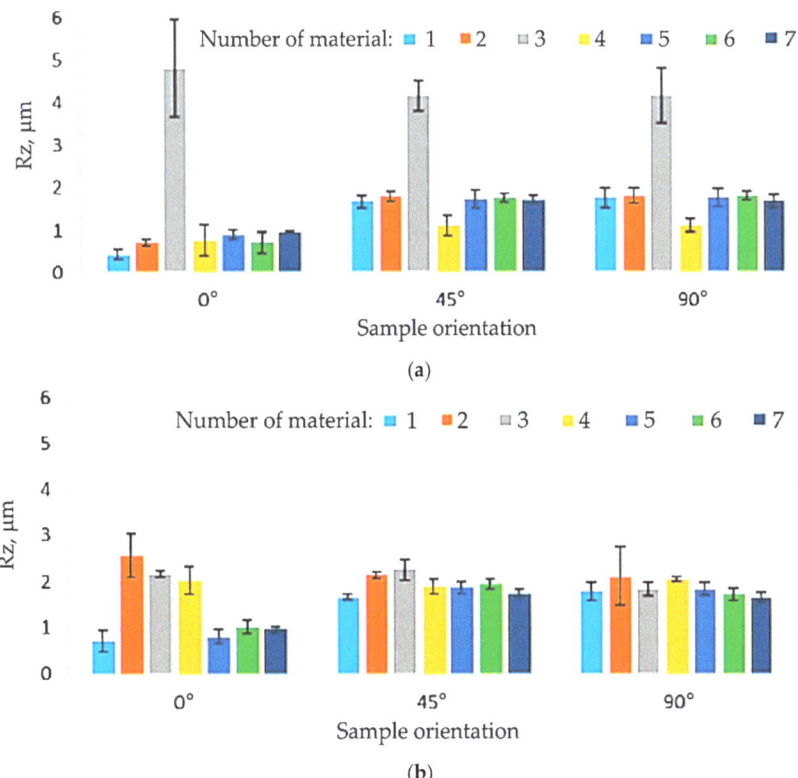

Figure 6. Values of the roughness parameter Rz depending on the orientation angle in relation to the sheet RD on (**a**) 'white' and 'white'/'yellow' sides.

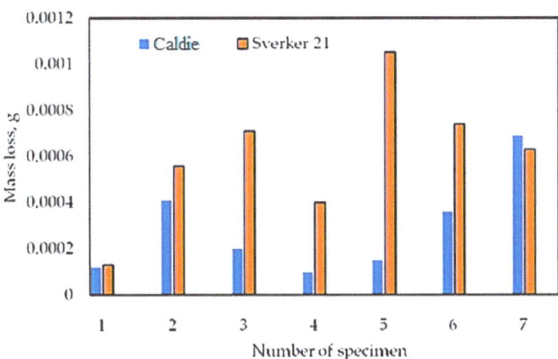

Figure 7. Mass loss of the countersample material.

The tested sheets, especially without the varnish layer (samples no. 1, 5–7), are characterised by different values of roughness parameters depending on the orientation relative to the sheet RD. The surface roughness of the lacquered sheets (samples no. 2–4) was greater than that of the non-lacquered surfaces, although the distribution of Ra and Rz values on their surface was more even. The transparent varnish (Salchi VE2028) causes a significant increase in the surface roughness of the tested material, regardless of the orientation angle in relation to the sheet RD. For this situation, the Ra and Rz parameter values are also the highest. To sum up, the application and curing of the varnish have a significant impact on the surface roughness of samples.

3.2. Wear Resistance and Coefficient of Friction (Roller-Block Test)

Table 4 shows the mass loss of the materials under a load of 20 N. In addition, the results of mass loss and average CoF are shown graphically in Figures 7 and 8, respectively. Analysing the presented results, for most of the tested sheets (samples 1–6), smaller mass losses were clearly recorded for the countersample of Caldie steel. Samples 1 and 5 without a varnish coating and sample 4 with a varnish coating are characterised by the lowest wear. The greatest difference in the CoF, approximately 50%, when testing with analysed countersamples, was observed for samples no. 5 and 6 (Figure 9). In turn, the smallest difference in the value of the CoF was observed for samples no. 4 and 7. Figures 9 and 10 show the variation of the friction force during friction testing. In the case of Caldie steel countersample the friction force is at the most stable level and reaches the lowest values for sample 4. On the other hand, the highest CoF occurs in the case of contact of samples no. 1 and 6 with the countersample made of Sverker 21 steel. Test results with the Sverker 21 steel countersample (Figure 10) appear to show a more stable friction process compared with the Caldie steel countersample (Figure 9). For most of the samples (no. 2, 3, 4 and 7), the value of the friction force varies slightly throughout the test period and is much lower than for the as-received sheet metal (sample no. 1). No clear relationship was observed between the surface roughness of the samples and the course of the friction process.

Table 4. Mass loss of test countersamples.

Sample Number	Countersample Material							
	Caldie				Sverker 21			
	m_1, g	m_2, g	Δm, g	Δm, %	m_1, g	m_2, g	Δm, g	Δm, %
1	0.03599	0.03587	0.00012	0.33	0.03940	0.03927	0.00013	0.33
2	0.04083	0.04042	0.00041	1.00	0.03879	0.03823	0.00056	1.44
3	0.03865	0.03845	0.0002	0.52	0.04063	0.03992	0.00071	1.75
4	0.04148	0.04138	0.0001	0.24	0.03993	0.03953	0.0004	1.00
5	0.03746	0.03731	0.00015	0.40	0.03548	0.03443	0.00105	2.96
6	0.03593	0.03557	0.00036	1.00	0.03388	0.03314	0.00074	2.18
7	0.03800	0.03731	0.00069	1.82	0.03693	0.03630	0.00063	1.71

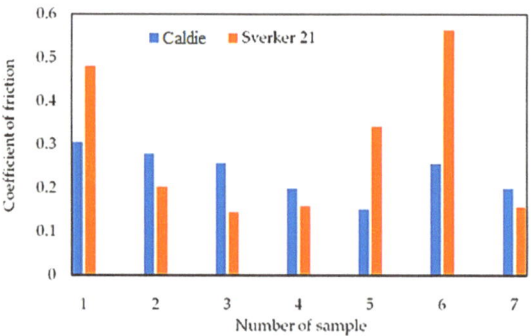

Figure 8. Average CoF of tested samples.

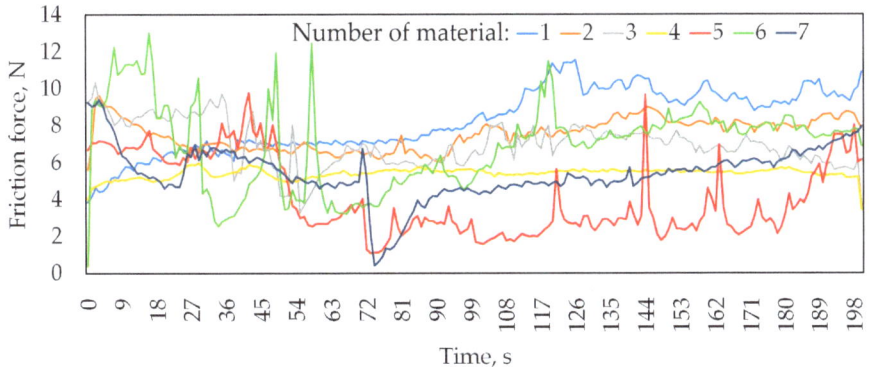

Figure 9. Changes in friction force as a function of test time for the Caldie steel countersample.

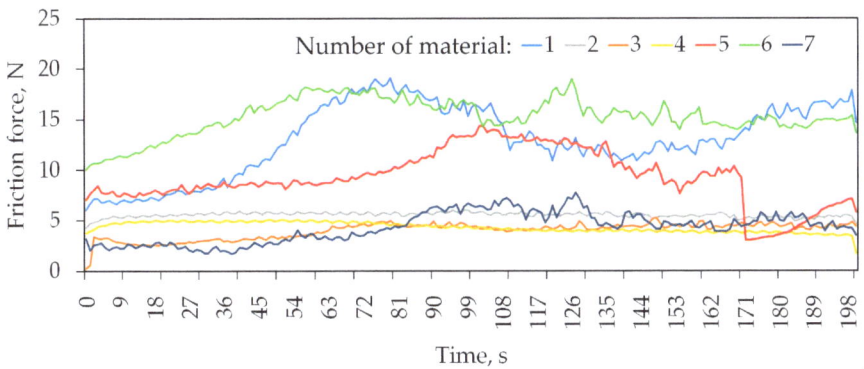

Figure 10. Changes in friction force as a function of test time for the Sverker 21 steel countersample.

The course of changes in frictional force is more unstable for the Caldie steel counterexample; however, the lapping time is shorter for this counterexample, with a maximum of 72 s, where for the Sverker 21 steel it is up to 100 s. The Caldie steel counterexample caused deep scratches on the surface of the sheet, especially on painted samples. It seems to show some affinity with the varnish and cause it to detach from the surface, as the friction force stabilises after rubbing through the varnish. In some of the curves showing changes in frictional force as a function of test time, we observe abrupt changes in force over very short time periods, especially for tests with a countersample made of Caldie steel (Figure 9). This is due to the presence of aluminium in the friction node, which causes adhesion to occur. This mechanism is related to the formation and breaking of adhesive joints formed at the friction node. A stable course after degradation of the varnish layer is observed for samples no. 3 and 4 (Figures 9 and 10). The behaviour of the samples in contact with Sverker 21 steel is completely different; only samples 1 and 5 show significant, abrupt changes in force during the entire tribological test. The surfaces of selected samples after the friction process, depending on the type of counterexample material, are shown in Figures 11–14.

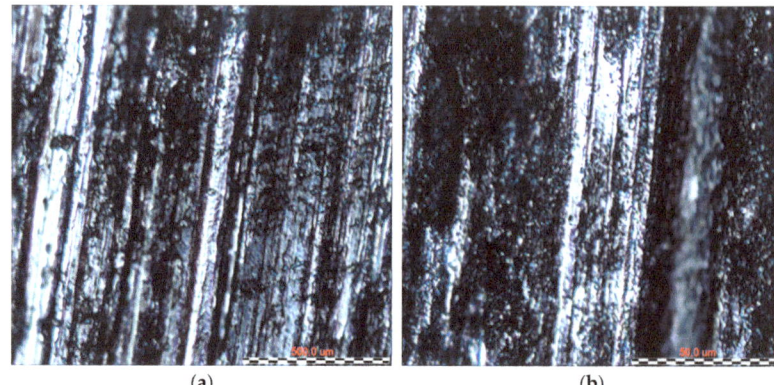

Figure 11. The surface of sample no. 1 after the friction process with countersample made of (**a**) Caldie and (**b**) Sverker 21 steel.

Figure 12. The surface of sample no. 2 after the friction process with countersample made of (**a**) Caldie and (**b**) Sverker 21 steel.

Figure 13. The surface of sample no. 3 after the friction process with countersample made of (**a**) Caldie and (**b**) Sverker 21 steel.

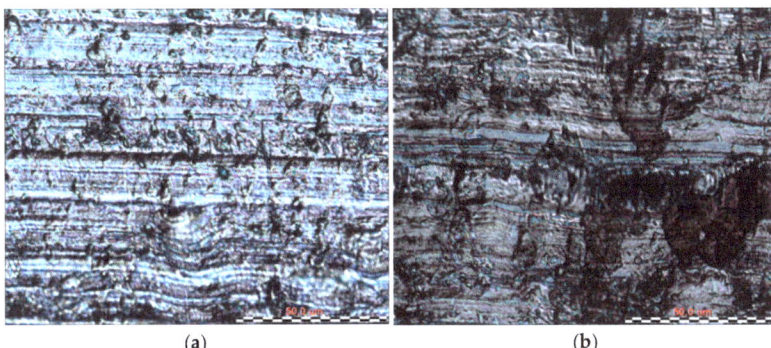

Figure 14. The surface of sample no. 4 after the friction process with countersample made of (**a**) Caldie and (**b**) Sverker 21 steel.

By analysing the friction surfaces shown in Figures 11–14, it can be concluded that the typical wear mechanism in these materials is abrasive wear, with scratching and ploughing identified. Locally, adhesive wear and the accumulation of wear products in the form of sticker patches are also observed (Figure 11b). For sample 1, the Sverker 21 counterexample results in the accumulation of more wear products on the surface, which consist of oxides chipped from the surface. The main differences for samples 2 and 3 can be identified in terms of the depth of the scratches and grooves on the surfaces after friction (Figures 12 and 13); deeper grooves were observed after using the counterexample made of Caldie steel. After using a counterexample made of Sverker 21 steel, the surfaces of the samples are 'smoother' and wear products adhering to the surface are also observed. Sample no. 4, for which the friction surfaces are summarised in Figure 14, differs from those presented in Figures 11–13. In addition to the scratches, areas that are locally deformed are visible. This may be related to the disruption and overspray of the painted layer. This phenomenon is more pronounced for countersamples made of Caldie steel. Similar wear mechanisms after analysing the surfaces of the specimens were observed for unpainted specimens no. 5–7, where the dominant wear mechanism was abrasion of the sheet surface through scratching and grooving.

To sum up, depending on the method of surface preparation of the samples and the grade of countersample material, fundamentally different results of mass loss and CoF were obtained. The course of changes in the friction force is less stable for the Caldie steel countersample, while a stable course after rubbing through the varnish layer was observed for samples no. 3 and 4. In the case of samples no. 1–6, smaller mass losses were recorded for the Caldie steel countersample. Non-lacquered samples no. 1 and 5 and sample no. 4 exhibited the lowest wear. The CoF is stable and reaches the lowest values for sample no. 4 for both countersample materials. The highest CoF occurs when samples no. 1 and 6 meet the Sverker 21 steel countersample.

3.3. Mechanical Properties

Figure 15 presents the results of the mechanical properties of sheets cut perpendicular, parallel and at an angle of 45° to the RD. Depending on the treatment method, the yield stress and ultimate tensile strength varied between 235.3 and 264 MPa, and between 272 and 301.3 MPa, respectively. However, all samples showed a decrease in these parameters compared to the as-received sample (sample no. 1). As-received material is characterised by the following mechanical parameters: yield stress $Rp_{0.2}$ = 279.3, ultimate tensile strength Rm = 315 MPa and elongation A = 5.4%. The highest mechanical properties are exhibited by sample no. 1, and the lowest by sample no. 4 which was soaked at 200 °C. So, soaking has been shown to reduce the strength of EN AW-5052-H28 aluminium alloy sheet. All sheets are characterised by high anisotropy, with the lowest tensile strength being shown by samples cut at an angle of 45° to the RD and the highest by sheets cut perpendicular to the RD.

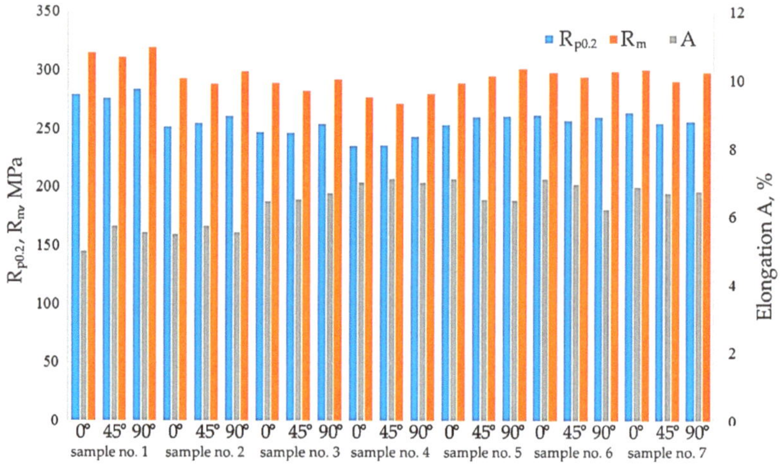

Figure 15. Average mechanical parameters obtained for all samples.

3.4. Hardness of Coatings

Coatings with high hardness and low modulus of elasticity can carry a significant load that does not plasticise the coating [14]. Ensuring both high hardness and hardness to Young's modulus ratio with adequate substrate stiffness is particularly important for thin coatings; however, in their tribological applications, the value of the CoF should also be considered.

For the assumed indenter loads of 20 and 50 mN, the penetration depth loads were significantly less than 1/10 of the coating thickness, which allows us to conclude that the obtained values are the properties of the tested coatings and the substrate had no effect on deformations. All coatings with the exception of MForce are characterised by greater hardness than the reference TiN coating. For the MPower shell, this value is more than twice as high. However, their Young's modulus is slightly higher. This indicates that the H/E ratio will be higher for the proposed coatings, and thus the expected wear resistance will also be higher. This is confirmed by the results of tribological tests presented in the further part of the work in Figure 16. For the applied loads of 20 and 50 mN, no coating cracks were observed, which is also evidenced by the lack of pop-ins on the indentation curves.

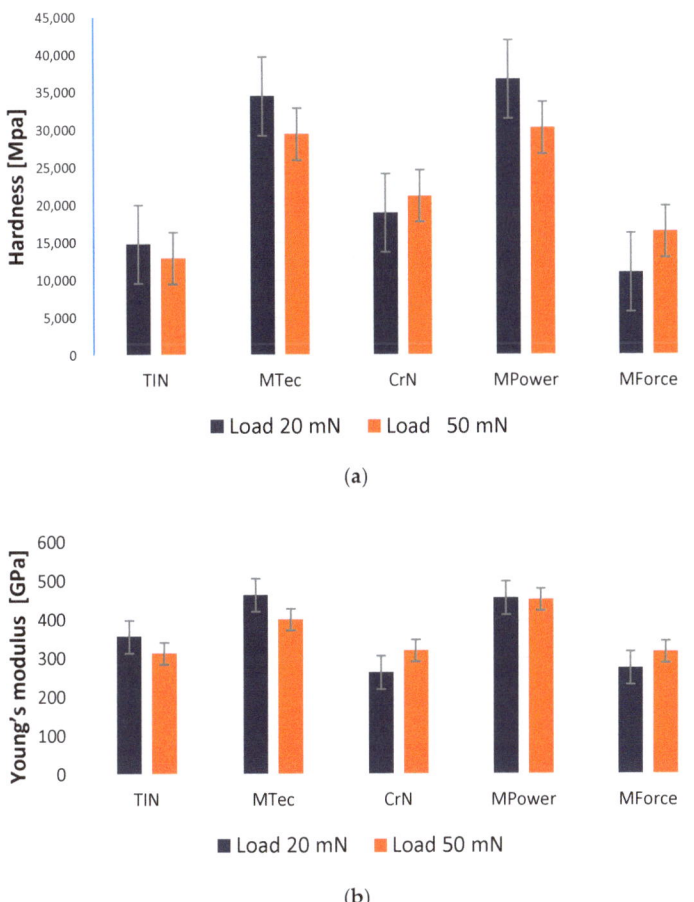

Figure 16. Hardness (**a**) and Young's modulus (**b**) of the coatings at the maximum loading force of 20 mN and 50 mN.

3.5. Coefficient of Friction and Wear Coefficient (Ball-on-Disc Test)

Figure 17 presents a summary of the average values of the CoF of the analysed coatings. The highest value of the average CoF was observed for samples coated with CVD-TiN and MTec coatings. The most favourable friction conditions are provided by the CrN coating, for which the value of the CoF is about 36% lower than that of the CVD-TiN coating. The low CoF of the CrN coating is also associated with the smallest coefficient of volumetric wear of the sample (Figure 18a) and countersample (Figure 18b).

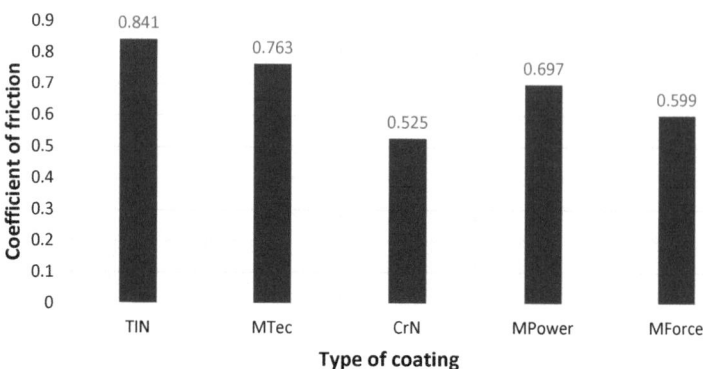

Figure 17. Average values of the CoF for analysed coatings.

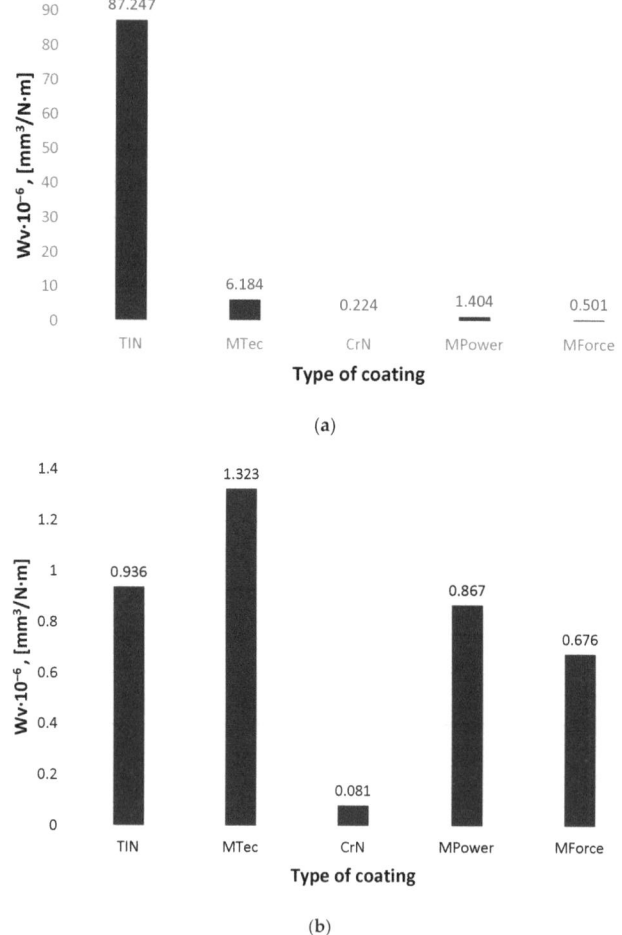

(a)

(b)

Figure 18. Coefficient of volumetric wear of the samples (a) and of the countersample (ball) (b).

The sample coated with CVD-TiN showed the highest coefficient of wear index (87.2476·10–6 mm^3/Nm). The coefficient of volumetric wear of the remaining coatings was less than 6.2·10–6 mm^3/Nm. The value of volumetric wear for MPower, MTec, MForce and CrN coatings corresponds well with the value of volumetric wear of countersample (ball). The greater the value of volumetric wear of the sample (Figure 18a), the greater the value of volumetric wear of the countersample (Figure 18b). This relation does not apply to the CVD-TiN countersample. Despite the high value of the volumetric wear of this countersample, the ball wears at a similar level as during the wear test with MPower coatings. This proves the low wear resistance of the CVD-TiN coating. The greatest wear of the countersample concerned cooperation with the sample coated with MTec (1.3233·10–6 mm^3/Nm).

Topographies of friction tracks of the coated samples are shown in Figures 19 and 20.

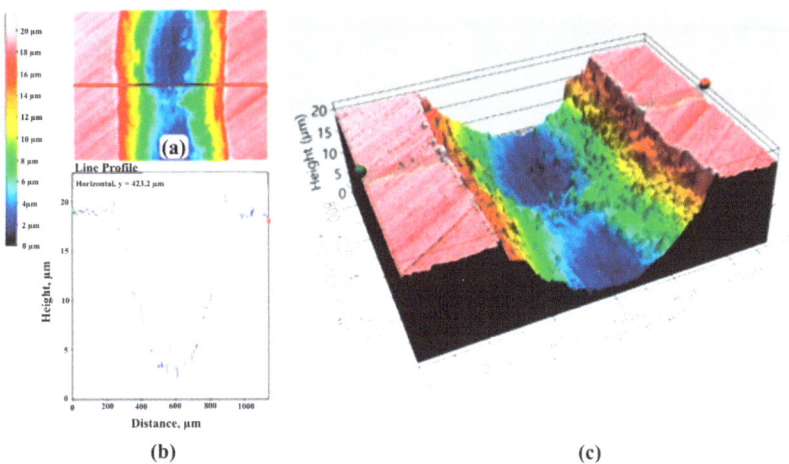

Figure 19. Topography of wear track (top view) (**a**), profile of friction track and (**b**) topography of friction track (isometric view) (**c**) for CVD-TiN sample.

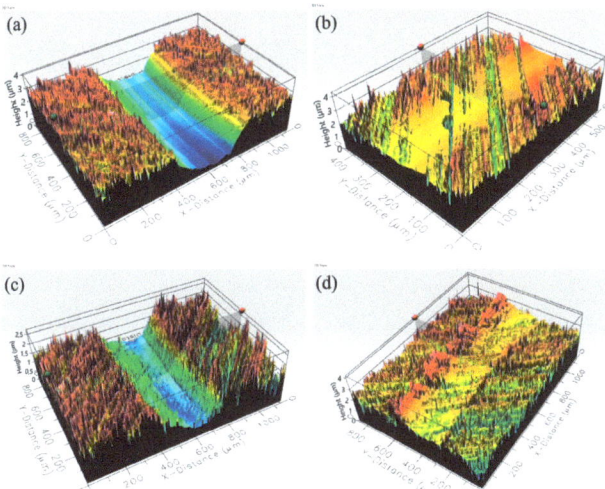

Figure 20. Three-dimensional topographies of friction track for samples coated with the following: (**a**) MTec, (**b**) CrN (**c**) Mpower, (**d**) Mforce.

3.6. Wear Resistance of Countersamples (Pin-on-Disc Tribometer)

Several countersamples with different properties in tribological contact with samples made of EN AW-5052-H28 aluminium alloy sheet were tested. The materials were analysed for their wear resistance, CoF and the morphology of worn surfaces in conditions of dry friction. The tests were carried out at ambient temperature (about 21 °C) and at about 30% humidity.

Tribological tests were carried out using countersamples with as-received surfaces. The highest mass loss (2.18%) was observed for the countersample with the MTec coating. The countersamples coated with MForce and CrN coatings turned out to be the most wear resistant (Figure 21a). Despite the lowest mass loss, the CrN-coated countersample showed the highest CoF value, μ = 0.398 (Figure 21b). Among all the coatings tested, the difference in the CoF value is less than 10%. Despite fluctuations in the friction force, the countersamples coated with MTec, MPower Nano, CrN and MPower exhibited a stable average friction force throughout the test. Only the MForce-coated countersample showed a continuous increase in the friction force over the entire test period. This may indicate premature rupture of the coating and intensification of the wear processes. On the other hand, the value of the friction force during the test with this countersample was the most stable without large fluctuations.

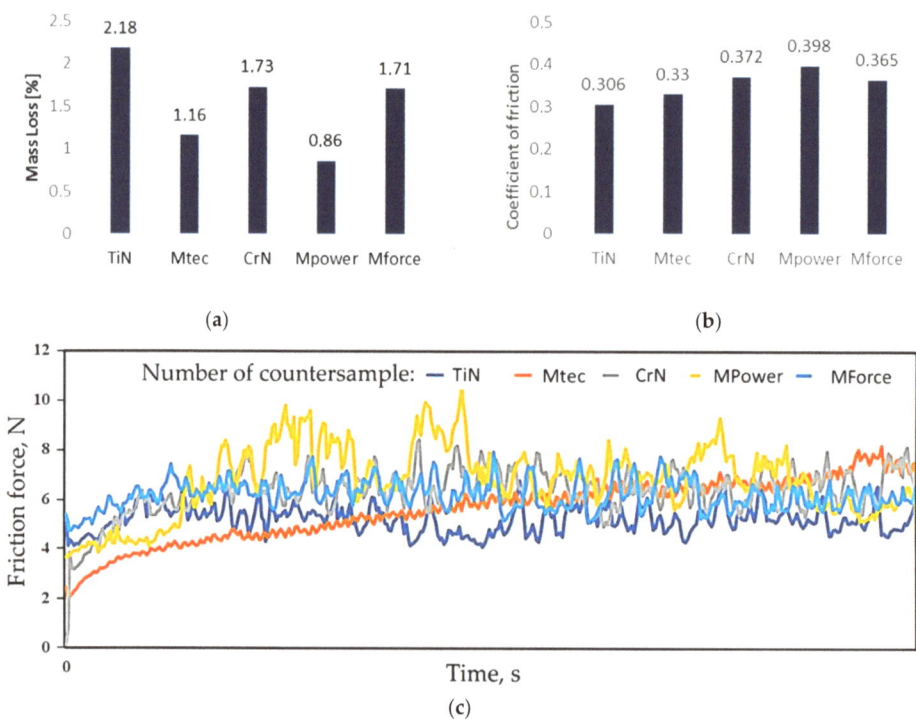

Figure 21. (a) Mass loss, (b) CoF and (c) variation of the friction force during the testing of coated countersamples against EN AW-5052-H28 aluminium alloy sheet.

The highest wear was recorded for the MTec-coated countersample at over 2% with a CoF of about 0.3. Practically no wear was observed on the surface of this countersample; there are trace scratches, while much greater abrasive wear from scratching and ploughing was found on the surface of the EN AW-5052-H28 aluminium alloy sheet. The course of changes in the friction force is quite unstable, and may be related to the accumulation of wear products on the surface of the sample. In the case of the MForce-coated countersample, its wear in contact with the EN AW-5052-H28 aluminium alloy sheet was insignificant and was within the measurement error limit, while almost the smallest mass loss of the countersample was recorded at over 1.16% with a slightly higher CoF (0.33) than in the case of the MTec-coated countersample.

A high coefficient of friction can increase wear on stamping tools, which increases production costs. Tools must be wear-resistant and durable to maintain the efficiency of the stamping process. In addition, the low coefficient of friction will improve the tool's guidance due to a reduction in the heat released during work.

3.7. Microstructure

The results of microstructural analyses of the tested coatings are summarised in Figures 22–25. Using metallographic microscopy, thin protective coatings were identified on the surface of the substrate. In the case of the first tested sample, the MTec coating is unevenly deposited on the surface of the substrate, and it occurs only in small fragments over the entire area of the tested material (Figure 22). The SEM micrographs confirmed the uneven deposition of the coating on the surface of the steel substrate. Non-coated areas and some fragments of coating were identified (Figure 23). It should be noted that where the coating is present, it is homogeneous and adheres well to the surface of the substrate.

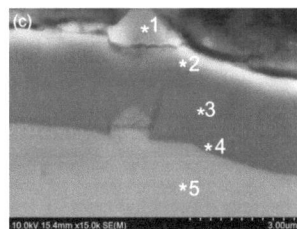

Point no.	N-K	Al-K	Ti-K	Cr-K	Fe-K	Mo-L
1	12.05	36	42.78	0.85	7.87	0.27
2	17.89	33.89	46.60	0.22	1.40	0.06
3	28.80	27.16	41.39	0.37	2.27	0.06
4	24.26	24.38	35.03	7.52	8.81	0.04
5	-	1.47	2.82	41.86	51.87	1.99

Figure 22. Microstructure of the MTec coating and substrate (SEM) (**a**), thickness of coating (**b**) and chemical composition (wt. %) of the MTec coating and the substrate (EDS-SEM) (**c**).

The thickness of the other two coatings, CrN and MPower, was, respectively, 0.8 and 2.5 μm (Figures 24 and 25). The CrN (Figure 23) and MPower (Figure 24) coatings researched have a flat and smooth structure. Funnel-like artefacts were observed for the MPower coating locally owing to the presence of a droplet phase embedded in the coating, which is connected with the nature of the employed coating deposition CVD process (Figure 25). No microstructural defects were observed for either CrN or MPower coatings. In a few areas, for the CrN coating, delamination between the coating and the substrate was visible (Figure 23). Other areas where the coatings adhered well to the substrate surface are characterised by tight adherence to the substrate material. Small areas of delamination should not affect the quality and strength of the coating. This can be confirmed by the results of the study of the coefficient of friction and wear coefficient (Figure 18).

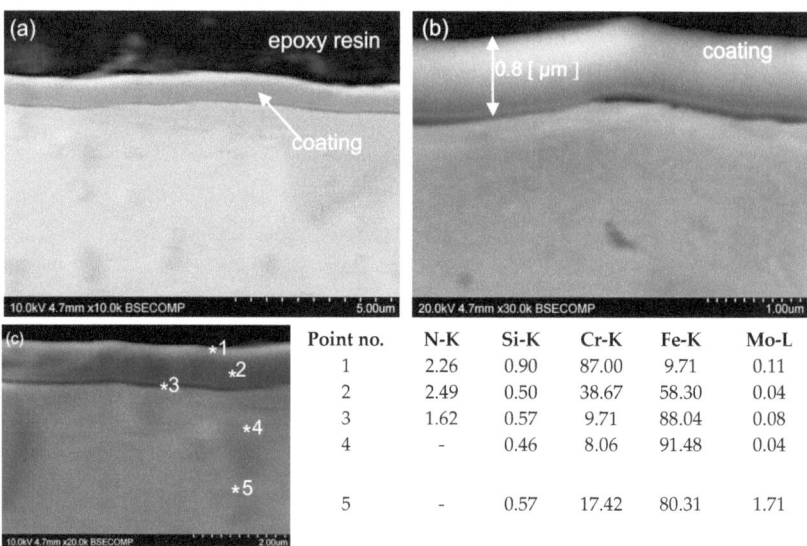

Point no.	N-K	Si-K	Cr-K	Fe-K	Mo-L
1	2.26	0.90	87.00	9.71	0.11
2	2.49	0.50	38.67	58.30	0.04
3	1.62	0.57	9.71	88.04	0.08
4	-	0.46	8.06	91.48	0.04
5	-	0.57	17.42	80.31	1.71

Figure 23. Microstructure of the CrN coating and substrate (SEM) (**a**), thickness of coating (**b**), chemical composition (wt. %) of the CrN coating and the substrate (EDS-SEM) (**c**).

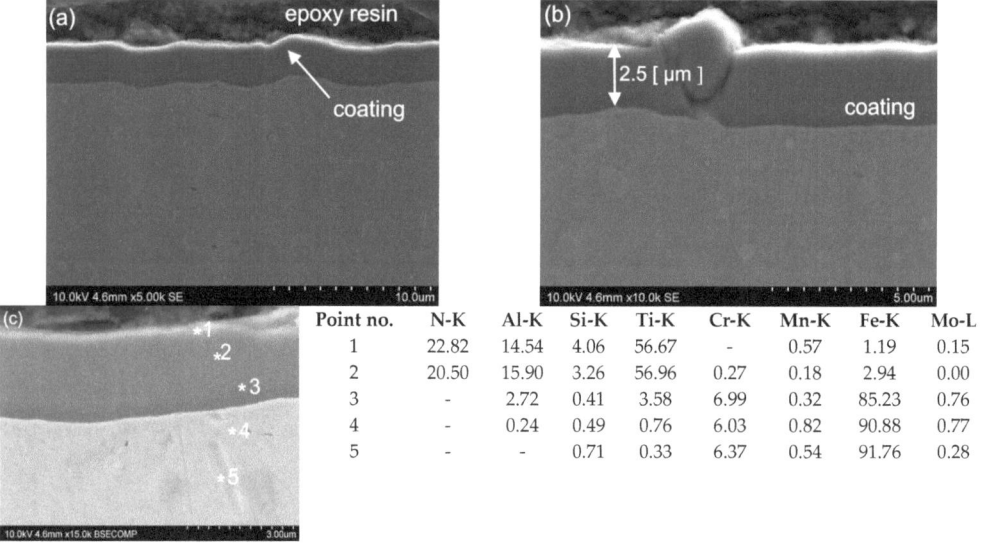

Point no.	N-K	Al-K	Si-K	Ti-K	Cr-K	Mn-K	Fe-K	Mo-L
1	22.82	14.54	4.06	56.67	-	0.57	1.19	0.15
2	20.50	15.90	3.26	56.96	0.27	0.18	2.94	0.00
3	-	2.72	0.41	3.58	6.99	0.32	85.23	0.76
4	-	0.24	0.49	0.76	6.03	0.82	90.88	0.77
5	-	-	0.71	0.33	6.37	0.54	91.76	0.28

Figure 24. Microstructure of the MPower coating and substrate (SEM) (**a**), thickness of coating (**b**), chemical composition (wt. %) of the MPower coating and the substrate (EDS-SEM) (**c**).

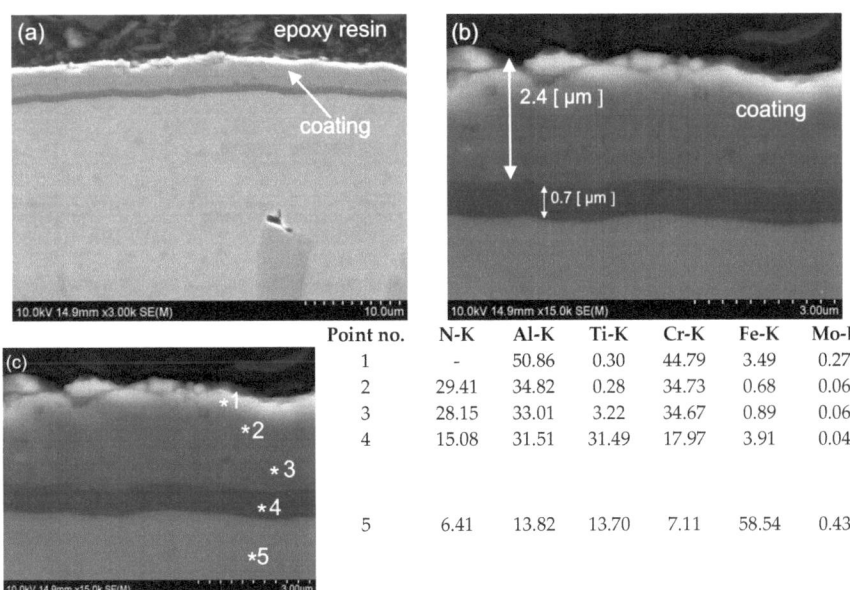

Figure 25. Microstructure of the MForce coating and substrate (SEM) (**a**), thickness of coating (**b**), chemical composition (wt. %) of the MForce coating and the substrate (EDS-SEM) (**c**).

The MForce multi-layer coatings reveal dense structures with not-visible delamination and defects. The coating was uniformly distributed over the entire width of the test sample. For a more accurate microstructural analysis, SEM measurements were performed, confirming the presence of two superimposed layers. The thickness of the main layers of the MForce (AlTi)N/(Al,Ti)N coating were equal 0.7 and 2.4 μm (Figure 25). In the tested areas of the coating, the point analysis using an X-ray energy dispersive spectrometer EDS indicated the presence of such elements as Al, Ti and N (Figure 22c, Figure 23c, Figure 24c, Figure 25c). The * symbol indicates the place of point analysis of chemical composition. Analysis of the chemical composition showed that the interlayer is rich in Ti, Al and N. The overlying coating contains Cr, Al and N. In the tested areas, the outer layer shows areas where microstructural defects in the form of cracks and material losses were identified. The coating shows good adhesion to the substrate surface.

4. Conclusions

This article presents the results of tribological and microstructural studies of coatings applied to tools for forming pull-off caps. As a test material, EN AW-5052-H28 aluminium alloy sheets with different surface treatments were used. Measurements of surface roughness of sheet metals and basic mechanical properties of sheet metals and hardness of coatings, as well as wear tests and formability tests using the Erichsen method, were carried out. Based on the comprehensive tribological and microstructural investigations, the following conclusions can be drawn:

- It was found that the Ra and Rz roughness parameters of the analysed samples depend on both the material and the sample orientation in relation to the sheet RD. The lowest values of roughness parameters were recorded for the orientation 0° in relation to the sheet RD.
- For most of the tested sheets (samples 1–6), smaller mass losses were recorded for the countersample made of Caldie steel.
- Lacquered and/or soaked samples showed a decrease in yield stress and ultimate tensile strength compared to as-received sample. Samples are characterised by high

- anisotropy, with the lowest tensile strength being shown by samples cut at an angle of 45° to the RD and the highest by sheets cut perpendicular to the RD.
- Based on SEM micrographs and EDS mapping, it was found that the microstructure of the test sheets is characterised by the existence of precipitates that are rich in Fe, Si, Mn, Mg and Si.
- The results of ball-on-disc tests concluded that the most favourable friction conditions are provided by the CrN coating. The low CoF of the CrN coating is associated with the smallest coefficient of volumetric wear of the sample and ball-shaped countersample.
- The tests of the chemical composition using the EDS method confirmed the composition of all applied coatings.
- The microstructure of the MForce coating is two-layered, and the analysis of the chemical composition confirmed the presence of (AlTi)N/(AlCr)N layers.
- The (AlTi)N interlayer in the MForce coating adheres to the substrate over the entire tested area and no detachment from its surface was observed, which proves good bonding at the substrate/coating interface.
- Metallographic tests showed that, in the case of the (AlTi)N coating, there are defects in the form of discontinuities.
- In the tested areas of the MForce (AlTi)N/(AlCr)N coating, defects in the form of cracks parallel to the substrate and material losses were identified.
- The tested CrN and MPower (AlTiN/TiAlSiXN) coating is characterised by a homogeneous, compact microstructure.
- The occurrence of funnel-shaped artefacts resulting from the existence of a droplet phase embedded in the coating was found. A few areas with delamination between the coating and the countersample substrate were also observed.
- Future work will include research in industrial conditions on the impact of the use of anti-wear coatings selected on the basis of tests of the geometric quality of products.

Author Contributions: Conceptualization, K.Ż. and K.C.; data curation, K.Ż. and K.C.; formal analysis, T.T.; investigation, K.Ż., K.C., M.K., I.N., M.M. and T.T.; methodology, K.Ż., K.C., M.K., I.N., M.M. and T.T.; validation, K.Ż., M.K., M.M. and T.T.; visualization, K.Ż., K.C., M.K., I.N., M.M. and T.T.; writing—original draft, K.Ż.; writing—review and editing, K.C., T.T., M.M. and M.K. All authors have read and agreed to the published version of the manuscript.

Funding: This research received no external funding.

Institutional Review Board Statement: Not applicable.

Informed Consent Statement: Not applicable.

Data Availability Statement: The data presented in this study are available on request from the corresponding author.

Conflicts of Interest: The authors declare no conflict of interest.

References

1. Clarysse, F.; Lauwerens, W.; Vermeulen, M. Tribological properties of PVD tool coatings in forming operations of steel sheet. *Wear* **2008**, *264*, 400–404. [CrossRef]
2. Aschauer, E.; Riedl, H.; Koller, C.M.; Bolvardi, H.; Arndt, M.; Polcik, P.; Mayrhofer, P.H. Adhesive wear formation on PVD coated tools applied in hot forming of Al-Si coated steel sheets. *Wear* **2019**, *430–431*, 309–316. [CrossRef]
3. Montgomery, S.; Kennedy, D.; O'Dowd, N. PVD and CVD Coatings for the Metal Forming Industry. In Proceedings of the Matrib 2010, Vela Luka, Croatia, 23–25 June 2010; pp. 1–13.
4. Espitia-Rico, M.J.; Caqsiano-Jiménez, G.; Ortega-López, C.; De la Espriella-Vélez, N.; Sánchez-Pacheco, L. A comparative study TiC/TiN and TiN/CN multilayers. *Dyna* **2014**, *81*, 160–165. [CrossRef]
5. Tkaczyk, W. PVD czy CVD—Metody Nanoszenia Powłok na Narzędzia. Available online: https://magazynprzemyslowy.pl/artykuly/pvd-czy-cvd-metody-nanoszenia-powlok-na-narzedzia (accessed on 22 July 2023).
6. Dobrzański, L.A.; Dobrzańska-Danikiewicz, A.D. Obróbka powierzchni materiałów inżynierskich. Zmiany struktury i własności powierzchni materiałów inżynierskich w wyniku eksploatacji. *Open Access Libr.* **2011**, *5*, 368–416.
7. Trzepieciński, T. Approaches for Preventing Tool Wear in Sheet Metal Forming Processes. *Machines* **2023**, *11*, 616.

8. Bohdal, Ł.; Kukiełka, L.; Patyk, R.; Kośka, K.; Chodór, J.; Czyżewski, K. Experimental and Numerical Studies of Tool Wear Processes in the Nibbling Process. *Materials* **2022**, *15*, 107. [CrossRef]
9. Bang, J.; Kim, M.; Bae, G.; Song, J.; Kim, H.G.; Lee, M.G. Quantitative Evaluation of Tool Wear in Cold Stamping of Ultra-High-Strength Steel Sheets. *Met. Mater. Int.* **2023**, *29*, 327–342. [CrossRef]
10. Klingenberg, W.; Singh, U.P. Principles for on-line Monitoring of Tool Wear During Sheet Metal Punching. In *Proceedings of the 34th International MATADOR Conference*; Hinduja, S., Ed.; Springer: London, UK, 2004.
11. Slota, J.; Kubit, A.; Gajdoš, I.; Trzepieciński, T.; Kaščák, Ľ. A Comparative Study of Hardfacing Deposits Using a Modified Tribological Testing Strategy. *Lubricants* **2022**, *10*, 187. [CrossRef]
12. Zhang, T.; Jiang, F.; Yan, L.; Jiang, Z.; Xiu, X. A novel ultrahigh-speed ball-on-disc tribometer. *Tribol. Int.* **2021**, *157*, 106901. [CrossRef]
13. Federici, M.; Straffelini, G.; Gialanella, S. Pin-on-Disc Testing of Low-Metallic Friction Material Sliding Against HVOF Coated Cast Iron: Modelling of the Contact Temperature Evolution. *Tribol. Lett.* **2017**, *65*, 121. [CrossRef]
14. Liu, R.; Yang, Q.; Gao, F. Tribological Behavior of Stellite 720 Coating under Block-on-Ring Wear Test. *Mater. Sci. Appl.* **2012**, *3*, 756–762. [CrossRef]
15. Wang, X.; Kwon, P.Y.; Schrock, D.; Kim, D. Friction coefficient and sliding wear of AlTiN coating under various lubrication condition. *Wear* **2013**, *304*, 67–76. [CrossRef]
16. Meng, Y.; Deng, J.; Zhang, Z.; Sun, Q. Enhanced wear resistance of AlTiN coatings by ultrasonic rolling substrate texturing. *Surf. Coat. Technol.* **2022**, *447*, 128841. [CrossRef]
17. Zhang, Q.; Wu, Z.; Xu, Y.X.; Wang, Q.; Chen, L.; Kim, K.H. Improving the mechanical and anti-wear properties of AlTiN coatings by the hybrid arc and sputtering deposition. *Surf. Coat. Technol.* **2019**, *378*, 125022. [CrossRef]
18. He, Q.; DePaiva, J.M.; Kohlscheen, J.; Veldhuis, S.C. Analysis of the performance of PVD AlTiN coating with five different Al/Ti ratios during the high-speed turning of stainless steel 304 under dry and wet cooling conditions. *Wear* **2022**, *492–493*, 204213. [CrossRef]
19. Zhang, C.; Gu, L.; Tang, G.; Mao, Y. Wear Transition of CrN Coated M50 Steel under High Temperature and Heavy Load. *Coatings* **2017**, *7*, 202. [CrossRef]
20. Biava, G.; Siqueira, I.B.A.F.; Vaz, R.F.; de Souza, G.B.; Jambo, H.C.M.; Szogyenyi, A.; Pukasiewicz, A.G.M. Evaluation of high temperature corrosion resistance of CrN, AlCrN, and TiAlN arc evaporation PVD coatings deposited on Waspaloy. *Surf. Coat. Technol.* **2022**, *438*, 128398. [CrossRef]
21. Navinšek, B.; Panjan, P. Novel applications of CrN (PVD) coatings deposited at 200 °C. *Surf. Coat. Technol.* **1995**, *74–75*, 919–926. [CrossRef]
22. Polok-Rubiniec, M.; Dobrzański, L.A.; Adamiak, M. The properties and wear resistance of the CrN PVD coatings. *J. Achiev. Mater. Manuf. Eng.* **2008**, *30*, 165–171.
23. Wu, L.; Qiu, L.; Du, Y.; Zeng, F.; Lu, Q.; Tan, Z.; Yin, L.; Chen, L.; Zhu, J. Structure and Mechanical Properties of PVD and CVD TiAlSiN Coatings Deposited on Cemented Carbide. *Crystals* **2021**, *11*, 598. [CrossRef]
24. Schulz, W.; Joukov, V.; Köhn, F.; Engelhart, W.; Schier, V.; Schubert, T.; Albrecht, J. The Behavior of TiAlN and TiAlCrSiN Films in Abrasive and Adhesive Tribological Contacts. *Coatings* **2023**, *13*, 1603. [CrossRef]
25. Merklein, M.; Schrader, T.; Engel, U. Wear Behavior of PVD-Coatings. *Tribol. Ind.* **2012**, *34*, 51–56.
26. Claver, A.; Randulfe, J.J.; Palacio, J.F.; Fernández de Ara, J.; Almandoz, E.; Montalá, F.; Colominas, C.; Cot, V.; García, J.A. Improved Adhesion and Tribological Properties of AlTiN-TiSiN Coatings Deposited by DCMS and HiPIMS on Nitrided Tool Steels. *Coatings* **2021**, *11*, 1175. [CrossRef]
27. Das, A.; Kamal, M.; Biswil, B.B. Comparative assessment between AlTiN and AlTiSiN coated carbide tools towards machinability improvement of AISI D6 steel in dry hard turning. *Proc. Inst. Mech. Eng. Part C J. Mech. Eng. Sci.* **2022**, *236*, 3174–3197. [CrossRef]
28. Liu, J.; Wang, Y.; Liu, G.; Hua, J.; Deng, X. Properties and Performance of TiAlSiN and AlCrN Monolayer and Multilayer Coatings for Turning Ti-6Al-4V. *Coatings* **2023**, *13*, 1229. [CrossRef]
29. Hu, C.; Chen, L.; Lou, Y.; Zhao, N.; Yue, J. Influence of Si content on the microstructure, thermal stability and oxidation resistance of TiAlSiN/CrAlN multilayers. *J. Alloy. Compd.* **2021**, *855*, 157441. [CrossRef]
30. Xiao, B.; Liu, J.; Liu, F.; Zhong, X.; Xiao, X.; Zhang, T.F.; Wang, Q. Effects of microstructure evolution on the oxidation behavior and high-temperature tribological properties of AlCrN/TiAlSiN multilayer coatings. *Ceram. Int.* **2018**, *44*, 23150–23161. [CrossRef]
31. Chen, W.; Lin, Y.; Zheng, J.; Zhang, S.; Liu, S.; Kwon, S. Preparation and characterization of CrAlN/TiAlSiN nano-multilayers by cathodic vacuum arc. *Surf. Coat. Technol.* **2015**, *265*, 205–211. [CrossRef]
32. Chen, L.; Wang, S.Q.; Du, Y.; Zhou, S.Z.; Gang, T.; Fen, J.C.; Chang, K.K.; Li, Y.W.; Xiong, X. Machining performance of Ti-Al-Si-N coated inserts. *Surf. Coat. Technol.* **2010**, *205*, 582–586. [CrossRef]
33. Chang, C.L.; Chen, W.C.; Tsai, P.C.; Ho, W.Y.; Wang, D.Y. Characteristics and performance of TiSiN/TiAlN multilayers coating synthesized by cathodic arc plasma evaporation. *Surf. Coat. Technol.* **2007**, *202*, 987–992. [CrossRef]
34. Xiao, B.; Zhang, T.F.; Guo, Z.; Li, Z.; Fan, B.; Chen, G.; Xiong, Z.; Wang, Q. Mechanical, oxidation, and cutting properties of AlCrN/AlTiSiN nano-multilayer coatings. *Surf. Coat. Technol.* **2022**, *433*, 128094. [CrossRef]
35. Fukumoto, N.; Ezura, H.; Suzuki, T. Synthesis and oxidation resistance of TiAlSiN and multilayer TiAlSiN/CrAlN coating. *Surf. Coat. Technol.* **2009**, *204*, 902–906. [CrossRef]
36. Sousa, V.F.C.; Da Silva, F.J.G.; Pinto, G.F.; Baptista, A.; Alexandre, R. Characteristics and Wear Mechanisms of TiAlN-Based Coatings for Machining Applications: A Comprehensive Review. *Metals* **2021**, *11*, 260. [CrossRef]
37. Endrino, J.L.; Fax-Rabinovich, G.S.; Gey, C. Hard AlTiN, AlCrN PVD coatings for machining of austenitic stainless steel. *Surf. Coat. Technol.* **2006**, *200*, 6840–6845. [CrossRef]

38. *ASTM B209/B209M-21a*; Standard Specification for Aluminum and Aluminum-Alloy Sheet and Plate. ASTM: West Conshehoken, PA, USA, 2022.
39. Deshwal, G.; Panjagari, N. Review on metal packaging: Materials, forms, food applications, safety and recyclability. *J. Food Sci. Technol.* **2020**, *57*, 2377–2392. [CrossRef]
40. *ISO 21920-1:2021*; Geometrical product specifications (GPS)—Surface texture: Profile—Part 1: Indication of Surface Texture. ISO: Geneva, Switzerland, 2021.
41. *EN ISO 6892-1:2022*; Metallic Materials—Tensile Testing—Part 1: Method of Test at Room Temperature. ISO: Geneva, Switzerland, 2022.
42. *ASTM-E2546-15*; Standard Practice for Instrumented Indentation Testing. ASTM: West Conshehoken, PA, USA, 2023.

Disclaimer/Publisher's Note: The statements, opinions and data contained in all publications are solely those of the individual author(s) and contributor(s) and not of MDPI and/or the editor(s). MDPI and/or the editor(s) disclaim responsibility for any injury to people or property resulting from any ideas, methods, instructions or products referred to in the content.

Article

Possibilities of a Hybrid Method for a Time-Scale-Frequency Analysis in the Aspect of Identifying Surface Topography Irregularities

Damian Gogolewski [1,*], Paweł Zmarzły [1], Tomasz Kozior [1] and Thomas G. Mathia [2]

1 Department of Mechanical Engineering and Metrology, Kielce University of Technology, al. Tysiąclecia Państwa Polskiego 7, 25-314 Kielce, Poland
2 Laboratoire de Tribologie et Dynamique des Systemes (LTDS), Ecole Centrale de Lyon, Centre National de la Recherche Scientifique, 69134 Lyon, France
* Correspondence: dgogolewski@tu.kielce.pl

Citation: Gogolewski, D.; Zmarzły, P.; Kozior, T.; Mathia, T.G. Possibilities of a Hybrid Method for a Time-Scale-Frequency Analysis in the Aspect of Identifying Surface Topography Irregularities. *Materials* 2023, *16*, 1228. https://doi.org/10.3390/ma16031228

Academic Editor: Przemysław Podulka

Received: 23 December 2022
Revised: 25 January 2023
Accepted: 30 January 2023
Published: 31 January 2023

Copyright: © 2023 by the authors. Licensee MDPI, Basel, Switzerland. This article is an open access article distributed under the terms and conditions of the Creative Commons Attribution (CC BY) license (https://creativecommons.org/licenses/by/4.0/).

Abstract: The article presents research results related to assessing the possibilities of applying modern filtration methods to diagnosing measurement signals. The Fourier transformation does not always provide full information about the signal. It is, therefore, appropriate to complement the methodology with a modern multiscale method: the wavelet transformation. A hybrid combination of two algorithms results in revealing additional signal components, which are invisible in the spectrum in the case of using only the harmonic analysis. The tests performed using both simulated signals and the measured roundness profiles of rollers in rolling bearings proved the advantages of using a complex approach. A combination of the Fourier and wavelet transformations resulted in the possibility to identify the components of the signal, which directly translates into better diagnostics. The tests fill a research gap in terms of complex diagnostics and assessment of profiles, which is very important from the standpoint of the precision industry.

Keywords: wavelet transformation; Fourier analysis; surface texture; rolling bearings; roundness; surface quality

1. Introduction

Metrology 4.0 is undoubtedly an important component of the fourth industrial revolution which is entering a dynamic phase. Measurements and a proper analysis of the acquired data constitute an integral part of the process and the functioning of systems. As part of Metrology 4.0, a number of measurement devices have been developed and improved, which can execute requested processes in an automatic manner in real-time using artificial intelligence and make decisions in an autonomous way, based on the acquired data. In recent years, new solutions were also introduced in the aspect of surface topography, allowing for a more detailed and comprehensive assessment and prediction [1,2]. Modern multiscale methods constitute an important contribution to the development of modern surface metrology, and they contribute to a considerable increase in the possibilities of its assessment [3,4]. When surfaces are observed on numerous scales, there is a possibility to observe and highlight their characteristic features. The surface texture of the produced elements is described as a set of periodic and aperiodic irregularities. This distribution depends on the procedures to which the surface was subjected, while taking into account the parameters of specific machining processes. There are a number of algorithms used to assess, filtrate, or detect certain important information included in such surfaces; however, the application of traditional algorithms often does not allow for the identification of all the components of a signal. It is, therefore, desirable to eliminate the shortcomings of the individual transforms and to analyze the measurement data in a more efficient way.

The hybrid application of several known methods in a proper order allows for a better, comprehensive evaluation of the signals.

The nature of irregularities of rotating elements is evaluated using a roundness profile transformation based on the Fourier transformation. In accordance with this approach, the individual profiles are described as a composition of periodic functions with a defined magnitude of the amplitude and period, which provide information on possible errors in the manufacturing process. Based on the magnitude of the individual harmonic components, it is possible to draw a conclusion about the prevalent deviations of the shape profile and about their potential impact on the subsequent work of an element. Nonetheless, in the case of local defects existing on the surface, a harmonic analysis does not provide a complete image. The values of the individual components of the signal are considerably disrupted by scratches or cracks. As a supplement to the Fourier transformation, one can suggest the concept of using an additional method developed at the turn of the 21st century: the wavelet transformation.

The wavelet transformation is one of the multiscale methods which allow for decomposing a signal and presenting it in individual ranges. Currently, it is also increasingly used to analyze 2D and 3D surface profiles. These profiles are a composition of periodic and aperiodic irregularities; therefore, they should be assessed as non-stationary signals and analyzed with the use of proper methods. Such methods include wavelet transformation, allowing for simultaneous localization in the time and frequency domain, which is its unquestionable advantage. However, this transformation also has disadvantages, such as, among other things, shift sensitivity, poor directionality, and the lack of phase information [5]. The discrete wavelet transform used in the study is based on the use of two types of filters, i.e., high-pass and low-pass. The particular combination of them allows for the evaluation of surface topography in a specific frequency band, the size of the scale. In addition, a wide spectrum of mother wavelets with different properties allows for their appropriate use, which translates into better diagnostics and detection of characteristic features of the signal which are invisible in the results of the analysis carried out by other methods. When analyzing the literature, it can be concluded that in surface metrology, a number of uses have already been found for the wavelet transformation, e.g., for correcting temperature errors [6], diagnosing, characterizing, and identifying characteristic features [7–9], decomposition the profile into components (roughness, waviness, form) [10–12], and estimating surface roughness parameters using image processing [13].

Therefore, it was desirable to compare the possibilities of using the developed wavelet approach and Fourier transformation in the aspect of analyzing various types of signals. Research pointing to the drawbacks and advantages of specific methods using the example of the surface profiles of rotating elements are described in [14]. In [15], the authors compare both methods in the aspect of assessing muscle fatigue, while in [16], the Fourier transform, the windowed Fourier transform, and the wavelet transform methods were applied to fringe pattern processing. Research intended to identify the error components influencing the measurement accuracy of the system using the Fourier transformation and the wavelet transformation is presented in the paper [17]. Based on the current state of the art, it should be concluded that both methods can complement each other; therefore, a proper combination of both methods can be used successfully for proper diagnostics of signals. A proper combination of both algorithms has been used to identify defects of bearings in [18–20]. Nonetheless, the use of a hybrid method based on the time-scale-frequency analysis to diagnose measurement signals provides potentially vast possibilities, and it requires a broader analysis. The novelty of the work is the application of a comprehensive approach to the evaluation of surfaces using the modern, multiscale, hybrid method including the analysis of its parameters. The tests fill a research gap in terms of complex diagnostics and assessment of profiles. Despite the many studies relating to the evaluation of wavelet transformation parameters and their impact on results, further comprehensive analysis of their applicability is required [21–24]. The relationship between the surface texture, which is influenced by the parameters of both the process and the measurement, and the methods of analysis, is very important in the context of a comprehensive analysis of surface quality, as shown in Figure 1. It should be pointed out that there is no research aimed at assessing the possibilities of using the wavelet transform to analyze the roundness profiles, in particular in the elements of rolling bearings. It should be added that this type of shape deviation has a considerable impact on the operating parameters of rolling bearings, and new methods of their assessment should be sought [25].

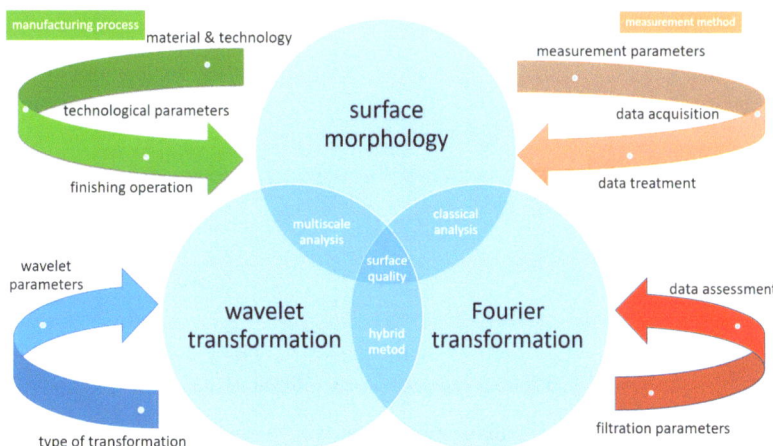

Figure 1. Relationship between surface morphology and methods for analyzing measurement data.

Figure 1 presents a schematic interpretation of the production and measurement process, taking into account the technological [26–31] and measuring process [32–39], with particular emphasis on the wavelet [40–43] and Fourier transformation [44–49].

2. Metrological Approach

This paper presents simulation tests intended to assess the possibilities of identifying signal disruptions using a time-scale-frequency analysis (Figure 2). The hybrid approach methodology is based on the performance of a spectral analysis of signals generated at the subsequent stages of the wavelet decomposition. Assessment of the signal, taking into account the scale of irregularities, will allow for the identification of disruptions that are invisible in the input signal spectrum.

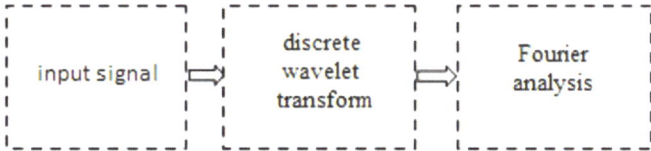

Figure 2. Diagram of the hybrid method of the time-scale-frequency analysis.

The tests were implemented using the concept of combining the Fourier transformation and the wavelet transformation. The paper presents a comparative analysis of various types of signals. To this end, simulation tests were performed for signals being a composition of periodic functions with various amplitude sizes and various periods (x_1, Equation (1), Figure 3), with noise from a periodic function with a relatively small amplitude and period (x_2, Equation (2), Figure 3), and a signal with noise from a series of coefficients with random values (x_3, Figure 3). The functions presented in Figure 3 are almost identical, differing only in additional relatively small disruptions, which cannot be distinguished in a visual manner.

$$x_1(t) = 2\sin(6\pi t) + 3\cos(10\pi t) + 0.3\cos(30\pi t) + 0.7\cos(60\pi t) + \cos(75\pi t) + 0.07\cos(90\pi t) \tag{1}$$

$$x_2(t) = x_1(t) + 0.018\cos(100\pi t) \tag{2}$$

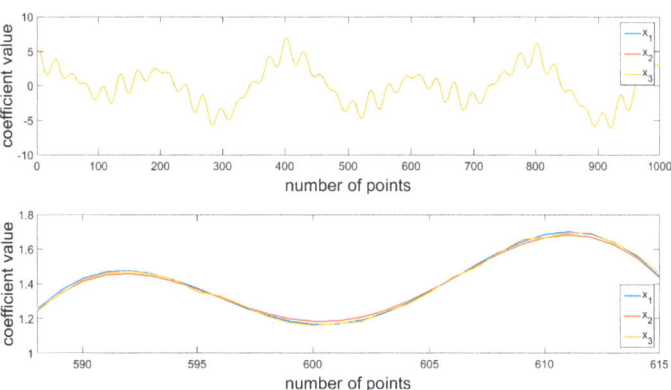

Figure 3. Analyzed surface profiles—simulated tests.

Experimental tests were performed to verify the results of simulation tests. Measurements of the roundness profiles of a series of rollers in rolling bearings were performed with the use of the Talyrond 365 measurement system. It is a precise measurement system; the operating principle of which is based on the method of measuring changes in the radius (radial method) with a rotary table. Talyrond 365 is commonly used in the bearing industry to analyze the quality of production of cylindrical elements [50,51]. On the other hand, the deviation of roundness and waviness of bearing rollers is of key significance, since its excess values cause the generation of additional vibrations of the rolling bearings. Moreover, they can also cause assembly problems. Figure 4a presents the measurement of a bearing roller using the example of Talyrond 365, while Figure 4b presents a sample-assessed roundness profile.

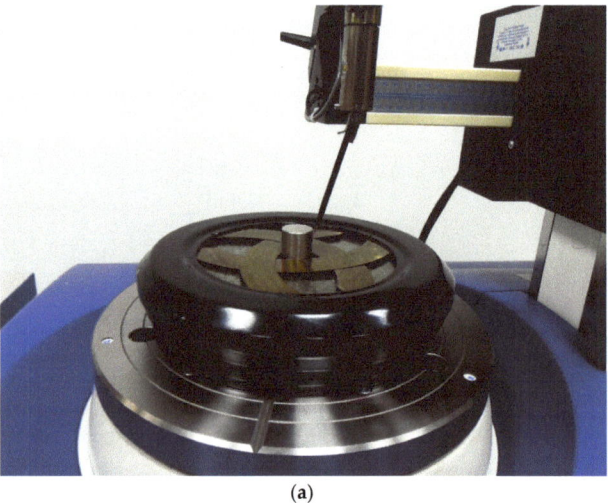

(a)

Figure 4. *Cont.*

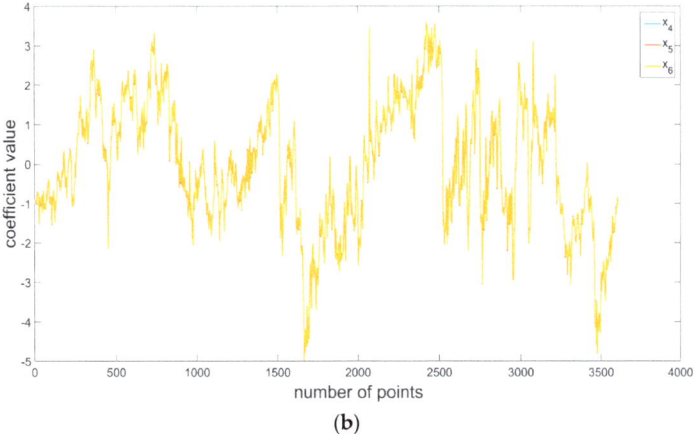

(b)

Figure 4. (a) A Talyrond 365 measurement of roundness profiles in a bearing roller and (b) a roundness profile with waviness components.

3. Results

3.1. Simulated Tests

A harmonic analysis was performed for the signals presented in Figure 3 (Figure 5). The resulting values of amplitudes for the individual harmonics resembling with each other, and they correspond to the modeled values of signal components. Based on either a visual assessment of the signal or the resulting values of its components, it is not possible to clearly distinguish between the individual signals. Based on the below figure, it cannot be concluded whether the signals are different, and optionally which ones of them contain additional information. Correct assessment of the signals is particularly important in terms of diagnosing a process or a product. At the initial stage of the destabilization of production processes in the individual signals received from machines and devices, there are indeed low resulting distortions, deviations from the nominal signal which should be identified as fast as possible in order to prevent deterioration of the process, which directly affects the quality of the manufactured elements. Therefore, it can be concluded that the use of the Fourier transformation only results in the lack of certain information about the signal in the produced results. The use of solely the Fourier transformation causes the loss of certain important features.

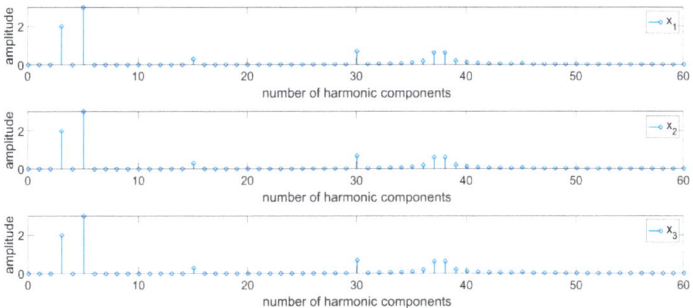

Figure 5. Amplitude spectral density.

In order to streamline the diagnostic process and obtain better, more precise, and comprehensive (with a broader range) information about the investigated signal, it was deemed purposeful to perform an analysis on many scales by means of the wavelet transformation.

The use of the discrete wavelet transformation followed by the Fourier transform will allow for analyzing the signal in its individual frequency bands. This will allow for focusing more on high-frequency information about small amplitudes which have been lost in the spectrum presented in Figure 5. In order to perform the analysis, the db2 mother wavelet was chosen, which due to its short support width, should be sensitive to small but rapid changes in the values of the coefficients describing the signal. Figure 6 presents the results of the analysis of details generated at the first level of decomposition of signals, while Figure 7 presents the results of filtration for seven levels. Red color indicates differences in the spectrum for particular signals.

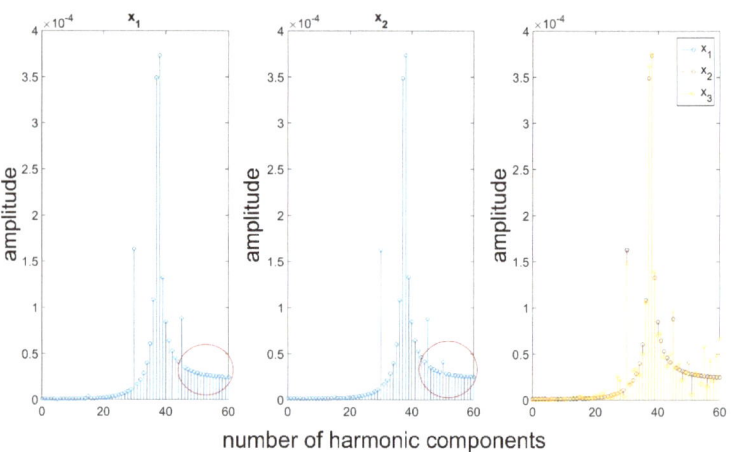

Figure 6. A spectrum of signals on the first level of decomposition.

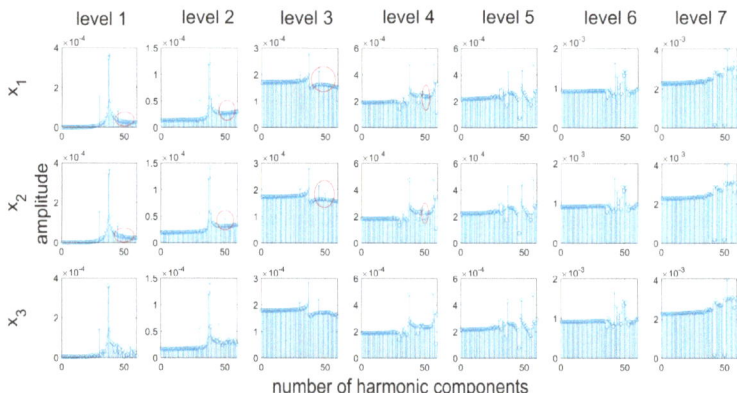

Figure 7. A spectrum of signals for seven levels of analysis.

An assessment of the resulting signal spectra indicated that there are components that have been made additionally visible in the signal spectra of details created after the performance of the wavelet analysis. For each of the assessed signals, at the initial stages of the analysis, different values were obtained for certain harmonic components. When assessing the possibilities of detecting a disturbance in the form of a periodic function, it should be concluded that a disturbance corresponding to the parameters of the additional component of the signal was noticed in high-frequency signals at the four initial levels of the analysis. Therefore, on this basis, it is possible to draw conclusions about its frequency. At the following levels of decomposition, the spectra for signals x_1 and x_2 do not exhibit any significant differences. The additional component has been filtered out. Analogical conclusions can be drawn based on an analysis of the noisy signal (x_3). In this case, differences between the nominal (x_1) and the noisy signal can be noticed in a broader range of harmonics in the assessed spectra. Nonetheless, the nature of the disruption is indicated by the fact that most disruptive components were filtered out at the first stage of the analysis. Moreover, there is a considerable difference in the resulting values of spectrum coefficients at the two initial levels of the analysis.

3.2. Analysis of Roundness Profiles

Analogical tests were also performed for the actual roundness profiles of rollers in a rolling bearing (x_4). The sample surface profile presented in Figure 4b is characterized by a completely different, irregular, and aperiodic distribution of irregularities. In order to assess the possibilities of the proposed approach, it was also deemed reasonable, analogically to the case of simulation tests, to add to the analyzed profile an additional periodic function with a relatively small amplitude and period (x_5), and a signal defined as a series of coefficients with random values (x_6). For a comprehensive assessment, thirty roundness profiles were selected for the tests.

A harmonic analysis was performed for the sample profile presented in Figure 3 (Figure 8). The introduction of additional components in the form of a periodic function or noise caused no noticeable changes in the values of amplitudes in the resulting spectra. Based on the resulting data, it is not possible to clearly distinguish between the individual signals. The detection of changes and any deviations from the nominal assumptions is particularly important in the case of the bearing industry. It should be noted that an erroneous interpretation of the predominant harmonic in the roundness profile of a bearing roller can result in the wrong assessment of the stability of the production process. In this case, unnecessary correction procedures can be introduced, which may result in the propagation of the already existing deviations. Moreover, the excess deviation values for roundness and surface waviness of bearing rollers cause the generation of additional vibrations and noises in the rolling bearings, which is an unpreferable phenomenon [52,53]. Due to this, in the case of surface profiles, a combination of two transformations was used in order to diagnose surface topography.

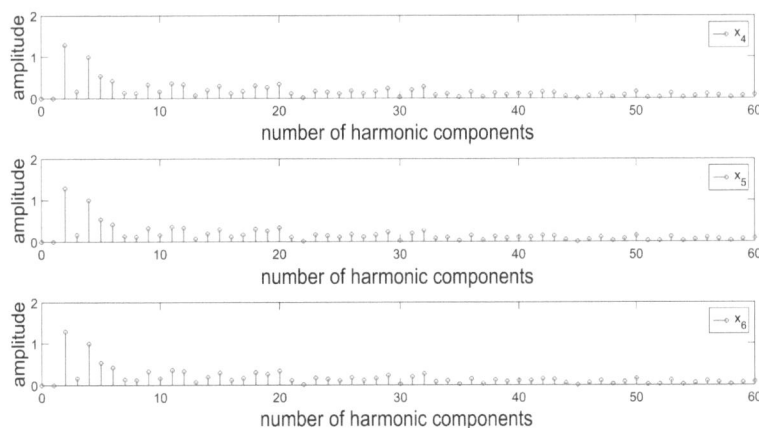

Figure 8. Amplitude spectral density of the roundness profiles of rollers in rolling bearings.

Surface profiles being a composition of periodic and aperiodic irregularities in the generated spectrum are defined by a significantly higher number of harmonic components. The assessment of irregularities in the individual bands will allow for narrowing down the scope of the analysis, which at the initial stages of decomposition results in an analysis of fast-changing information corresponding to the introduced disruptions. Figure 8 presents the results of the analysis of details generated at the first level of decomposition of signals, performed by means of the db2 wavelet. In this case, differences in the amplitude values were recorded for a single harmonic corresponding to the value of the modeled function. In the case of noise, a similar distribution of amplitude values can be observed; however, the values are proportionally higher (Figure 9). At the subsequent stages of the analysis; the resulting spectra contain no clear information about the signal. It was, therefore, deemed reasonable to use a different mother wavelet, the support width of which is even shorter, i.e., db1. The dbN family of wavelets (Daubechies), where N refers to the number of vanishing moments and wavelet order, is characterized by 2 N-1 compact support [54]. For the db1 wavelet, information in the detail signals was acquired in a clearer manner (Figure 10). Therefore, it can be concluded that for non-stationary signals, the use of wavelets with possibly short support width results in better diagnostics of the signal. In the upper part of Figure 10, there are noticeable differences for the 50th harmonic between the surface profile resulting from the measurements and the signal resulting from the addition of an additional function. There is also a noticeable level of decomposition, for which the additional component has been removed. In the case of noise, it is also possible to record the level of decomposition for which it has been filtered out. Since the wavelet analysis is a lossless transformation [55], the signal can be reconstructed, and the possible causes of the resulting defect can be analyzed to a given level.

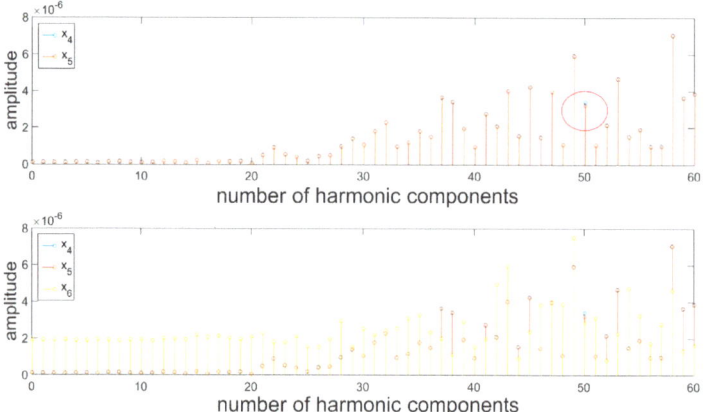

Figure 9. A spectrum of surface profiles on the first level of decomposition—the db2 wavelet.

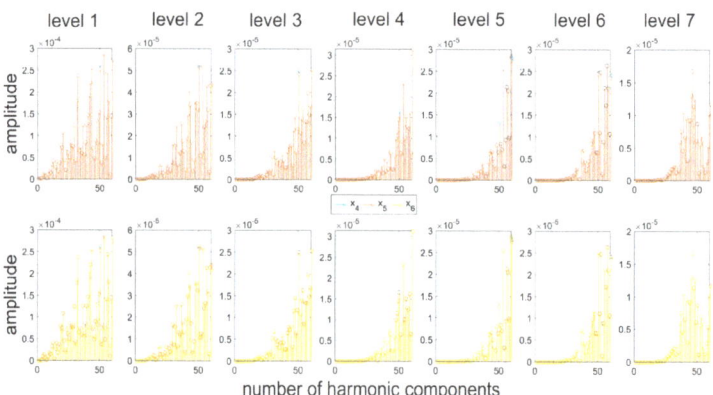

Figure 10. A Spectrum of surface profiles for seven levels of analysis—the db1 wavelet.

4. Discussion

The continuing industrial revolution and the dynamically developing industry necessitate an increasingly accurate and comprehensive assessment of the process and the quality control of the final elements. Modern production technologies entail modern methods for analyzing measurement data. The currently widely developed multiscale methods constitute significant supplementation of the traditional algorithms used for the metrological assessment of measurement data. One of these methods is the wavelet transformation. The combined use of the wavelet and Fourier transformations allowed for extending the range of information about the assessed roundness profiles. The method can be useful in diagnostic processes for fast and comprehensive evaluation of surface topography irregularities. Feature extraction, detection of small signal changes, and additional relatively small components in signals/profiles, which in the further operation of the mechanisms can cause a real reduction in efficiency, durability, and applicability of particular elements, is a real benefit of this solution. The tests fill a research gap in terms of diagnostics and assessment of profiles, which is very important from the standpoint of the precision industry.

Understanding the possibilities of detecting individual morphological features and determining the scale on which the irregularities occur required a comprehensive assessment and verification of the assumed procedure. The combination of both algorithms resulted in the identification and the possibility of a detailed analysis of the important aspects of geometry, which could influence the subsequent operation of the elements of machine parts. In this aspect, the traditional perception of roundness profiles seems insufficient.

The tests indicated vast potential application possibilities for multiscale methods in order to supplement the currently used traditional methods. A multiscale analysis using discrete one-dimensional wavelet transformation indicated both the prevalent components of surface irregularities, as well as additional components which had not been properly filtered out based on the Fourier transformation. An assessment of the distribution of irregularities on multiple scales allowed for better diagnostics of the roundness profiles, taking into account the scale of the individual features. The tests performed for a wide spectrum of mother wavelets and roundness profiles allowed for verifying the impact of the mother wavelet and its properties on the filtration process, positively verifying the adopted concept. Moreover, no significant diversity of the resulting values of the spectrum was obtained as a function of profile change.

The conducted tests also came with certain limitations, to be analyzed as part of research in the future. In particular, future studies should focus on a greater diversity of surface types, in particular functional surfaces, and on taking into account a broader range of types of wavelet transformations, since the individual approaches and types of the mother wavelet are characterized by different properties. This results in highlighting other characteristic morphological features on the surface [56–58]. In future research, it is planned to include machine parts that will contain certain surface irregularities created after a certain work cycle. The research will help to develop modern diagnostic systems allowing for fast control of the key produced industrial elements.

5. Conclusions

The paper assesses the possibilities of utilizing modern filtration methods for diagnosing signals. The tests were performed both for simulated irregularities distributions, as well as for the actual roundness profiles of rollers in rolling bearings. The application of an algorithm is a combination of two currently used signal analysis methods, i.e., the Fourier transformation and the wavelet transformation, which results in a better, more comprehensive analysis of data. The proposed hybrid algorithm allows for the detection of signal components that are invisible in the spectrum created in the case of performing the analysis using the Fourier transformation only. The combination of the two methods allows for providing additional information about the characteristic harmonics in the signal. There is a possibility to identify the disruptions and the occurrence of noise, as well as the parameters which describe them, which are hindered when using the Fourier transformation only. The individual values can be noted in the figures presented in the paper, where the resulting differences in spectrums are marked. Differences in amplitude values were noted for individual numbers of harmonic components, as shown, for example, in the values of the fiftieth harmonic component. Moreover, it has been noticed that the use of wavelets with a shorter support width result in better, clearer filtration at the initial stages of decomposition of additional high-frequency disrupting components. The tests performed for numerous profiles with a diverse distribution of irregularities indicated that the resulting spectra contained important additional modeled components, which were properly identified, which proves the effectiveness of the proposed approach.

Author Contributions: Conceptualization, D.G.; methodology, D.G.; software, D.G., P.Z. and T.K.; validation, D.G. and T.G.M.; formal analysis, D.G. and P.Z.; investigation, D.G.; resources, D.G. and P.Z.; data curation, D.G. and P.Z.; writing—original draft preparation, D.G., P.Z., T.K. and T.G.M.; writing—review and editing, D.G.; visualization, D.G.; supervision, D.G.; project administration, D.G.; funding acquisition, D.G. All authors have read and agreed to the published version of the manuscript.

Funding: This research received no external funding.

Institutional Review Board Statement: Not applicable.

Informed Consent Statement: Not applicable.

Data Availability Statement: The data presented in this study are available upon request from the corresponding author.

Conflicts of Interest: The authors declare no conflict of interest.

Nomenclature

x_1	A function that is a composition of periodic functions with various amplitude sizes and various periods.
x_2	A function that is a composition of periodic functions with various amplitude sizes and various periods, with additional noise from a periodic function with a relatively small amplitude and period.
x_3	A function that is a composition of periodic functions with various amplitude sizes and various periods, with additional noise from a series of coefficients with random values.
x_4	The roundness profile of rollers in a rolling bearing.
x_5	The roundness profile of rollers in a rolling bearing, with additional noise from a periodic function with a relatively small amplitude and period.
x_6	The roundness profile of rollers in a rolling bearing, with additional noise from a series of coefficients with random values.
N	The number of vanishing moments.

References

1. Panda, A.; Sahoo, A.K.; Panigrahi, I.; Rout, A.K. Investigating Machinability in Hard Turning of AISI 52100 Bearing Steel through Performance Measurement: QR, ANN and GRA Study. *Int. J. Automot. Mech. Eng.* **2018**, *15*, 4935–4961. [CrossRef]
2. Panda, A.; Sahoo, A.K.; Panigrahi, I.; Rout, A.K. Prediction models for on-line cutting tool and machined surface condition monitoring during hard turning considering vibration signal. *Mech. Ind.* **2020**, *21*, 520. [CrossRef]
3. Gogolewski, D.; Bartkowiak, T.; Kozior, T.; Zmarzły, P. Multiscale Analysis of Surface Texture Quality of Models Manufactured by Laser Powder-Bed Fusion Technology and Machining from 316L Steel. *Materials* **2021**, *14*, 2794. [CrossRef]
4. Gogolewski, D. Multiscale assessment of additively manufactured free-form surfaces. *Metrol. Meas. Syst.* **2023**, *30*.
5. Fernandes, F.; van Spaendonck, R.; Burrus, C. A new framework for complex wavelet transforms. *IEEE Trans. Signal Process.* **2003**, *51*, 1825–1837. [CrossRef]
6. Makieła, W.; Świderski, J.; Gogolewski, D.; Makieła, K. Compensation of temperature errors when measuring surface textures by applying a two-dimensional wavelet transform. *Eng. Mech.* **2018**, *2018*, 537–540.
7. Bruzzone, A.; Montanaro, J.; Ferrando, A.; Lonardo, P. Wavelet Analysis for Surface Characterisation: An Experimental Assessment. *CIRP Ann.* **2004**, *53*, 479–482. [CrossRef]
8. Edjeou, W.; Cerezo, V.; Zahouani, H.; Salvatore, F. Multiscale analyses of pavement texture during polishing. *Surf. Topogr. Metrol. Prop.* **2020**, *8*, 024008. [CrossRef]

9. Méndez, A.; Reyes, Y.; Trejo, G.; StȨpień, K.; Țălu, Ş. Micromorphological characterization of zinc/silver particle composite coatings. *Microsc. Res. Tech.* **2015**, *78*, 1082–1089. [CrossRef] [PubMed]
10. Blateyron, F. Good practices for the use of areal filters. In Proceedings of the 3rd Seminar on Surface Metrology of the Americas (SSMA), Albuquerque, NM, USA, 12–13 May 2014.
11. Lucas, K.; Sanz-Lobera, A.; Antón-Acedos, P.; Amatriain, A. A Survey of Bidimensional Wavelet Filtering in Surface Texture Characterization. *Procedia Manuf.* **2019**, *41*, 811–818. [CrossRef]
12. Wang, X.; Shi, T.; Liao, G.; Zhang, Y.; Hong, Y.; Chen, K. Using Wavelet Packet Transform for Surface Roughness Evaluation and Texture Extraction. *Sensors* **2017**, *17*, 933. [CrossRef] [PubMed]
13. Ali, J.M.; Jailani, H.S.; Murugan, M. Surface roughness evaluation of electrical discharge machined surfaces using wavelet transform of speckle line images. *Measurement* **2020**, *149*, 107029. [CrossRef]
14. Adamczak, S.; Makieła, W.; Stepien, K. Investigating Advantages and Disadvantages of the Analysis of a Geometrical Surface Structure with the Use of Fourier and Wavelet Transform. *Metrol. Meas. Syst.* **2010**, *17*, 233–244. [CrossRef]
15. Chowdhury, S.K.; Nimbarte, A.D. Comparison of Fourier and wavelet analysis for fatigue assessment during repetitive dynamic exertion. *J. Electromyogr. Kinesiol.* **2015**, *25*, 205–213. [CrossRef]
16. Huang, L.; Kemao, Q.; Pan, B.; Asundi, A.K. Comparison of Fourier transform, windowed Fourier transform, and wavelet transform methods for phase extraction from a single fringe pattern in fringe projection profilometry. *Opt. Lasers Eng.* **2010**, *48*, 141–148. [CrossRef]
17. Xu, B.; Shimizu, Y.; Takeishi, T.; Ito, S.; Gao, W. Surface Form Measurement and Analysis of a Cylindrical Workpiece with Microstructures. *J. Adv. Mech. Des. Syst. Manuf.* **2012**, *6*, 936–948. [CrossRef]
18. Yan, R.; Gao, R.X.; Chen, X. Wavelets for fault diagnosis of rotary machines: A review with applications. *Signal Process.* **2014**, *96*, 1–15. [CrossRef]
19. Gao, R.X.; Yan, R. *Wavelets: Theory and Applications for Manufacturing*; Springer: Boston, MA, USA, 2010. ISBN 978-1-4419-1545-0.
20. Wang, C.; Gao, R.X.; Yan, R. Unified time–scale–frequency analysis for machine defect signature extraction: Theoretical framework. *Mech. Syst. Signal Process.* **2009**, *23*, 226–235. [CrossRef]
21. Grzesik, W.; Brol, S. Wavelet and fractal approach to surface roughness characterization after finish turning of different workpiece materials. *J. Mater. Process. Technol.* **2009**, *209*, 2522–2531. [CrossRef]
22. Karolczak, P.; Kowalski, M.; Wiśniewska, M. Analysis of the Possibility of Using Wavelet Transform to Assess the Condition of the Surface Layer of Elements with Flat-Top Structures. *Machines* **2020**, *8*, 65. [CrossRef]
23. Guo, T.; Zhang, T.; Lim, E.; Lopez-Benitez, M.; Ma, F.; Yu, L. A Review of Wavelet Analysis and Its Applications: Challenges and Opportunities. *IEEE Access* **2022**, *10*, 58869–58903. [CrossRef]
24. Trivizakis, E.; Ioannidis, G.S.; Souglakos, I.; Karantanas, A.H.; Tzardi, M.; Marias, K. A neural pathomics framework for classifying colorectal cancer histopathology images based on wavelet multi-scale texture analysis. *Sci. Rep.* **2021**, *11*, 15546. [CrossRef] [PubMed]
25. Zmarzły, P. Multi-Dimensional Mathematical Wear Models of Vibration Generated by Rolling Ball Bearings Made of AISI 52100 Bearing Steel. *Materials* **2020**, *13*, 5440. [CrossRef] [PubMed]
26. Gogolewski, D.; Kozior, T.; Zmarzły, P.; Mathia, T.G. Morphology of Models Manufactured by SLM Technology and the Ti6Al4V Titanium Alloy Designed for Medical Applications. *Materials* **2021**, *14*, 6249. [CrossRef]
27. Skrzyniarz, M.; Nowakowski, L.; Blasiak, S. Geometry, Structure and Surface Quality of a Maraging Steel Milling Cutter Printed by Direct Metal Laser Melting. *Materials* **2022**, *15*, 773. [CrossRef]
28. Zárybnická, L.; Petrů, J.; Krpec, P.; Pagáč, M. Effect of Additives and Print Orientation on the Properties of Laser Sintering-Printed Polyamide 12 Components. *Polymers* **2022**, *14*, 1172. [CrossRef] [PubMed]
29. Mesicek, J.; Ma, Q.-P.; Hajnys, J.; Zelinka, J.; Pagac, M.; Petru, J.; Mizera, O. Abrasive Surface Finishing on SLM 316L Parts Fabricated with Recycled Powder. *Appl. Sci.* **2021**, *11*, 2869. [CrossRef]
30. Panda, A.; Sahoo, A.K.; Rout, A.K. Investigations on surface quality characteristics with multi-response parametric optimization and correlations. *Alex. Eng. J.* **2016**, *55*, 1625–1633. [CrossRef]
31. Sahoo, A.K. Application of Taguchi and regression analysis on surface roughness in machining hardened AISI D2 steel. *Int. J. Ind. Eng. Comput.* **2014**, *5*, 295–304. [CrossRef]
32. Grochalski, K.; Wieczorowski, M.; Jakubek, B. Influence of thermal disturbances on profilometric measurements of surface asperities. *Measurement* **2022**, *190*, 110694. [CrossRef]
33. Wieczorowski, M. Spiral sampling as a fast way of data acquisition in surface topography. *Int. J. Mach. Tools Manuf.* **2001**, *41*, 2017–2022. [CrossRef]
34. Eifler, M.; Klauer, K.; Kirsch, B.; Aurich, J.C.; Seewig, J. Performance verification of areal surface texture measuring instruments with the Sk-parameters. *Measurement* **2021**, *173*, 8257–8283. [CrossRef]
35. Zmarzły, P.; Kozior, T.; Gogolewski, D. The Effect of Non-Measured Points on the Accuracy of the Surface Topography Assessment of Elements 3D Printed Using Selected Additive Technologies. *Materials* **2023**, *16*, 460. [CrossRef] [PubMed]
36. Adamczak, S.; Kundera, C.; Swiderski, J. Assessment of the state of the geometrical surface texture of seal rings by various measuring methods. *IOP Conf. Series: Mater. Sci. Eng.* **2017**, *233*, 012031. [CrossRef]
37. Podulka, P. Thresholding Methods for Reduction in Data Processing Errors in the Laser-Textured Surface Topography Measurements. *Materials* **2022**, *15*, 5137. [CrossRef]

38. Pawlus, P.; Reizer, R.; Wieczorowski, M. Comparison of Results of Surface Texture Measurement Obtained with Stylus Methods and Optical Methods. *Metrol. Meas. Syst.* **2018**, *25*, 589–602.
39. Pawlus, P.; Reizer, R.; Wieczorowski, M. Problem of Non-Measured Points in Surface Texture Measurements. *Metrol. Meas. Syst.* **2017**, *24*, 525–536. [CrossRef]
40. Jiang, X.; Scott, P.; Whitehouse, D. Wavelets and their applications for surface metrology. *CIRP Ann.* **2008**, *57*, 555–558. [CrossRef]
41. Li, G.; Zhang, K.; Gong, J.; Jin, X. Calculation method for fractal characteristics of machining topography surface based on wavelet transform. *Procedia CIRP* **2019**, *79*, 500–504. [CrossRef]
42. Sun, J.; Song, Z.; He, G.; Sang, Y. An improved signal determination method on machined surface topography. *Precis. Eng.* **2018**, *51*, 338–347. [CrossRef]
43. Abdul-Rahman, H.; Jiang, X.; Scott, P. Freeform surface filtering using the lifting wavelet transform. *Precis. Eng.* **2013**, *37*, 187–202. [CrossRef]
44. Lucido, M. Helmholtz–Galerkin Regularizing Technique for the Analysis of the THz-Range Surface-Plasmon-Mode Resonances of a Graphene Microdisk Stack. *Micro* **2022**, *2*, 295–312. [CrossRef]
45. Hirschberg, V.; Wilhelm, M.; Rodrigue, D. Combining mechanical and thermal surface fourier transform analysis to follow the dynamic fatigue behavior of polymers. *Polym. Test.* **2021**, *96*, 107070. [CrossRef]
46. Markovska, I.; Pavlov, S.; Yovkova, F.; Michalev, T.; Yaneva, S. Estimation of ceramics surface roughness by fourier analysis. *J. Chem. Technol. Metall.* **2018**, *53*, 1139–1143.
47. Liu, M.; Dong, L.; Zhang, Z.; Li, L.; Wang, L.; Song, Z.; Weng, Z.; Han, X.; Zhou, L.; Wang, Z. Fast Fourier transport analysis of surface structures fabricated by laser interference lithography. *Appl. Phys. Express* **2014**, *12*, 096503. [CrossRef]
48. Chen, H.-Y.; Lee, C.-H. Deep Learning Approach for Vibration Signals Applications. *Sensors* **2021**, *21*, 3929. [CrossRef]
49. Singru, P.; Krishnakumar, V.; Natarajan, D.; Raizada, A. Bearing failure prediction using Wigner-Ville distribution, modified Poincare mapping and fast Fourier transform. *J. Vibroeng.* **2018**, *20*, 127–137. [CrossRef]
50. Holub, M.; Jankovych, R.; Vetiska, J.; Sramek, J.; Blecha, P.; Smolik, J.; Heinrich, P. Experimental Study of the Volumetric Error Effect on the Resulting Working Accuracy—Roundness. *Appl. Sci.* **2020**, *10*, 6233. [CrossRef]
51. Adamczak, S.; Zmarzły, P.; Janecki, D. Theoretical And Practical Investigations Of V-Block Waviness Measurement Of Cylindrical Parts. *Metrol. Meas. Syst.* **2015**, *22*, 181–192. [CrossRef]
52. Liu, J.; Shao, Y. Vibration modelling of nonuniform surface waviness in a lubricated roller bearing. *J. Vib. Control.* **2015**, *23*, 1115–1132. [CrossRef]
53. Wrzochal, M.; Adamczak, S. The problems of mathematical modelling of rolling bearing vibrations. *Bull. Pol. Acad. Sci. Tech. Sci.* **2020**, *68*, 1363–1372.
54. Misiti, M.; Misiti, Y.; Oppenheim, G.; Poggi, J.-M. *MathWorks Wavelet Toolbox User's Guide*; MathWorks Inc.: Natick, MA, USA, 2015.
55. Kumar, R.N.; Jagadale, B.N.; Bhat, J.S. A lossless image compression algorithm using wavelets and fractional Fourier transform. *SN Appl. Sci.* **2019**, *1*, 266. [CrossRef]
56. Gogolewski, D. Influence of the edge effect on the wavelet analysis process. *Measurement* **2020**, *152*, 107314. [CrossRef]
57. Gogolewski, D. Fractional spline wavelets within the surface texture analysis. *Measurement* **2021**, *179*, 109435. [CrossRef]
58. Herrmann, F.J. A Scaling Medium Representation a Discussion on Well-Logs, Fractals and Waves. Ph.D Thesis, Beeld en Gra sch Centrum, Technische Universiteit Delft, Delft, The Netherlands, 1997.

Disclaimer/Publisher's Note: The statements, opinions and data contained in all publications are solely those of the individual author(s) and contributor(s) and not of MDPI and/or the editor(s). MDPI and/or the editor(s) disclaim responsibility for any injury to people or property resulting from any ideas, methods, instructions or products referred to in the content.

Article

The Effect of Non-Measured Points on the Accuracy of the Surface Topography Assessment of Elements 3D Printed Using Selected Additive Technologies

Paweł Zmarzły *, Tomasz Kozior and Damian Gogolewski

Faculty of Mechatronics and Mechanical Engineering, Kielce University of Technology, al. Tysiąclecia Państwa Polskiego 7, 25-314 Kielce, Poland
* Correspondence: pzmarzly@tu.kielce.pl; Tel.: +48-41-342-44-53

Citation: Zmarzły, P.; Kozior, T.; Gogolewski, D. The Effect of Non-Measured Points on the Accuracy of the Surface Topography Assessment of Elements 3D Printed Using Selected Additive Technologies. *Materials* **2023**, *16*, 460. https://doi.org/10.3390/ma16010460

Academic Editor: Przemysław Podulka

Received: 16 December 2022
Revised: 29 December 2022
Accepted: 30 December 2022
Published: 3 January 2023

Copyright: © 2023 by the authors. Licensee MDPI, Basel, Switzerland. This article is an open access article distributed under the terms and conditions of the Creative Commons Attribution (CC BY) license (https:// creativecommons.org/licenses/by/ 4.0/).

Abstract: The paper presents the results of research aimed at evaluating the surface topography including the analysis of the number of unmeasured points of the samples 3D printed using four additive technologies (i.e., PolyJet Matrix, fused deposition modeling, selective laser sintering, and selective laser melting). The samples were made in three variants of location on the printing platform of 3D printers. Measurements of the samples' surface topography were carried out using a Talysurf CCI Lite optical profilometer and a Talysurf PGI 1230 contact profilometer. The percentage of non-measured points for each sample and the parameters of the surface topography were determined. Then, the non-measured points were complemented and the topography parameters for the corrected surface were recalculated. In addition, to perform comparative measurements, each surface was measured using a contact profilometer Talysurf PGI 1230. Preliminary results of the research showed that the measurement of the surface topography of the samples made using selective laser sintering technology with the Taysurf CCI optical measuring system is very unreliable, as the number of non-measured points for the analyzed samples was higher than 98%. The highest accuracy of optical measurement was obtained for PJM technology and three variants of location on the printing platform of the 3D printer.

Keywords: non-measured points; surface topography; 3D printers; additive technology; optical measurement

1. Introduction

Additive technologies are becoming increasingly popular in the ongoing Industry 4.0 revolution [1]. In the beginning, they were mainly used for the construction of prototypes and models. However, currently, due to the decrease in the prices of 3D printers and printing materials, they are used in various sectors of the economy (i.e., in foundries, automotive, space, textiles, and medicine [2–7]. Along with the dynamic development of 3D printers, the dimensional and shape accuracy and mechanical strength of the elements printed using additive technologies are constantly improving [8,9]. This allows for the production of finished (precise) products without the need for further machining [10] as well as using the tools of reverse engineering [11]. Therefore, there is a need for a comprehensive analysis of the surface topography of such elements using contact or optical profilometers [12].

The main materials used in additive manufacturing technologies for the printing of functional elements are plastics, but recently also metal alloys (steel, aluminum alloys) [13–15]. In the case of some types of additive technologies and the materials used for 3D printing, there is a problem with the measurement and analysis of the surface topography. Additionally, the use of functional coatings can pose measurement problems [16,17]. For example, the elements made using SLS technology (selective laser sintering) have a low yield point [18]. The surface topography of such elements should not be analyzed using contact methods due to the risk of scratching (destruction) the measured surface [19]. In such cases, the use of non-contact, optical methods of measurement is recommended [20]. However, due to the layered nature of the 3D print and anisotropy of the surface, there is a very large number of non-measured points. This problem technically makes it impossible to use this type of measuring instrument for the analysis of surface stereometry. Therefore, other methods of evaluating the surface topography of elements made using some additive technologies should be sought.

The main purpose of the research presented in this paper was to analyze the number of non-measured points for selected additive technologies by taking into account the 3D printing direction, and to assess the impact of non-measured points on the accuracy of the surface topography parameters.

There have been some research projects dealing with the problem of measuring the surface topography of some materials using optical profilometers, however, these projects have mainly been based on elements made using conventional methods of production [21]. The study in [22] presented research aimed at assessing the impact of undetected points on the accuracy of the measurement of the surface of elements subjected to various surface machining methods (i.e., milling, honing, and polishing). In addition, the effect of the lighting intensity setting of the Talysurf CCI measuring device on the number of undetected points was tested. The research has shown that even a small number of undetected points significantly affects the accuracy of the 3D roughness parameter determination. The authors in [23] analyzed the surface topography of the faces of seal rings made of silicon carbide and carbon graphite by using three different measurement systems. A contact profilometer, an optical profilometer, and an atomic force microscope were used for comparative analysis. The research showed that for the silicon carbide ring, surface texture measurements using atomic force microscopy and optical instruments more accurately represented the actual topography than the measurements determined by the stylus profilometer. A research project was also performed to determine the optimal method for measuring the surface roughness in medical applications (i.e., assessing the surface roughness of a femoral head) [24]. The test results proved significant discrepancies between the values of the 3D roughness parameters (St, Ssk, and Sku) by using optical and contact profilometers. In [25], practical issues related to optical measurements were presented. In [26], they presented the experimental research of a preloaded asymmetric multi-bolted connection.

The analysis of the literature showed that there are many works related to the analysis of contact and optical methods of surface topography measurement. However, there have been no research projects on the selection of the optimal method for evaluating the surface topography of elements 3D printed using commonly used additive technologies. Moreover, most works on optical measurements do not mention the number of non-measured points, which is a critical issue when it comes to the accuracy of determining the surface topography parameters. It should be noted that the so-called "good measurement practice" is not always adhered to and metrologists "blindly" trust the software, which does not indicate the percentage of the measured points. Sometimes, especially in industrial applications, previously prepared measuring programs (templates) are used to control the surface of mass-produced elements. Then, the non-measured points are filled in automatically (without the operator's involvement). This is an important problem that has been overlooked in many publications.

Therefore, the present paper presents studies aimed at the evaluation of the percentage of non-measured points depending on the type of additive technology, printing material, and the way the samples are located on the 3D printing platform in the XZ axis. In addition, an assessment of the impact of supplementing non-measured measurement points using DigitalSurf's MountainsMap® software on the values of selected types of surface topography parameters was performed. The types of tested parameters were selected in such a way as to define in detail the nature of the surface irregularities. To obtain reference values for the analyzed surface topography parameters, each surface was tested using a Talysurf PGI 1230 contact profilometer. The contact measurement details were compared with the results of the optical measurements.

2. Materials and Methods

The experimental research conducted in relation to this study was carried out in two stages. The first stage involved making samples using four additive technologies (PJM, FDM, SLS, SLM), taking into account three directions of the models' location on the printing platform in the XZ axis (i.e., Pd = 0°, Pd = 45° and Pd = 90°). The second stage involved surface topography measurements using two measuring systems: a Talysurf CCI optical profilometer and a Talysurf PGI 1230 contact profilometer.

2.1. Sample Preparation

The below CAD model was used to make samples for all of the analyzed additive technologies. Then, the CAD model was approximated using a triangle mesh into an STL (stereolithography language) file. The approximation parameters were selected in such a way that the accuracy of the STL model was greater than the accuracy of the most accurate printer used to print the sample (PJM printer). Due to the geometry of the samples, each possible notation approximated the 3D models with 12 triangles. The STL model prepared in this way was sent to each 3D printer. Samples located on the printing platform were made for each printing technology in three variants in the XZ axis (Figure 1).

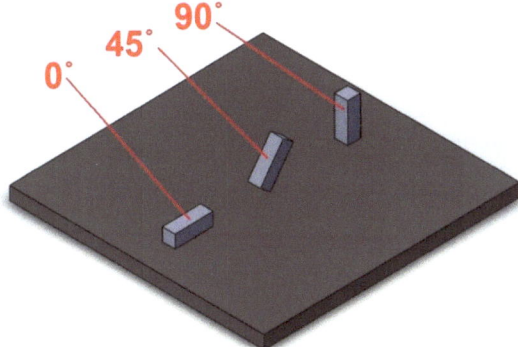

Figure 1. Location of the samples on the 3D printing platform.

A different material and a different method of connecting individual layers of the material were used for each of the researched printing technologies. Samples were printed using PolyJet Matrix (PJM) Technology with a Stratasys Connex 350 machine and liquid resin—FullCure 720 material. The layer thickness used was Lt = 0.016 mm. To implement fused deposition modeling (FDM) technology, a 3D printer Dimension 1200es (Stratasys) and ABS P430 material were used. The thickness of a single layer of the printing material was Lt = 0.254 mm. A Formiga P100 printer (0.1 mm layer thickness) was used to print the sample using selective laser sintering (SLS) technology. To make samples with this technology, polyamide powder PA2200 based on polyamide PA12 was used. A Concept laser M2 printer was used to print the sample using powder laser sintering (SLM) technology. The printing material was a 316L tool steel powder. The thickness of a single layer of the material applied using SLM technology was Lt = 0.060 mm. Figure 2 shows the 3D printers used for the research.

Due to the different chemical and physical properties of the materials used to print the samples, the selected mechanical properties of the tested materials are given in Table 1.

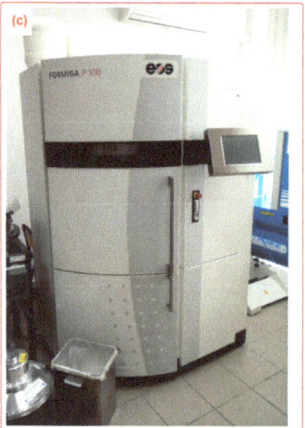

Figure 2. The 3D printers used to 3D print the samples. (**a**) Connex 350, (**b**) 3D Dimension 1200es, (**c**) Formiga P100, (**d**) Concept laser M2.

Table 1. Selected mechanical properties of the materials used to print the samples [18,27,28].

Type of Additive Technology	Material	Tensile Strength, MPa	Young's Modulus, MPa	Hardness
PJM	FullCure 720	60	2870	83 (Shore scale D)
FDM	ABS P430	37	2320	76 (Shore scale D)
SLS	PA2200	48	1700	75 (Shore scale D)
SLM	316 L grade steel	615	$200 \cdot 10^3$	20 HRC

2.2. Metrological Measurements

The first step of the research procedure was the measurement of the surface topography of the samples using a Talysurf CCI optical profilometer. This measurement system is equipped with a set of lenses ($\times 2.5$; $\times 10$; $\times 20$; $\times 50$) and enables the analysis of the surface texture with a vertical resolution of up to 0.01 nm. The horizontal resolution achieved by the device was of the order of 0.33 μm (for a $\times 50$ lens). The measuring range in the Z axis was 2.2 mm. Measurements of samples were made on an area of 1.66 mm $\times$ 1.66 mm. The sampling interval was as follows: $\Delta X = 1.62$ μm, $\Delta Y = 1.62$ μm. Based on the measured surfaces, the parameters of the 3D surface topography (i.e., Sa, Sq, Sz, Ssk, Sku) were determined. Parameter Sa is known as the arithmetical mean height and it expresses, as an absolute value, the difference in height of each point compared to the arithmetical mean of the surface. Parameter Sq is the root mean square height and it represents the root mean square value of ordinate values within the definition area. The parameter Sz indicates the maximum height. This parameter is defined as the sum of the largest peak height value and the largest pit depth value within the definition area. Ssk (skewness) values represent the degree of bias of the roughness shape (asperity). The kurtosis parameter (Sku) value is a measure of the sharpness of the roughness profile.

The non-measured surface points were then filled using MountainsMap® software options called "smooth shape calculated from the neighbors—optimize for space points" to assess the effect of the procedure of filling the non-measured points on the values of the 3D roughness parameters. The parameters of the surface topography for the same surfaces were determined again once the procedure of filling the points of the non-measured surfaces was complete. In addition, to obtain reference results (reference values), each surface at the same locations was measured using the Talysurf PGI 1230 contact profilometer. Due to the laser interferometer mounted in the "z" axis of the Talysurf PGI 1230 measurement system, the measurement resolution in the z axis was up to 0.8 nm. The measuring range in the "z" axis of Talysurf PGI 1230 was 12.5 mm. The measurement parameters were selected in such a way as to correspond to the measurements carried out using the Talysurf CCI optical system.

3. Results and Discussion

Table 2 presents the measurement results obtained using the Talysurf CCI optical device without filling the non-measured points (before filling) and after applying the filling procedure called "smooth shape calculated from the neighbors". In addition, the table includes the type of additive technology used and the 3D printing direction (Pd).

Table 2. Measurement results obtained for the Talysurf CCI optical profilometer.

Technology		SLS			PJM			FDM			SLM		
Printing direction, Pd		0°	45°	90°	0°	45°	90°	0°	45°	90°	0°	45°	90°
Non-measured points,%		99.6	99.6	98.8	9.91	12.2	1.5	88.8	89.8	60.4	72.9	30.1	40.7
Sa, µm	Before filling	22.2	21.46	20.50	3.23	3.96	2.03	9.5	2.91	6.17	10.55	6.08	5.27
	After filling	16.18	17.46	19.03	3.22	4.02	2.04	8.69	3.19	7.46	11.81	6.36	5.30
Sq, µm	Before filling	29.00	26.66	26.72	3.59	4.51	2.47	12.15	3.59	7.87	14.43	8.24	6.67
	After filling	20.93	22.46	23.82	3.6	4.56	2.47	11.06	4.18	9.79	15.54	8.56	6.75
Sz, µm	Before filling	200.7	217.4	181.0	41.88	30.72	46.44	72.36	54.65	76.82	111.8	94.51	106.1
	After filling	198.1	207.2	178.7	41.88	30.77	46.41	69.01	54.99	77.01	112.6	94.82	106.2
Ssk	Before filling	−1.01	−0.45	1.07	0.19	0.65	0.23	−0.19	−0.19	−0.38	1.42	0.94	−0.02
	After filling	−0.72	−0.44	0.52	0.25	0.54	0.21	0.15	−0.77	−0.87	1.08	0.89	−0.06
Sku	Before filling	4.67	3.58	3.58	1.73	2.01	2.85	2.63	4.04	4.00	6.87	5.37	3.37
	After filling	4.79	3.96	2.9	2.06	1.92	2.81	2.76	5.47	4.05	4.85	5.15	3.30

When analyzing the results presented in Table 2, one can note that some surfaces of the elements printed using selected additive technologies were "immeasurable" using the Talysurf CCI optical profilometer (about 1% of the measured surface). For example, for selective laser sintering (SLS) technology, depending on the selected print direction, the number of non-measured points was in the range of 98.8–99.6%. This is closely related to the nature of the surface of the elements printed using SLS technology and the properties of the material in terms of light absorption. The sintered powder creates a surface that scatters light to a large extent, which makes it much more difficult to carry out optical measurements. This means that the selected surfaces technically cannot be measured using optical devices and other measurement methods (e.g., contact measurements) should be sought. The same is true for the most popular additive technology, namely, FDM technology. Here, for the printing direction Pd = 45°, the number of non-measured points was 89.8%, while for the 3D printing direction of Pd = 90° (i.e., perpendicular to the surface of the printing platform), the number of non-measured points was 60.4%. Considering the results obtained for selective laser melting (SLM) technology, relatively divergent results were obtained depending on the printing direction. The best results were obtained for the printing angle of Pd = 45° (30.1%), and the worst for Pd = 0° (72.9%). Due to the method of laying successive layers of material by the printer, the direction of Pd = 45° is the most recommended. The best results were recorded when using photo-curing of liquid polymer resin technology—PJM. Using this technology, for the print angle Pd = 90°, only 1.5% of the points were not measured. It should be added that the surface of the elements printed using PJM technology is highly reflective, which greatly facilitates the measurements performed with the Talysurf CCI profilometer. Therefore, it can be concluded that optical measurements can be used to analyze the topography of the surface of elements printed using PJM technology.

Considering the impact of the procedure of filling the non-measured points, one can note a clear change in the values of the surface topography parameters depending on the percentage of unmeasured points. For the SLS technology, where the level of undetected points was the highest and fluctuated around 99%, a clear decrease in the surface topography parameters Sa, Sqm, and Sz was observed. This was clearly visible, especially for the print angle Pd = 0° and the parameter Sa, where the reduction was about 27%. Similar trends were noted for FDM technology and the printing direction Pd = 0°, where the number of non-measured points was 88.8%; however, in this technology, the reduction in the surface topography parameters was much lower. This may be due to the fact that clear and regular paths of the printing material are visible when using FDM technology, which resulted in a higher accuracy of the mapping of the non-measured points.

For the other technologies, where the number of undetected points was significantly smaller, the values of parameters Sa, Sq, and Sz increased. It should be noted that for the PolyJet Matrix Technology, where the surfaces were measured with the highest accuracy (the smallest number of measured points), the procedure of filling the non-measured points technically did not affect the values of the surface topography parameters. When analyzing the parameters of surface skewness Ssk and kurtosis Sku, it can be concluded that as a result of the procedure of filling the non-measured points, similar tendencies (trends) were observed as for the parameters of height (i.e., Sa, Sq, and Sz). However, the interpretation of the surface based on the Ssk and Sku parameters, despite a slight change in their values, did not change significantly.

When analyzing the results presented in Table 2 as a whole, it can be concluded that in the case of hard-to-measure surfaces, where the number of non-measured points is high, the procedure of filling them significantly reduces the values of the surface topography parameters. If this fact is not taken into account when analyzing the surface topography of 3D printed elements, erroneous conclusions concerning the nature of the surface can be drawn.

However, referring to the results of the surface topography measurements using the Talysurf PGI 1230 contact profilometer (see: Table 3), one can conclude that the highest values of Sa and Sq roughness parameters were obtained for FDM technology and the printing direction of Pd = 45°. These results are inconsistent with the results of the roughness measurement using the Talysurf CCI profilometer, where the highest values of the Sa and Sq parameters were measured when SLS technology was used. The lowest values of the Sa and Sq parameters were recorded for the technology of PolyJet Matrix (PJM). This is consistent with the results obtained when using the optical profilometer. It is important to say that for this technology, the highest number of points was detected. This was confirmed by the fact that the number of non-detected points directly affects the quality of the surface topography assessment, and consequently, the values of the roughness parameters. When analyzing the parameter determining the maximum height of the surface topography (maximum height), it should be stated that for both the optical and contact measurements, the maximum value of the Sz parameter was obtained for samples printed using SLS technology at the sample position of Pd = 0°.

It should be added that for all of the analyzed additive technologies, the smallest values of the amplitude parameters (Sa and Sq) were obtained for the printing angle of Pd = 90°. This is closely related to the way the successive layers of the printing material are formed. Therefore, this way of positioning the models on the printing platform is recommended to lower the roughness of the surface. A certain disadvantage of such a location of the element on the printing platform is the long printing time.

Table 3. Measurement results obtained for the Talysurf PGI optical profilometer 1230.

Technology Name	SLS			PJM			FDM			SLM		
Printing direction	0°	45°	90°	0°	45°	90°	0°	45°	90°	0°	45°	90°
Sa, μm	15.75	19.36	15.73	1.89	2.65	1.07	21.17	30.45	8.68	5.68	11.1	5.59
Sq, μm	19.93	23.92	19.92	2.52	3.08	1.32	25.2	35.65	10.73	7.23	14.78	7.13
Sz, μm	167.2	190	153.1	18.57	19.05	9.18	116.9	148.1	63.20	58.10	113.4	65.54
Ssk	−0.24	0.09	0.03	0.21	0.14	0.00	−0.41	−0.69	−0.55	0.64	1.16	−0.05
Sku	3.29	2.74	2.91	3.68	1.98	2.68	2.25	2.27	3.07	3.94	5.56	3.39

When analyzing the values of the skewness, namely, the Ssk parameter measured by the contact method, it can be concluded that in SLS technology and the print directions Pd = 45° and Pd = 90°, the values of the Ssk parameter were close to zero. This means that the considered surfaces are symmetrical (i.e., there is a large number of individual peaks and valleys). This can be proven by the surface topography views given in Figures 3–5. A similar effect was visible when using PJM Technology for the Pd = 90° printing directions. In the case of FDM technology, the values of the Ssk parameter were negative for all the analyzed print directions. This indicates that during printing, plateau-shaped surfaces were created (i.e., with single depressions and a concentration of material around the peaks, see Figures 6–8). A similar concentration of material around individual peaks was noted for SLS technology and the printing direction of Pd = 0°. However, when considering the surface of the elements printed of metal powder using SLM technology for the printing directions Pd = 0° and Pd = 45°, it can be concluded that the surfaces had few peaks and the 3D printing material was concentrated around the valleys. This was also confirmed by the kurtosis parameter interpretation (i.e., Sku). For SLM technology and the printing directions Pd = 0° and Pd = 45°, the Sku parameter values significantly exceeded 3. As a result, the surface was full of sharp peaks.

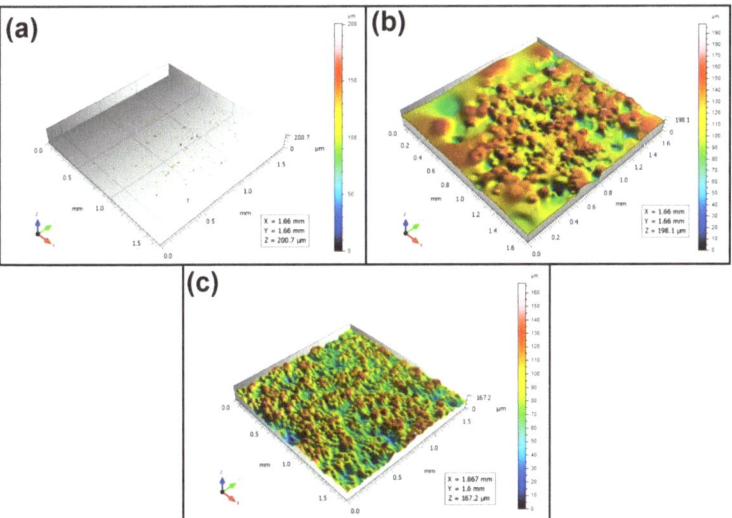

Figure 3. Surface topography of the element 3D printed using SLS technology (Pd = 0°). (a) Talysurf CCI—without after filling non-measured points, (b) Talysurf CCI—after filling non-measured points, (c) Talysurf PGI 1230.

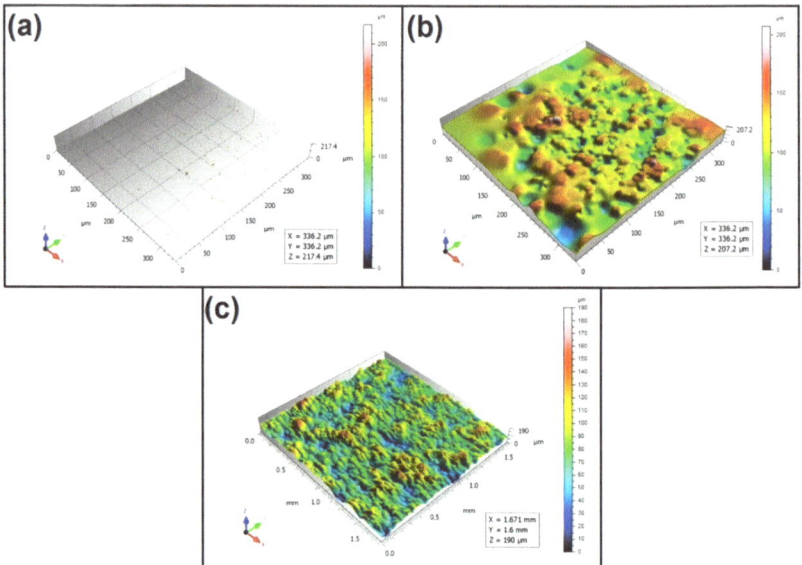

Figure 4. Surface topography of the element 3D printed using SLS technology (Pd = 45°). (**a**) Talysurf CCI—without filling non-measured points, (**b**) Talysurf CCI—after filling non-measured points, (**c**) Talysurf PGI 1230.

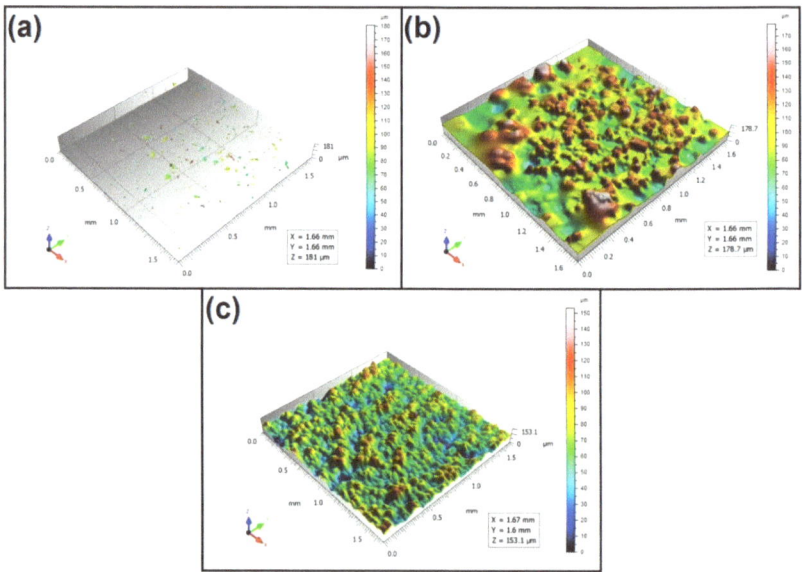

Figure 5. Surface topography of the element 3D printed using SLS technology (Pd = 90°). (**a**) Talysurf CCI—without filling non-measured points, (**b**) Talysurf CCI—after filling non-measured points, (**c**) Talysurf PGI 1230.

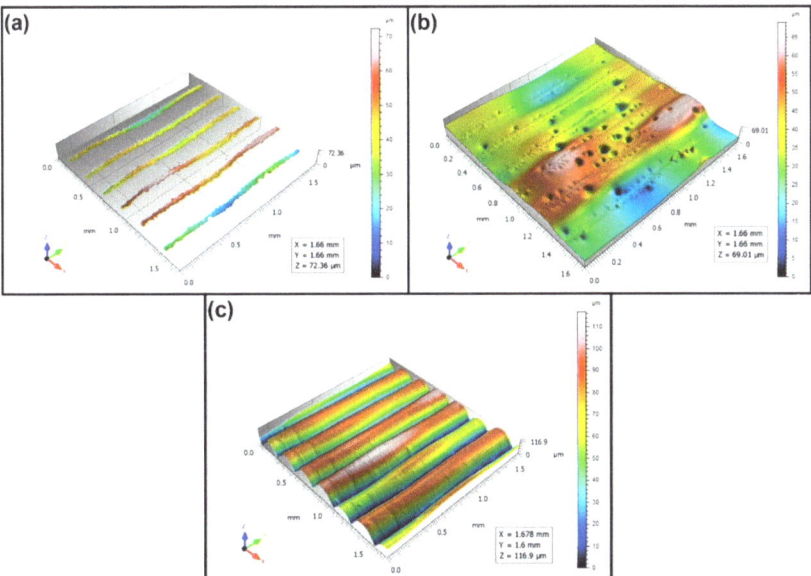

Figure 6. Surface topography of the element 3D printed using FDM technology (Pd = 0°). (**a**) Talysurf CCI—without filling non-measured points, (**b**) Talysurf CCI—after filling non-measured points, (**c**) Talysurf PGI 1230.

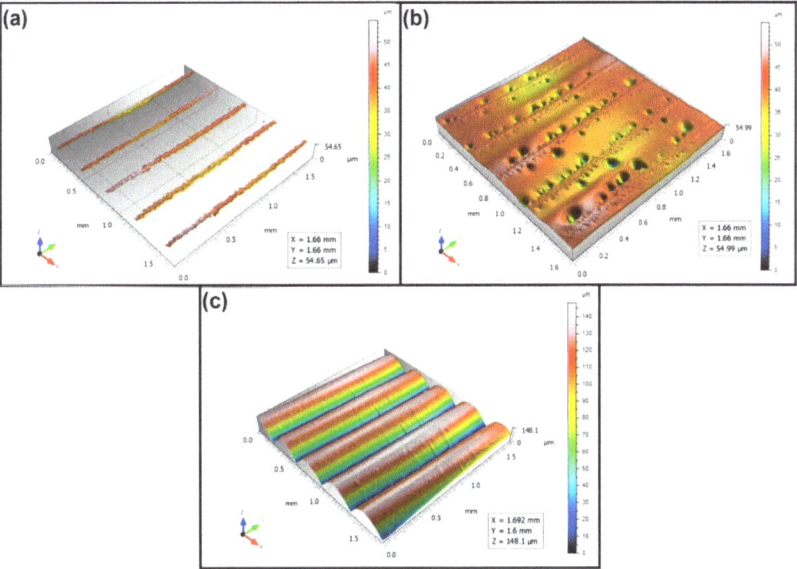

Figure 7. Surface topography of the element 3D printed using FDM technology (Pd = 45°). (**a**) Talysurf CCI—without filling non-measured points, (**b**) Talysurf CCI—after filling non-measured points, (**c**) Talysurf PGI 1230.

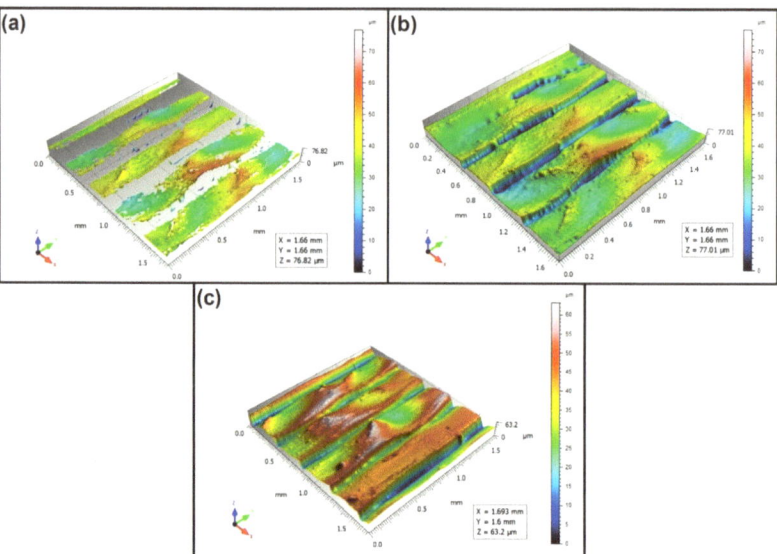

Figure 8. Surface topography of the element 3D printed using FDM technology (Pd = 90°). (**a**) Talysurf CCI—without filling non-measured points, (**b**) Talysurf CCI—after filling non-measured points, (**c**) Talysurf PGI 1230.

To visually assess the surface topography of elements printed with selected additive technologies, views of the surface topography obtained using two measuring systems (i.e., Talysurf CCI and Talysurf PGI 1230 are presented in the below figures, Figures 3–14). It should be noted that the Talysurf PGI 1230 measurement system utilizes the contact measurement method and, in this case, 100% of the measured values were assumed.

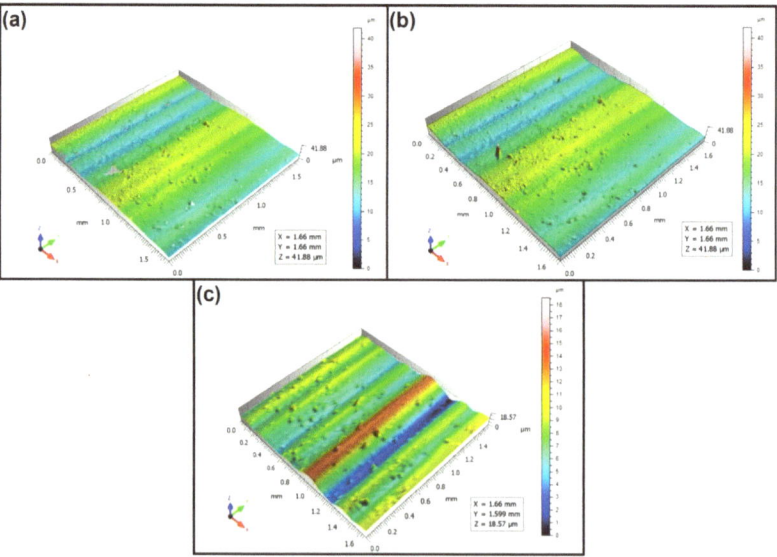

Figure 9. Surface topography of the element 3D printed using PJM technology (Pd = 0°). (**a**) Talysurf CCI—without filling non-measured points, (**b**) Talysurf CCI—after filling non-measured points, (**c**) Talysurf PGI 1230.

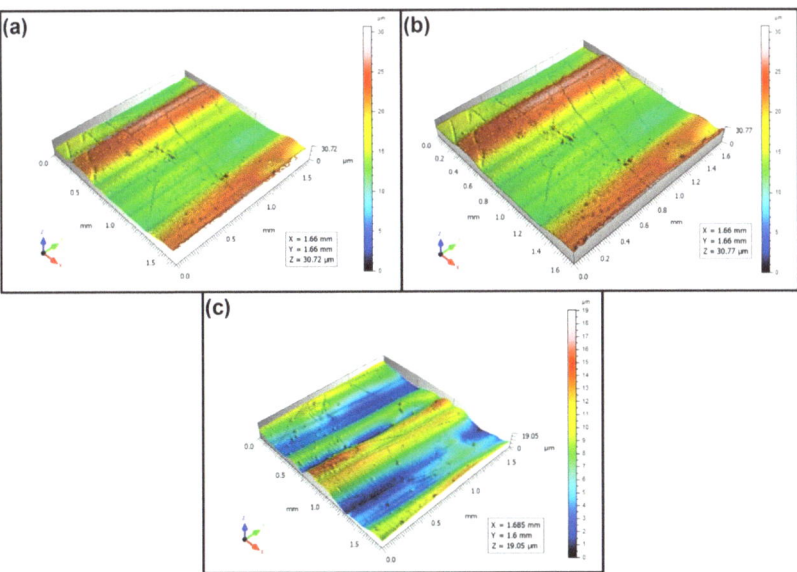

Figure 10. Surface topography of the element 3D printed using PJM technology (Pd = 45°). (**a**) Talysurf CCI—without filling non-measured points, (**b**) Talysurf CCI—after filling non-measured points, (**c**) Talysurf PGI 1230.

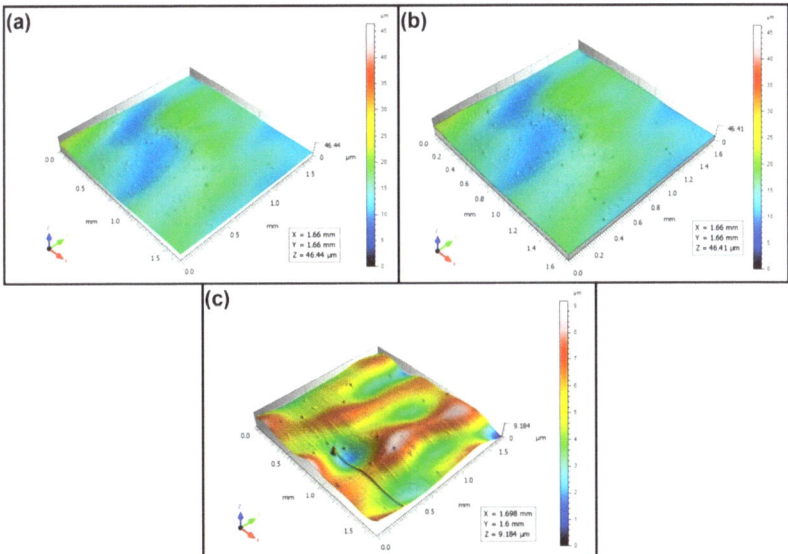

Figure 11. Surface topography of the element 3D printed using PJM technology (Pd = 90°). (**a**) Talysurf CCI—without filling non-measured points, (**b**) Talysurf CCI—after filling non-measured points, (**c**) Talysurf PGI 1230.

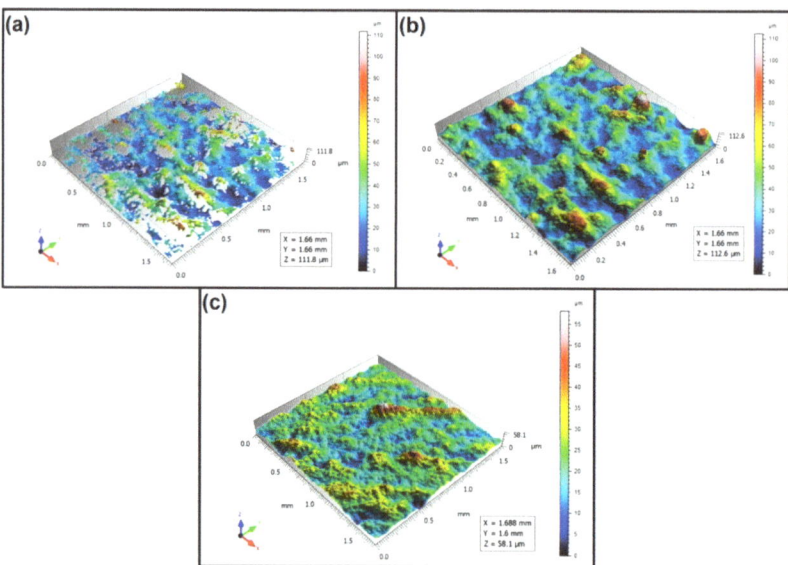

Figure 12. Surface topography of the element 3D printed using SLM technology (Pd = 0°). (**a**) Talysurf CCI—without filling non-measured points, (**b**) Talysurf CCI—after filling non-measured points, (**c**) Talysurf PGI 1230.

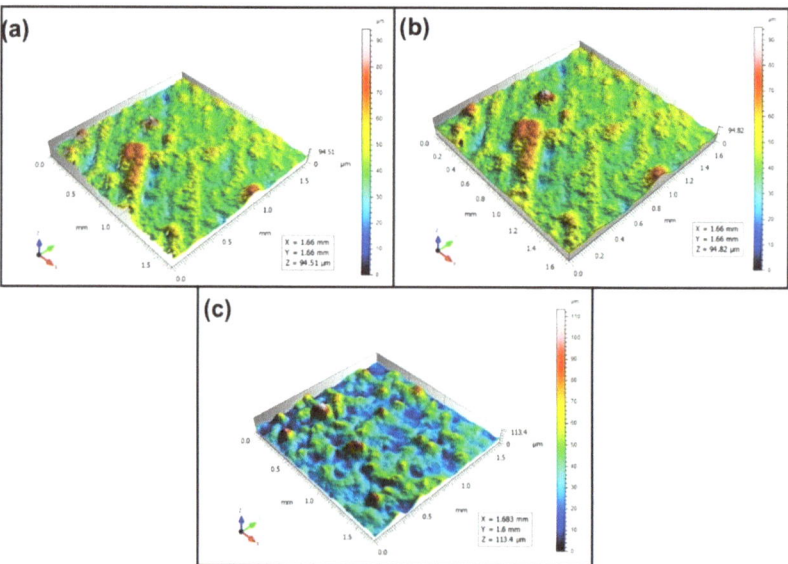

Figure 13. Surface topography of the element 3D printed using SLM technology (Pd = 45°). (**a**) Talysurf CCI—without filling non-measured points, (**b**) Talysurf CCI—after filling non-measured points, (**c**) Talysurf PGI 1230.

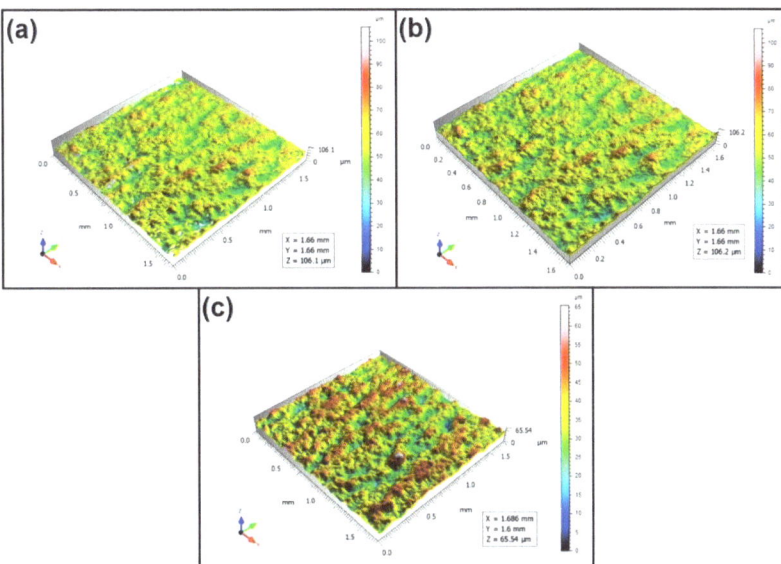

Figure 14. Surface topography of the element 3D printed using SLM technology (Pd = 90°). (**a**) Talysurf CCI—without filling non-measured points, (**b**) Talysurf CCI—after filling non-measured points, (**c**) Talysurf PGI 1230.

When analyzing the views of the surface topography obtained using the Talysurf CCI optical profilometer (Figures 3a, 4a and 5a), one can see that the surfaces were not measured. This was confirmed by the results presented in Table 2. The procedure of filling unmeasured points gives a misleading idea of the measured area (see Figures 3b, 4b and 5b). However, when analyzing the isometric images of the surface topography obtained by using the Talysurf PGI 1230 contact profilometer, one can note a clear structure distinguished by a large number of individual peaks and valleys. For each analyzed printing direction, the characteristics of the surface irregularities were similar. This is due to the way the SLS technology is implemented by the 3D printer. For each printing direction, a layer of material (powder) is sintered by a laser beam.

When analyzing the views of the surface topography of elements printed using FDM technology, one can periodically see occurring hills, related to the direction of the printing material application. When analyzing Figures 6a, 7a and 8a, we can see that only the top surface of the peaks was measured. Detailed measurement of the surface of the valleys was difficult. However, the structure obtained for the angle of Pd = 90° (see Figure 8) differed from the surface topography of the elements printed at the angles Pd = 0° and Pd = 45°. In this case, more "flattened" surfaces were obtained. It should be added that the largest number of measured points was obtained for this printing direction.

When analyzing the surface topography view for the samples printed of liquid resin using PJM Technology, one can see a clear directionality of the structure depending on the printing direction. Similarly to FDM technology, periodically occurring peaks can also be seen here (Figures 9–11). In addition, in the case of PJM Technology, in the print direction Pd = 90° (Figure 11), for example, local traces of material curing initiated by UV light were visible.

When analyzing the views of the surface topography presented in Figures 12–14, one can see that the nature of the surface irregularities of the metal samples printed using SLM technology was similar to the surface of the elements printed using SLS technology. This was due to the similarity of the operation of the printer in both of the analyzed technologies. In SLS technology, the plastic powder is sintered by a laser, while in SLM technology, metal powder is sintered by a laser. However, the surfaces made of metal powders were better mapped than those made of polyamide. This is due to the fact that metal surfaces are more reflective, while the polyamide surface diffuses mode light, which directly affects the number of unmeasured points.

4. Conclusions

The intensive development of additive technologies makes them more and more popular, and in the future, they may replace conventional manufacturing technologies. The current state-of-the-art of 3D printers allows for the production of functional machine parts. Therefore, the accuracy of the 3D printing of such elements should be examined in detail. However, in the case of responsible machine parts, the quality of the surface layer should also be tested by measuring and analyzing the surface stereometry parameters.

Due to the different methods of 3D printing as well as the different types of material used to print the elements, sometimes, the assessment of surface topography using non-contact measuring systems is difficult to perform or even impossible. Therefore, the main purpose of the research described in this paper was to analyze the selected measurement factors (i.e., the number of non-measured points to the values of surface topography parameters).

The results of the research presented in this paper prove that interferometric methods may not always be useful for assessing the surface topography of elements printed using some methods of additive technologies. Among all of the analyzed additive technologies, namely, SLS, FDM, PJM, and SLM, only in the case of PJM Technology and the printing angle of Pd = 90° was the surface measured with relatively high accuracy. The number of non-measured points was only 1.5%. For the remaining additive technologies analyzed, the number of non-measured points ranged from 30.1% to 99.6%. The values were significant. The worst results were obtained when using the technology of the selective laser sintering of metal powders, where, depending on the printing direction, the number of non-measured points was in the range of 98.8–99.6%, which excludes the analyzed measurement method for analyzing the surface topography of elements printed using SLS technology.

By analyzing the values of topography parameters before and after applying the procedure of filling non-measured points, it can be concluded that this function reduces the values of the tested surface topography parameters. This reduction depends on the type of additive technology used, the technological parameters, and the number of non-measured points.

Therefore, in the case of a large number of non-measured points, the use of the MountainsMap® software option called "smooth shape calculated from the neighbors—optimize for space points" may provide an unreliable estimate of the surface topography.

When analyzing the values of the surface topography parameters as a whole, it can be concluded that the best results were obtained for PJM Technology and the printing direction of Pd = 90°, where the value of the parameter Sa = 1.07 μm was obtained. Moreover, for the analyzed additive technologies, the lowest values of the topography parameters were obtained for the Pd = 90° printing direction. However, the use of this direction is associated with a longer manufacturing time, which in some cases is unprofitable.

As the aim of further research, the authors will analyze other methods of measuring the surface topography of elements printed using additive technologies. Among other things, atomic force microscopy (AFM) will be used for these types of measurements. In addition, we plan to analyze other factors that may affect the quality of the measured surface mapping (e.g., the type of light used, the method of removing support material from the 3D printed samples, etc.).

Author Contributions: Conceptualization, P.Z., T.K. and D.G.; Methodology, P.Z.; Software, P.Z. and D.G.; Validation, P.Z., T.K. and D.G.; Formal analysis, P.Z., T.K. and D.G.; Investigation, P.Z., T.K. and D.G.; Resources, P.Z.; Data curation, P.Z.; Writing—original draft preparation, P.Z.; Writing—review and editing, P.Z., T.K. and D.G.; Visualization, P.Z.; Supervision, P.Z., T.K. and D.G.; Project administration, P.Z.; Funding acquisition, P.Z. All authors have read and agreed to the published version of the manuscript.

Funding: The research was financed by the National Center for Research and Development as part of the Lider XI project, number LIDER/44/0146/L-11/19/NCBR/2020, under the title "An analysis of application possibilities of the additive technologies to rapid fabrication of casting patterns".

Institutional Review Board Statement: Not applicable.

Informed Consent Statement: Not applicable.

Data Availability Statement: Not applicable.

Conflicts of Interest: The authors declare no conflict of interest.

References

1. Wieczorowski, M.; Oliwa, R.; Woźniak, J.; Sobolewski, B.; Paszkiewicz, A.; Przeszłowski, Ł.; Oleksy, M.; Budzik, G. The Place of 3D Printing in the Manufacturing and Operational Process Based on the Industry 4.0 Structure. *Teh. Glas.* **2022**, *16*, 252–257. [CrossRef]
2. Kozior, T.; Döpke, C.; Grimmelsmann, N.; Juhász Junger, I.; Ehrmann, A. Influence of Fabric Pretreatment on Adhesion of Three-Dimensional Printed Material on Textile Substrates. *Adv. Mech. Eng.* **2018**, *10*, 1687814018792316. [CrossRef]
3. Yan, Q.; Dong, H.; Su, J.; Han, J.; Song, B.; Wei, Q.; Shi, Y. Additive Manufacturing—Review A Review of 3D Printing Technology for Medical Applications. *Engineering* **2018**, *4*, 729–742. [CrossRef]
4. Kozior, T.; Bochnia, J.; Gogolewski, D.; Zmarzły, P.; Rudnik, M.; Szot, W.; Szczygieł, P.; Musiałek, M. Analysis of Metrological Quality and Mechanical Properties of Models Manufactured with Photo-Curing PolyJet Matrix Technology for Medical Applications. *Polymers* **2022**, *14*, 408. [CrossRef]
5. Adamczak, S.; Zmarzly, P.; Kozior, T.; Gogolewski, D. Analysis of the dimensional accuracy of casting models manufactured by fused deposition modeling technology. In Proceedings of the 23rd International Conference on Engineering Mechanics 2017, Svratka, Czech Republic, 15–18 May 2017.
6. Blasiak, S.; Laski, P.A.; Takosoglu, J.E. Rapid Prototyping of Pneumatic Directional Control Valves. *Polymers* **2021**, *13*, 1458. [CrossRef]
7. Borawski, A. Impact of Operating Time on Selected Tribological Properties of the Friction Material in the Brake Pads of Passenger Cars. *Materials* **2021**, *14*, 884. [CrossRef]
8. Bochnia, J.; Blasiak, M.; Kozior, T. Tensile Strength Analysis of Thin-Walled Polymer Glass Fiber Reinforced Samples Manufactured by 3d Printing Technology. *Polymers* **2020**, *12*, 2783. [CrossRef]
9. Kozior, T. Rheological Properties of Polyamide Pa 2200 in Sls Technology. *Teh. Vjesn.* **2020**, *27*, 242308. [CrossRef]
10. Hatz, C.R.; Msallem, B.; Aghlmandi, S.; Brantner, P.; Can, F.M.T. Can an Entry-Level 3D Printer Create High-Quality Anatomical Models? Accuracy Assessment of Mandibular Models Printed by a Desktop 3D Printer and a Professional Device. *Int. J. Oral Maxillofac. Surg.* **2019**, *49*, 142–148. [CrossRef]
11. Łukaszewicz, A.; Miatliuk, K. Reverse Engineering Approach for Object with Free-Form Surfaces Using Standard Surface-Solid Parametric CAD System. *Solid State Phenom.* **2009**, *147–149*, 706–711. [CrossRef]
12. Mathia, T.G.; Pawlus, P.; Wieczorowski, M. Recent Trends in Surface Metrology. *Wear* **2011**, *271*, 494–508. [CrossRef]
13. Galy, C.; Le Guen, E.; Lacoste, E.; Arvieu, C. Main Defects Observed in Aluminum Alloy Parts Produced by SLM: From Causes to Consequences. *Addit. Manuf.* **2018**, *22*, 165–175. [CrossRef]
14. Skrzyniarz, M.; Nowakowski, L.; Blasiak, S. Geometry, Structure and Surface Quality of a Maraging Steel Milling Cutter Printed by Direct Metal Laser Melting. *Materials* **2022**, *15*, 773. [CrossRef]
15. Ma, Q.P.; Mesicek, J.; Fojtik, F.; Hajnys, J.; Krpec, P.; Pagac, M.; Petru, J. Residual Stress Build-Up in Aluminum Parts Fabricated with SLM Technology Using the Bridge Curvature Method. *Materials* **2022**, *15*, 6057. [CrossRef]
16. Podulka, P. Advances in Measurement and Data Analysis of Surfaces with Functionalized Coatings. *Coatings* **2022**, *12*, 1331. [CrossRef]
17. Feng, X.; Senin, N.; Su, R.; Ramasamy, S.; Leach, R. Optical Measurement of Surface Topographies with Transparent Coatings. *Opt. Lasers Eng.* **2019**, *121*, 261–270. [CrossRef]
18. Adamczak, S.; Zmarzly, P.; Kozior, T.; Gogolewski, D. Assessment of roundness and waviness deviations of elements produced by selective laser sintering technology. In Proceedings of the 23rd International Conference on Engineering Mechanics 2017, Svratka, Czech Republic, 15–18 May 2017; pp. 70–73.
19. Zmarzły, P.; Adamczak, S. The Evaluation of a Contact Profilometer Measuring Tip Movement on the Surface Texture of the Sample. In Proceedings of the International Symposium for Production Research, Vienna, Austria, 28–30 August 2018.
20. Podulka, P. Thresholding Methods for Reduction in Data Processing Errors in the Laser-Textured Surface Topography Measurements. *Materials* **2022**, *15*, 5137. [CrossRef]
21. *Optical Measurement of Surface Topography*; Leach, R.K. (Ed.) Springer: New York, NY, USA, 2011.
22. Pawlus, P.; Reizer, R.; Wieczorowski, M. Problem of Non-Measured Points in Surface Texture Measurements. *Metrol. Meas. Syst.* **2017**, *24*, 525–536. [CrossRef]
23. Adamczak, S.; Kundera, C.; Swiderski, J. Assessment of the State of the Geometrical Surface Texture of Seal Rings by Various Measuring Methods. *IOP Conf. Ser. Mater. Sci. Eng.* **2017**, *233*, 012031. [CrossRef]

24. Merola, M.; Ruggiero, A.; Salvatore, J.; Mattia, D.; Affatato, S. On the Tribological Behavior of Retrieved Hip Femoral Heads Affected by Metallic Debris. A Comparative Investigation by Stylus and Optical Profilometer for a New Roughness Measurement Protocol. *Measurement* **2016**, *90*, 365–371. [CrossRef]
25. Koizumi, H.; Maki, A.; Yamamoto, T. Optical Topography: Practical Problems and Novel Applications. In Proceedings of the Optics InfoBase Conference Papers, Miami Beach, FL, USA, 7–10 April 2002.
26. Grzejda, R.; Parus, A.; Kwiatkowski, K. Experimental Studies of an Asymmetric Multi-Bolted Connection under Monotonic Loads. *Materials* **2021**, *14*, 2353. [CrossRef] [PubMed]
27. Bochnia, J. Evaluation of Relaxation Properties of Digital Materials Obtained by Means of PolyJet Matrix Technology. *Bull. Polish Acad. Sci. Tech. Sci.* **2018**, *66*, 891–897. [CrossRef]
28. Hardes, C.; Pöhl, F.; Röttger, A.; Thiele, M.; Theisen, W.; Esen, C. Cavitation Erosion Resistance of 316L Austenitic Steel Processed by Selective Laser Melting (SLM). *Addit. Manuf.* **2019**, *29*, 100786. [CrossRef]

Disclaimer/Publisher's Note: The statements, opinions and data contained in all publications are solely those of the individual author(s) and contributor(s) and not of MDPI and/or the editor(s). MDPI and/or the editor(s) disclaim responsibility for any injury to people or property resulting from any ideas, methods, instructions or products referred to in the content.

Article

Multiscale Analysis of Functional Surfaces Produced by L-PBF Additive Technology and Titanium Powder Ti6Al4V

Damian Gogolewski *[ ], Paweł Zmarzły [ ] and Tomasz Kozior [ ]

Faculty of Mechatronics and Mechanical Engineering, Kielce University of Technology, al. Tysiąclecia Państwa Polskiego 7, 25-314 Kielce, Poland
* Correspondence: dgogolewski@tu.kielce.pl

Abstract: The article discusses experimental studies assessing the possibility of mapping surfaces with a characteristic distribution of irregularities. Tests involved surfaces produced using the L-PBF additive technology, using titanium-powder-based material (Ti6Al4V). An evaluation of the resulting surface texture was extended to cover the application of a modern, multiscale analysis, i.e., wavelet transformation. The conducted analysis that involved using selected mother wavelet enabled production process errors and involved determining the size of resulting surface irregularities. The tests provide guidelines and enable a better understanding of the possibility of producing fully functional elements on surfaces, where morphological surface features are distributed in a characteristic way. Conducted statistical studies showed the advantages and disadvantages of the applied solution.

Keywords: wavelet transformation; multiscale analysis; surface texture; L-PBF; additive manufacturing

Citation: Gogolewski, D.; Zmarzły, P.; Kozior, T. Multiscale Analysis of Functional Surfaces Produced by L-PBF Additive Technology and Titanium Powder Ti6Al4V. *Materials* 2023, *16*, 3167. https://doi.org/10.3390/ma16083167

Academic Editor: Przemysław Podulka

Received: 29 March 2023
Revised: 11 April 2023
Accepted: 16 April 2023
Published: 17 April 2023

Copyright: © 2023 by the authors. Licensee MDPI, Basel, Switzerland. This article is an open access article distributed under the terms and conditions of the Creative Commons Attribution (CC BY) license (https:// creativecommons.org/licenses/by/ 4.0/).

1. Introduction

The fourth industrial revolution is a concept that covers the technological and organizational transformation process. Its particularly important aspects are modern manufacturing techniques, especially 3D printing technologies, which enable rapid production of prototypes and models with a complex geometry [1]. The development of additive technologies that allow the production of elements of any complex shape determines the applicability of these methods over an ever-wider spectrum. The production of fully functional components using 3D printing has been implemented in many industrial areas: founding [2], automotive [3], aerospace [4], or pneumatic and hydraulic industries [5], as well as a basis for the production of elements with specific properties or as medicinal aspects [6]. Despite its numerous advantages, these technologies also exhibit disadvantages, which often make the mapping of a CAD-designed model seem problematic. It is particularly evident in the case of free-form and rough, irregular surfaces with specific morphological features. Process limitations, such as minimum layer thickness or aspects of approximating a model with a triangle mesh (most common STL file), translate directly to the quality of produced features and their shapes and sizes [7]. The resulting geometrical surface structure is defined by a number of process parameters (e.g., material, layer thickness, printing direction, laser power and speed for contour and infill parameters, laser beam diameter and path parameters, gas atmosphere, support material placement, further thermal processes, etc.), but also by material parameters including chemical composition and powder parameters like grain distribution and size, which can reach values greater than the layer thickness, depending on the material [8]. A comprehensive analysis of additively manufactured parts also requires an assessment of the potential existence of internal defects in the material [9,10].

The development of modern technologies has also determined the need to research measuring techniques and evaluation methods [11]. It is a common belief that traditional perception and evaluation of a surface structure through Gaussian transformations (roughness or waviness assessment) is insufficient and does not provide a complex of information about morphological surface features [12–15]. Therefore, new methods were developed [16], as well as hybrid methods that use both classical and multiscale approaches in their data evaluation [17,18]. Multiscale procedures provide a wider spectrum of information on the studied surfaces and enable presenting them on many scales, depending on the type and size of individual surface features [19]. There are currently ongoing studies on the adaptation of multiscale methods for surface texture assessment. Various types of transformations are developed, including sliding bandpass filters, structural functions [20], geometric methods [21,22], or wavelet transformations. Wavelet transformations are used in an increasing number of cases of surface metrology [23–28]. The properties of individual wavelets enable an effective and comprehensive assessment of non-periodic irregularities [29], assessment, diagnostics, and indication of the place of occurrence for individual features [30,31], evaluation of manufacturing process parameters [32], tool wear and damage [33], surface texture extraction [34], engineering surface separation [35], or the estimation of surface roughness parameters based on surface images [36].

Based on the current state-of-the-art, it should be concluded that wavelet analysis is an appropriate tool that could be successfully developed to verify the applicability of modern additive technologies in terms of producing characteristic surface features (surface with a characteristic irregularity distribution). It potentially provides great opportunities in terms of measurement signal diagnostics and requires a more in-depth analysis. Please note that there are no studies aimed at evaluating the applicability of wavelet transformations to verify surfaces with a characteristic irregularity distribution for diagnosing the production process, and to assess process errors and irregularity distribution. The previously used, classical filtration methods exhibit limitations and often do not emphasize significant irregularities of components, which are crucial for additive processes. The studies fill the research gap and improve the applicability of modern multiscale methods, which are part of the Fourth Industrial Revolution, Metrology 4.0.

2. Materials and Methods

Test samples used to model surfaces with a specific distribution of irregularities were designed in the NX software (Siemens, Plano, TX, USA). Six samples with a surface defined by specific period and amplitude values were executed. Surfaces No. 1–3 were defined using a period function with a period equal to 0.2 mm and an amplitude of 0.34 mm. Surfaces No. 4–6 were defined using a composition of four periodic functions with periods of 0.4; 0.3; 0.25; and 0.2 mm, and amplitudes of 0.5; 0.14; 0.01; and 0.34 mm, respectively. The samples were saved as .stl files using the SolidWorks software (Dassault Systèmes SolidWorks Corp., Waltham, MA, USA), with a linear and angular accuracy of +/− 0.01 mm. Figure 1 shows a visualization of produced surfaces.

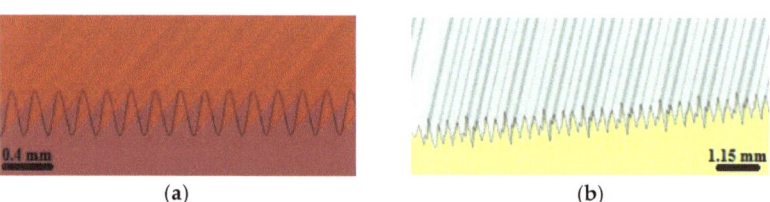

(a) (b)

Figure 1. Visualization of produced surfaces (a) No. 1–3, (b) No. 4–6.

Tests involved surfaces produced using the L-PBF additive technology. The samples were made from a titanium-powder-based material (Ti6Al4V), produced by EOS (EOS GmbH, Krailling, Germany) [37]. A 3D printer EOS M290 machine was used to build the sample. Samples No. 1–3 and No. 4–6 were built as an angle increment function relative to the building platform (20°, 45°, 70°). The samples were made with the following technological parameters: Inskin laser power—340 W, laser spot size—100 µm, laser speed—1250 mm/s, hatch distance—0.12 mm, layer thickness—60 µm. The platform temperature was set at a value of 35 °C, argon was used as a shielding, the powder fulfilled ASTM F1472 and ASTM F2924 standards, and samples were heat treated (necessary to stress-relieve treatment) at 800 °C for 2 h in an argon inert atmosphere as instructed by EOS. A surface view of sample No. 6 is shown in Figure 2a.

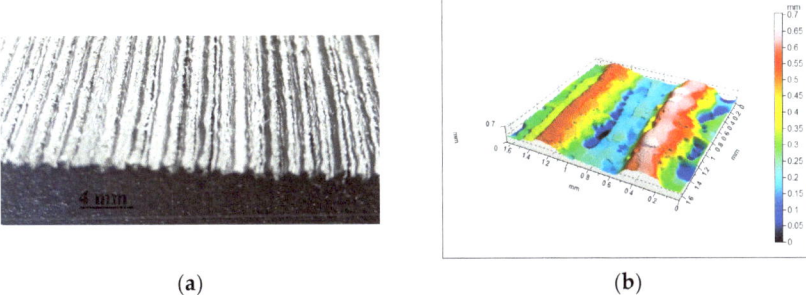

Figure 2. Sample No. 6 (**a**) surface view; (**b**) measured surface isometric image.

The measurements of the modelled surface irregularities distribution were conducted using an optical profilometer Talysurf CCI Lite (Taylor Hobson, Leicester, UK) with a vertical resolution of up to 0.01 nm. A magnification equal to ×10 was used for measurements, resulting in a surface size of 1.64 × 1.64 mm, which was represented by a point matrix of 1024 × 1024. TalyMap Platinum 6 (Digital Surf, Besançon, France) and Matlab software (The MathWorks, Natick, MA, USA) were used in the study. An isometric view of the measured sample No. 6 is shown in Figure 2b.

In addition, in order to provide a comprehensive analysis of research samples, the study was enhanced by analysing the samples using SEM (scanning electron microscope) and micro-CT (microfocus computed tomography). Microstructure studies were conducted using a scanning electron microscope JEOL JSM-7100F (JEOL Ltd. Akishima, Tokio, Japan) with different magnifications. The CT scanning and analysis were carried out using a computed tomography system (NIKON M2 LES SYSTEM (Nikon, Minato, Tokio, Japan)) that combines three radiation sources, i.e., two micro- and one minifocus X-ray sources (225 kV, 450 kV, and 450 kV, respectively). The examinations were conducted using a 225 kV X-ray tube with a 2 mm thick copper filter. The scanning data were then processed and visualized using VG Studio 3.5.2. software (Volume Graphics GmbH Heidelberg, Germany). The images were segmented using gray-scale thresholding. The 3D geometry was obtained using a 3 × 3 median filter. In addition, to remove small voids and inclusions, remove options were applied for objects up to 2 voxels in size. Measurements were made with these set parameters: voltage 210 kV, current 195 µA, power 41.0 W, voxel size 30.01 µm, exposure total 1.42 s.

3. Results

The first analysed aspect involved experimental studies focusing on assessing the feasibility of mapping surfaces with a characteristic distribution of irregularities using additive technology. Series with thirty surface profiles perpendicular to the modelled irregularities distribution were assessed for each sample surface produced at a different angle. Successively, each of the surface profiles was approximated, i.e., surfaces No. 1–3 with one periodic function (Figure 3), while for samples No. 4–6, there were four periodic functions (Figure 4). In addition, the modeled CAD function profile is provided in Figures 3 and 4. In the figures below, the abscissa axis shows the measurement section while the ordinate axis shows the height of the irregularity.

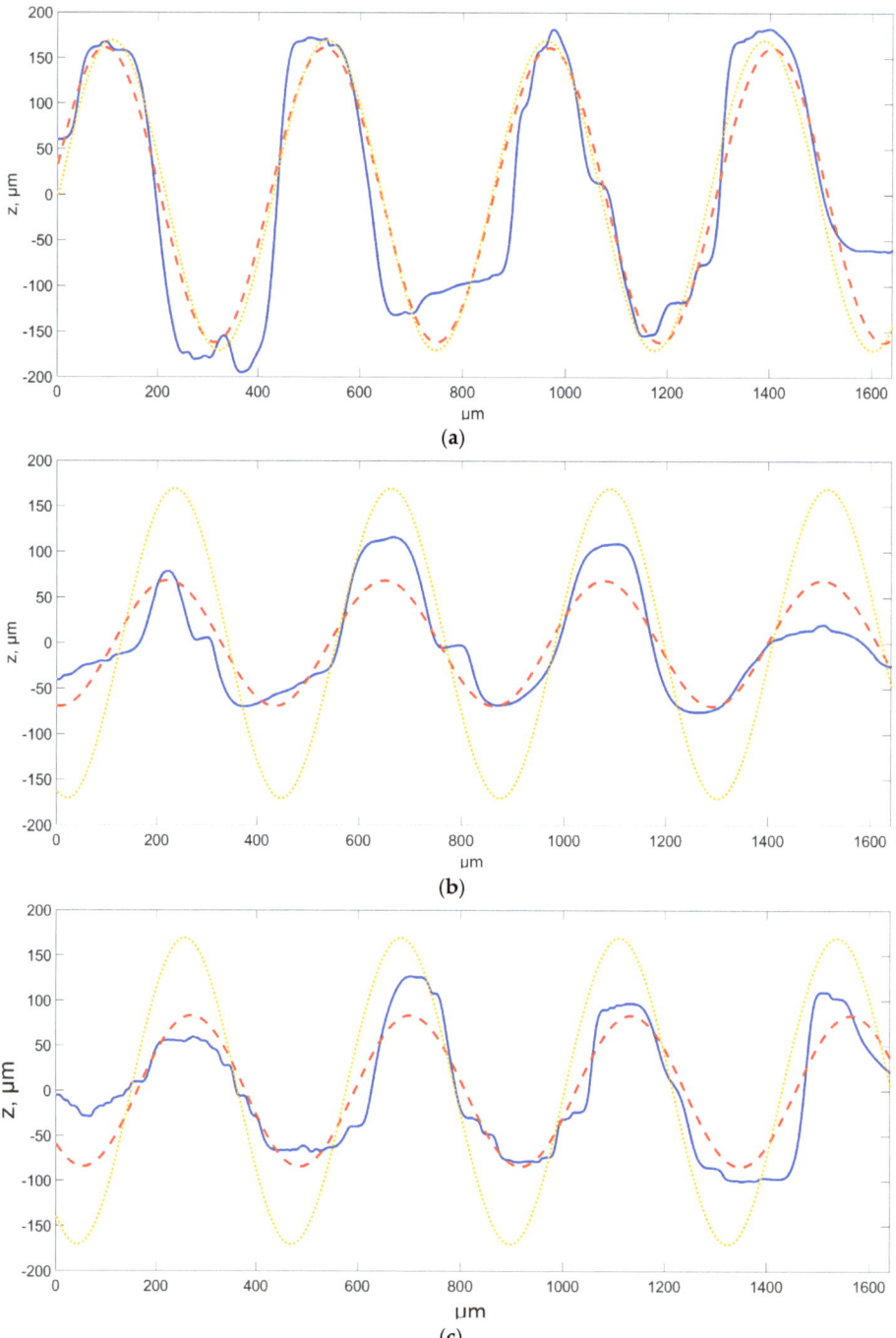

Figure 3. Example of a surface profile for samples No. 1–3 with an approximating function, respectively, as a function of building angle (**a**) 20°, (**b**) 45°, (**c**) 70°. The blue color indicates the measured profile, the red color indicates approximation, and the green color indicates the CAD model.

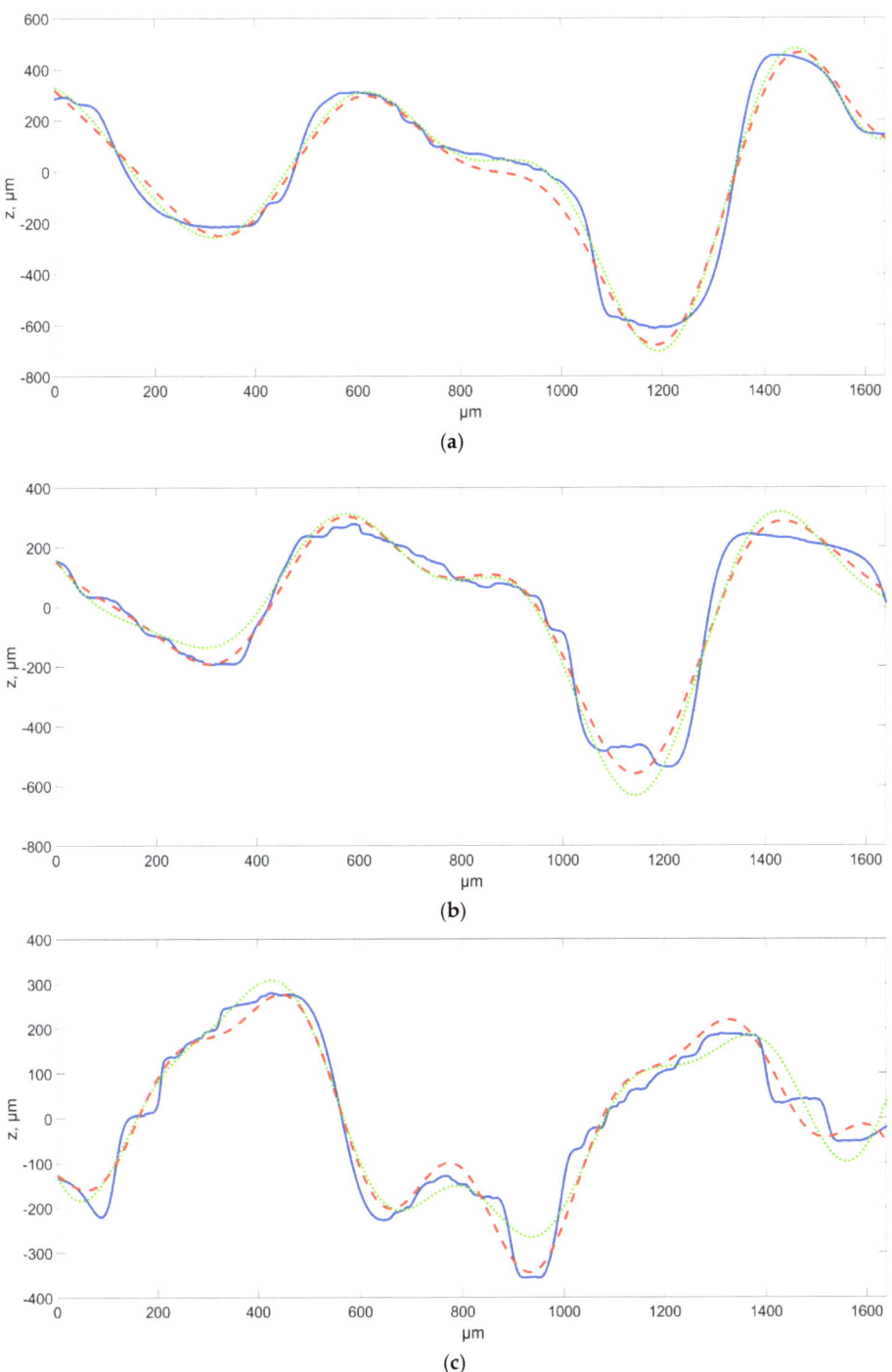

Figure 4. Example of a surface profile for samples No. 4–6 with an approximating function, respectively, as a function of construction angle, (**a**) 20°, (**b**) 45°, (**c**) 70°. The blue color indicates the measured profile, the red color indicates approximation, and the green color indicates CAD model.

Tests showed that the distribution of irregularities on the evaluated surfaces was close to nominal. However, the presence of morphological surface features was recorded due to, among other factors, spreading of the material between individual irregularities or the incomplete formation and melting of individual irregularities. The occurrence intensity of such features was variable and depended on the location on the sample. Approximating surface profiles with periodic functions enabled estimating the possibility of producing a surface of characteristic irregularities distribution. The studies showed that, for samples No. 1–3 defined by one period function, the R^2 factor for matching the approximating function to the measured, assessed profile ranged from about 0.7 to about 0.85. It should also be noted that the coefficient value decreased as a function of the building angle increasing. The differences may have been caused by, among other factors, difficulties in accurate model mapping and incomplete melting of individual peaks, which can be seen in, e.g., Figure 3b for the end profile. The causes also included limitations to the production process in terms of layer thickness and model approximation, which led to the formation of additional patterns on individual sinusoid waves, for which the height difference corresponded to the assumed layer thickness. It was also noted that the values of the defined approximating function were not fully consistent with the theoretical model. Amplitude values differed relatively by approximately twenty percent on average, depending on the profile. Additionally, in this case, the value decreased when the angle increased. However, it should be noted that periodic function values were convergent with theoretical ones. A relative difference in the values for the assessed profiles was around a few percent. No significant impact of the positioning angle in the case of assessed surface profile was recorded for this parameter.

A similar analysis was conducted for the surfaces of samples 4–6, which were defined by four periodic functions defined by different amplitudes and periods. The R^2 matching coefficient value for these samples was more than 0.95. Similarly like in the case of surfaces defined by one periodic function, the matching coefficient values decreased together when the building angle increased; however, these changes were insignificant. The studies involved assessing amplitude and period values for each of the four functions. However, please note the presence of one amplitude with a value close to assumed accuracy. It can be presumed that it will not be correctly mapped on the surface; however, it is important to approximate the profile based on four sine curve functions, due to the assumed period of a given function. When analysing the obtained result, it should be concluded that the relative amplitude for the assessed profiles was about thirty percent on average, while the average relative difference of the period value was approximately several percent. For a certain group of surface profiles, the indicated lowest amplitude value was not recorded for approximated functions, which directly translated to the value of other signal amplitudes. It cannot be clearly concluded whether the application of more functions resulted in better or worse mapping of the surface. Analysing the results of surface profile measurements, and taking into account the influence of powder particles on the profile parameters, it can be assumed that the chaotic character of surface irregularities of individual valleys and peaks is determined by the nature of the technological process in which, among other things, there are areas of not fully melted powder. The analysis carried out in relation to the measured and approximated profile showed that the function approximates the tested profile in a very effective way, and the differences between individual points of the profiles are less than 10 μm for the location of the model on the platform at an angle of 20°. For the other angles, the differences reached up to 100 μm. It seems that the location of manufactured models for smaller angles in relation to the building platform allows for the manufacture of a much more precise modeled surface with a noticeably smaller number of technological defects.

In terms of the resulting errors of the manufacturing process, for the samples in terms of the surface texture based on the analysis carried out using a scanning electron microscope, several characteristic morphological features can be distinguished. No unmelted powder grains were found on the tested surfaces (Figure 5a). Only minor impurities were found, which, due to their frequency of occurrence and their size, cannot be identified as roughness or waviness during optical measurements (please see the arrows in the Figure 5b). In addition, on the surface, there were single areas of melting of agglomerated powder grains (marked area in the Figure 5c,d), but this phenomenon was very rare (found only at 70° samples), such that its influence on the measurement is negligibly small. In this case, irregularities of up to 50–60 μm in height and 100–110 μm in diameter appeared on the surface. Due to the rarity of the discussed technological errors on the tested surfaces and evaluated profiles, the influence of the above features on the profiles of the samples was not noted.

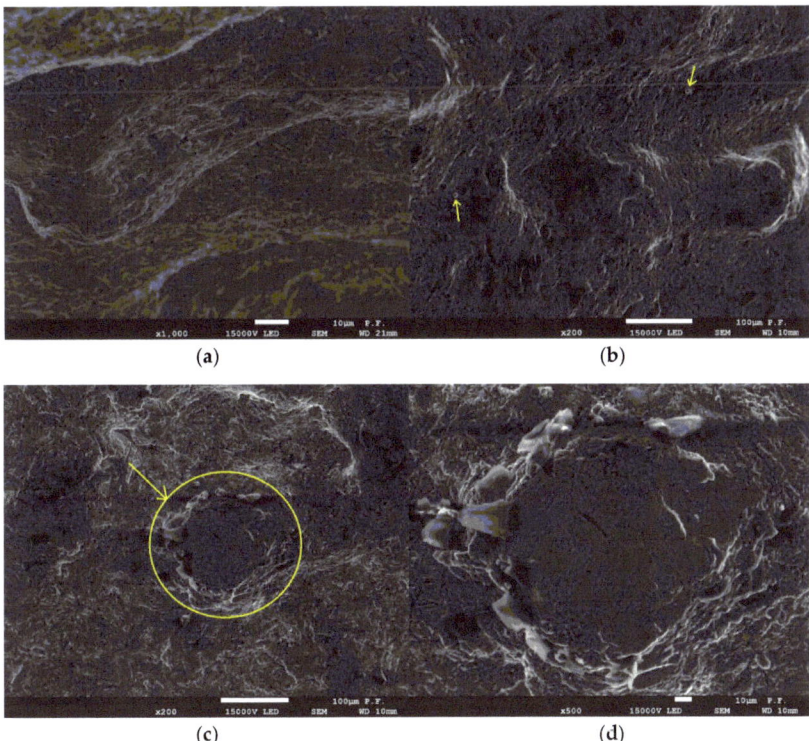

Figure 5. The microstructure of an example test sample: (**a**) surface view, (**b**) minor impurities, (**c**) agglomerated powder grains ×200, (**d**) agglomerated powder grains ×500.

Additional tests conducted using computer tomography showed that the geometric structure of the surface and the material on the analysed cross-sections of all measured profiles did not reveal any material defects related to the technological process of melting metal powders. No cavities, discontinuities, or inclusions were found. Observations near the measured surface layer do not indicate a possible influence of technological defects on the shape of the modeled irregularities. The visualization of the measurement results using CT for the selected sample is presented in Figure 6.

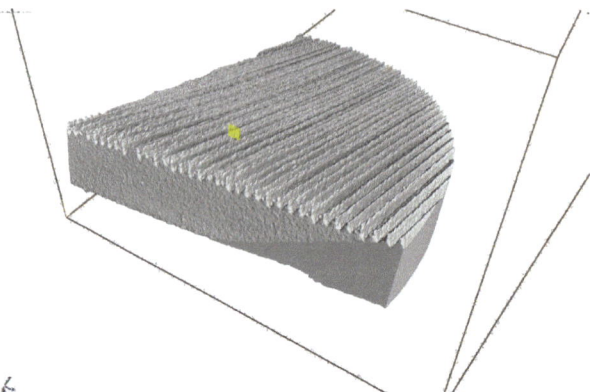

Figure 6. Visualization of the measurement results using CT for the selected sample.

Tests focusing on evaluating the possibility of mapping characteristic morphological surface features using additive technologies indicated the validity of comprehensive assessment of the resulting surface texture. Therefore, it was concluded that the application of modern, multiscale analysis, i.e., wavelet transformations, was justified. The research was aimed at assessing the possibility for noise reduction and study production process errors, as well as determining the size of resulting surface irregularities. A one-dimensional, discrete wavelet transformation was used for this purpose. A number of mother wavelet forms with different characteristics and properties were selected for the analysis. The following wavelets were used: db2, db12, db20, coif5, sym2, sym8, bior1.5, bior2.4, bior3.9, bior5.5. Figure 7 shows examples of surface irregularities distribution for samples No. 1 and 4, resulting from the application of coif5 and sym8 mother wavelets at the sixth analysis level. In the figures below, the abscissa axis shows the measurement section while the ordinate axis shows the height of the irregularity.

When assessing the obtained profile, it can be observed that the matching values for the coefficients describing the profile and approximating function were much better than for the profile before filtration. Therefore, it can be inferred that the dominating components of errors in a profile with a regular feature distribution are small, high-frequency pieces of information resulting from production process errors and corresponding to surface pores. Similar tests were conducted for all one hundred and eighty assessed profiles. The studies showed that, for profiles of samples No. 1–3, the authors obtained a matching coefficient value that grew with decomposition progress. However, this tendency can be observed up to the sixth decomposition level. Filtration of further levels for this mother wavelet leads to profile distortion. The tests covered all selected wavelets. Period and amplitude parameter values were greatly dependent on the mother wavelet support length. It was observed that better profile matching coefficients were obtained for mother wavelets with a longer support and that the obtained relative period difference values for individual wavelets were lower than nominal, compared to the parameters determined for the non-filtered profile. Furthermore, signal smoothing and filtering high-frequency information resulted in a change of the approximating function amplitude value. These values improved by several percentage points, depending on the applied wavelet.

Similarly, matching coefficient values for samples 4–6 were determined along with the progress of decomposition. Information that the filtration level leads to profile distortion from the sixth level for this mother wavelet was obtained for surfaces described by the sum of periodic functions. In the case of the indicated level, the obtained values for the studied surfaces were most similar to nominal ones, analogous to samples modelled using a single periodic function. However, for the assessed mother wavelet forms, these values slightly decreased together with wavelet support width increase.

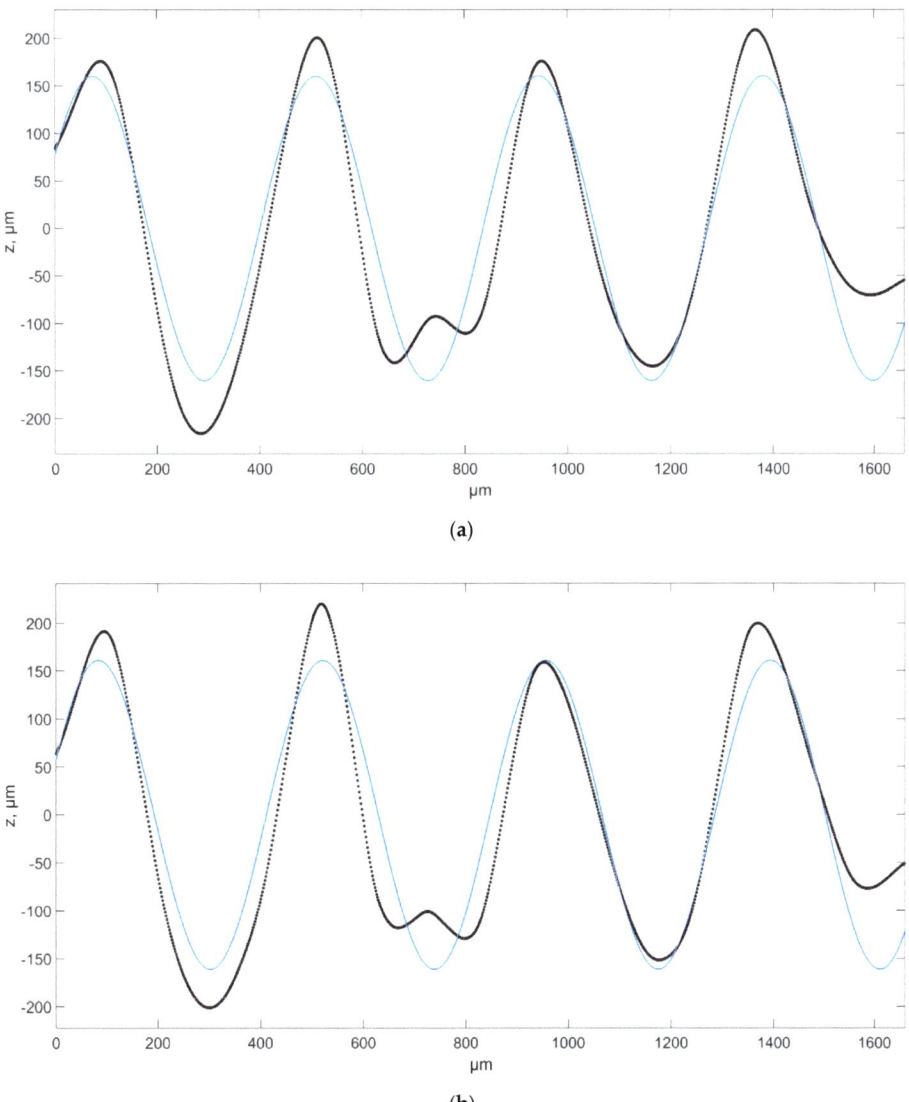

Figure 7. *Cont.*

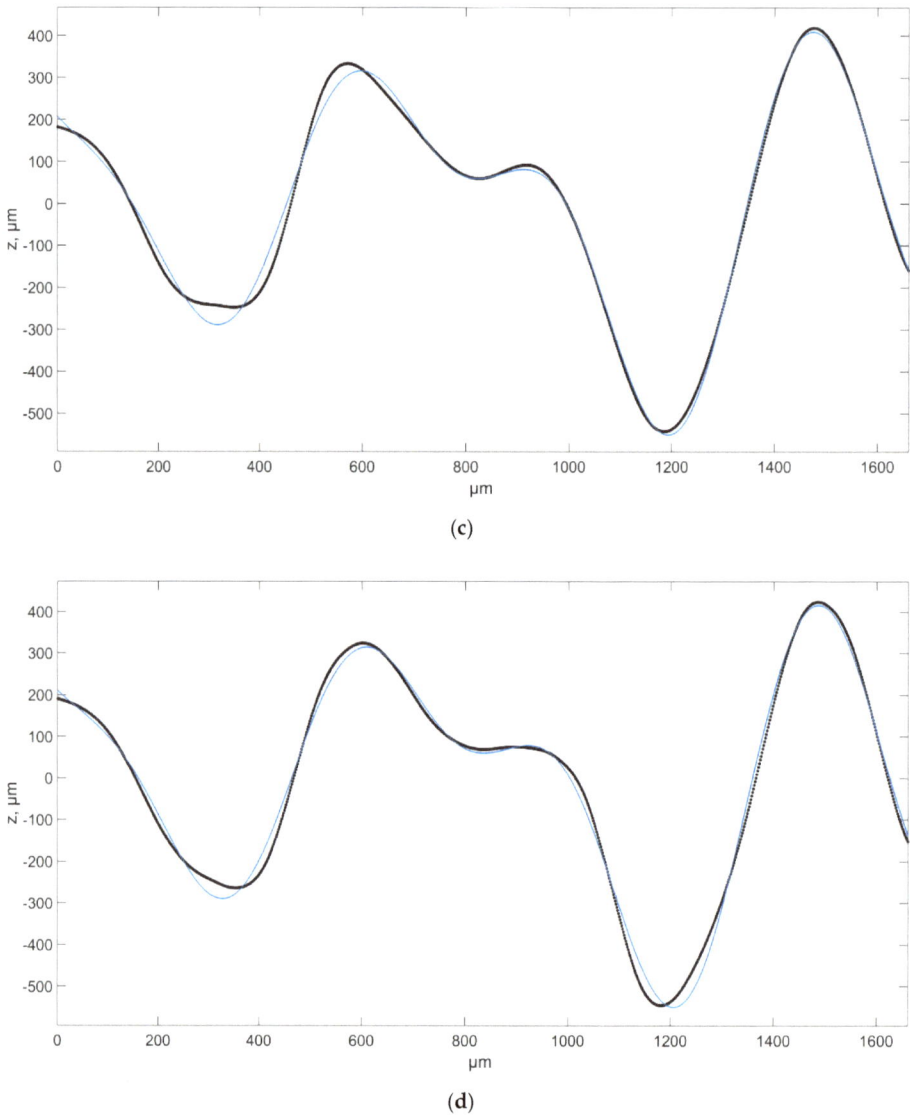

Figure 7. Example of a surface profile together with a sixth analysis level approximation function: (**a**) sample No. 1 coif5 wavelet, (**b**) sample no. 1 sym8 wavelet, (**c**) sample no. 4 coif5 wavelet, (**d**) sample no. 4 sym8 wavelet. The black color indicates the profile obtained by wavelet analysis, the blue color indicates approximation.

4. Discussion

Modern additive technologies enable producing fully functional models. However, the key issue is assessing the possibility of producing complex, characteristic morphological features on the surface of elements, since they directly impact the operation of individual machine parts at a later stage. Research was focused on evaluating the feasibility of producing characteristic irregularities distribution on the surface and process control. The studies were expanded with multiscale assessment of the resulting surface texture, based on discrete, one-dimensional wavelet transformation. The conducted analysis enabled expanding surface diagnostics or process capabilities through a broad and comprehensive assessment of individual surface features. The research provides hints in terms of producing elements, as well as indicates possible process errors, filling the research gap in the field of process diagnostics through assessing additively produced surfaces.

Research carried out using a scanning electron microscope did not indicate the regular, systematic occurrence of unmelted powder grains or impurities. This is a significant advantage over other materials where, in the case of additive technologies, the occurrence of unmelted particles and other defects causing deterioration of surface quality and misinterpretation of test results are observed.

Understanding the manner and scale in which a production process impacted morphological features required a comprehensive evaluation of individual processes, in order to analyse all important geometry aspects that resulted from it. A classic ISO-based perception of surface textures seems to be insufficient in these aspects [38], due to high process complexity [39].

Studies assessing the possibility of producing characteristic surface features indicated high potential applicability of additive technologies. An impact of the building angle on the resulting feature distribution was observed. Both the value of the matching coefficient reduced as a building angle function and the value of parameters describing individual function on the surfaces reached values that differed relatively from nominal values by approximately several percent on average. At the same time, it was noted that increasing the number of functions describing surfaces led to an ambiguous change in the aforementioned parameters.

Multiscale analysis using discrete, one-dimensional wavelet transformations showed dominant surface irregularities components. Assessing irregularities distribution on many scales enabled evaluating the production process in terms of porosity and additional features formed on the surface. The studies showed that filtering out high-frequency components at the initial analysis levels resulted in an improvement of the assessed parameters. Therefore, it can be inferred that the initial differences in the values were caused by random micro-roughness. The research covering a wide spectrum of mother wavelets enabled verifying the impact of a mother wavelet and its properties on the process of filtering individual surface profiles. It also provided hints on the potential diagnostic possibilities associated with the wavelet method.

The conducted tests also came with certain limitations, to be analysed as part of research in the future. In particular, future studies should focus on the greater differentiation of building angles, materials, process parameters like layer thickness or surface types, among others, through adding more functions or analysis free-form surfaces and specified, characteristic locations on individual surfaces. The research will help find functional dependencies of the production process and will translate to its in-depth diagnostics and understanding of the production process for individual morphological features.

5. Conclusions

The article assesses the applicability of additive technologies for shaping characteristic irregularities distribution on surfaces. The study involved using a modern approach based on wavelet transformation. An analysis of the results presented above led to the following conclusions:

1. It is possible to manufacture precise models with characteristic morphological features of various sizes and shapes using additive technologies. Based on the scanning electron microscope and computed tomography analysis, it can be noted that there are no defects caused by the technological process and no unmelted powder grains on the tested surfaces. The production of surfaces with a much more complicated shape should not be problematic for additive technologies compared to the limitations known for conventional technologies such as machining. The research has shown that it is a clear advantage compared to conventional methods, where shaping

such irregularities and defined morphological features on the surface is hindered or sometimes impossible.
2. It cannot be clearly concluded whether the application of more surface modelling functions resulted in better or worse mapping of the model surface. In the case of surfaces described by a single function (compared approximation function and measured profile), the differences in the parameters differed relatively by an average of approximately twenty percent in terms of the amplitude and several percent in terms of the period, for a matching value of 0.7–0.85, depending on the profile, which means a correlation dependence according to J.P. Guildford's classification. In the case of a surface defined by several periodic functions, these parameters differed by thirty and several percent, respectively, which for a matching of more than 0.95 proves a very clear correlation dependence. Moreover, comparing the profile specified in the CAD model with the profile of the approximating function, there are differences in the accuracy of the fit depending on the printing direction of the sample models. The most favorable variant due to amplitude and periodic differences is to place the models at the smallest possible angle to the building platform: for the assessed samples, it was an angle equal to $20°$. In this case, the amplitude differences reached only a few micrometers. A reduction in the surface irregularities mapping quality was observed with increasing building angle (printing direction), which has a negative effect on the building time, layer number, and stair-step effect.
3. When analysing the data obtained through wavelet filtration, it can be concluded that the dominant error component was high-frequency information resulting from production process errors and corresponding to surface pores. An assessment of the resulting signals leads to a conclusion that signals from the sixth level upwards do not contain such information.
4. In the case of mother wavelets with a large support, the obtained profile-matching coefficients, as well as the approximating function period and amplitude values, were better. However, they slightly decreased when support increased. This tendency could be observed up to the sixth decomposition level. Unnatural distortion of the resulting signals was observed at further levels.
5. The research showed that wavelet transformation can be successfully applied as a diagnostic tool in surface texture assessment and used as a base to diagnose the production process. It seems that a significant limitation of the technological process is that the layer thickness is determined, among other things, by the size of the powder grains, and in future research, it will be possible to analyse much more precisely manufactured models using multiscale analysis.

Author Contributions: Conceptualization, D.G.; methodology, D.G.; software, D.G. and T.K.; validation D.G., P.Z. and T.K.; formal analysis, D.G., P.Z. and T.K.; investigation, D.G.; resources, D.G.; data curation, D.G.; writing—original draft preparation, D.G., P.Z. and T.K.; writing—review and editing, D.G., P.Z. and T.K.; visualization, D.G.; supervision, D.G.; project administration, D.G.; funding acquisition, D.G. All authors have read and agreed to the published version of the manuscript.

Funding: The research presented in this paper was supported by the National Science Centre of Poland under the scientific work No. 2020/04/X/ST2/00352 "Multiscale analysis of free-form and functional surfaces manufactured by additive technology".

Institutional Review Board Statement: Not applicable.

Informed Consent Statement: Not applicable.

Data Availability Statement: The data presented in this study are available upon request from the corresponding author.

Conflicts of Interest: The authors declare no conflict of interest.

References

1. Snopiński, P.; Król, M.; Pagáč, M.; Petrů, J.; Hajnyš, J.; Mikuszewski, T.; Tański, T. Effects of equal channel angular pressing and heat treatments on the microstructures and mechanical properties of selective laser melted and cast AlSi10Mg alloys. *Arch. Civ. Mech. Eng.* **2021**, *21*, 92. [CrossRef]
2. Piekło, J.; Garbacz-Klempka, A. Use of selective laser melting (SLM) as a replacement for pressure die casting technology for the production of automotive casting. *Arch. Foundry Eng.* **2021**, *21*, 9–16.
3. Budzik, G.; Przeszlowski, L.; Wieczorowski, M.; Rzucidlo, A.; Gapinski, B.; Krolczyk, G. Analysis of 3D printing parameters of gears for hybrid manufacturing. *AIP Conf. Proc.* **2018**, *1960*, 140005.
4. Diaz, A. Surface texture characterization and optimization of metal additive manufacturing-produced components for aerospace applications. In *Additive Manufacturing for the Aerospace Industry*; Elsevier: Amsterdam, The Netherlands, 2019; pp. 341–374. [CrossRef]
5. Wang, Y.; Peng, T.; Zhu, Y.; Yang, Y.; Tang, R. A comparative life cycle assessment of a selective-laser-melting-produced hydraulic valve body using design for Property. *Procedia CIRP* **2020**, *90*, 220–225. [CrossRef]
6. Edelmann, A.; Dubis, M.; Hellmann, R. Selective laser melting of patient individualized osteosynthesis plates—Digital to physical process chain. *Materials* **2020**, *13*, 5786. [CrossRef]
7. Gogolewski, D.; Kozior, T.; Zmarzły, P.; Mathia, T.G. Morphology of Models Manufactured by SLM Technology and the Ti6Al4V Titanium Alloy Designed for Medical Applications. *Materials* **2021**, *14*, 6249. [CrossRef]
8. Triantaphyllou, A.; Giusca, C.L.; Macaulay, G.D.; Roerig, F.; Hoebel, M.; Leach, R.K.; Tomita, B.; Milne, K.A. Surface texture measurement for additive manufacturing. *Surf. Topogr. Metrol. Prop.* **2015**, *3*, 024002. [CrossRef]
9. Qian, W.; Wu, S.; Wu, Z.; Ahmed, S.; Zhang, W.; Qian, G.; Withers, P.J. In situ X-ray imaging of fatigue crack growth from multiple defects in additively manufactured AlSi10Mg alloy. *Int. J. Fatigue* **2022**, *155*, 106616. [CrossRef]
10. Hu, Y.N.; Wu, S.C.; Wu, Z.K.; Zhong, X.L.; Ahmed, S.; Karabal, S.; Xiao, X.H.; Zhang, H.O.; Withers, P.J. A new approach to correlate the defect population with the fatigue life of selective laser melted Ti-6Al-4V alloy. *Int. J. Fatigue* **2020**, *136*, 105584. [CrossRef]
11. Khaemba, D.N.; Azam, A.; See, T.L.; Neville, A.; Salehi, F.M. Understanding the role of surface textures in improving the performance of boundary additives, part I: Experimental. *Tribol. Int.* **2020**, *146*, 106243. [CrossRef]
12. Zmarzły, P. Multi-dimensional mathematical wear models of vibration generated by rolling ball bearings made of aisi 52100 bearing steel. *Materials* **2020**, *13*, 5440. [CrossRef]
13. Gogolewski, D. Multiscale assessment of additively manufactured free-form surfaces. *Metrol. Meas. Syst.* **2023**, *30*. [CrossRef]
14. Pagani, L.; Townsend, A.; Zeng, W.; Lou, S.; Blunt, L.; Jiang, X.Q.; Scott, P.J. Towards a new definition of areal surface texture parameters on freeform surface: Re-entrant features and functional parameters. *Meas. J. Int. Meas. Confed.* **2019**, *141*, 442–459. [CrossRef]
15. Podulka, P. Thresholding Methods for Reduction in Data Processing Errors in the Laser-Textured Surface Topography Measurements. *Materials* **2022**, *15*, 5137. [CrossRef]
16. Townsend, A.; Senin, N.; Blunt, L.; Leach, R.K.; Taylor, J.S. Surface texture metrology for metal additive manufacturing: A review. *Precis. Eng.* **2016**, *46*, 34–47. [CrossRef]
17. Gogolewski, D.; Zmarzły, P.; Kozior, T.; Mathia, T.G. Possibilities of a Hybrid Method for a Time-Scale-Frequency Analysis in the Aspect of Identifying Surface Topography Irregularities. *Materials* **2023**, *16*, 1228. [CrossRef] [PubMed]
18. Prabhakar, D.V.N.; Sreenivasa Kumar, M.; Gopala Krishna, A. A Novel Hybrid Transform approach with integration of Fast Fourier, Discrete Wavelet and Discrete Shearlet Transforms for prediction of surface roughness on machined surfaces. *Meas. J. Int. Meas. Confed.* **2020**, *164*, 108011. [CrossRef]
19. Brown, C.A.; Hansen, H.N.; Jiang, X.J.; Blateyron, F.; Berglund, J.; Senin, N.; Bartkowiak, T.; Dixon, B.; Le Goïc, G.; Quinsat, Y.; et al. Multiscale analyses and characterizations of surface topographies. *CIRP Ann.* **2018**, *67*, 839–862. [CrossRef]
20. Zare, M.; Solaymani, S.; Shafiekhani, A.; Kulesza, S.; Ţălu, Ş.; Bramowicz, M. Evolution of rough-surface geometry and crystalline structures of aligned TiO2 nanotubes for photoelectrochemical water splitting. *Sci. Rep.* **2018**, *8*, 10870. [CrossRef]
21. Bartkowiak, T.; Brown, C.A. Multiscale 3D curvature analysis of processed surface textures of aluminum alloy 6061 T6. *Materials* **2019**, *12*, 257. [CrossRef] [PubMed]
22. Maleki, I.; Wolski, M.; Woloszynski, T.; Podsiadlo, P.; Stachowiak, G. A comparison of multiscale surface curvature characterization methods for tribological surfaces. *Tribol. Online* **2019**, *14*, 8–17. [CrossRef]
23. Li, G.; Zhang, K.; Gong, J.; Jin, X. Calculation method for fractal characteristics of machining topography surface based on wavelet transform. *Procedia CIRP* **2019**, *79*, 500–504. [CrossRef]
24. Sun, J.; Song, Z.; He, G.; Sang, Y. An improved signal determination method on machined surface topography. *Precis. Eng.* **2018**, *51*, 338–347. [CrossRef]
25. Gogolewski, D. Fractional spline wavelets within the surface texture analysis. *Meas. J. Int. Meas. Confed.* **2021**, *179*, 109435. [CrossRef]
26. Pahuja, R.; Ramulu, M. Study of surface topography in Abrasive Water Jet machining of carbon foam and morphological characterization using Discrete Wavelet Transform. *J. Mater. Process. Technol.* **2019**, *273*, 116249. [CrossRef]
27. Abdul-Rahman, H.S.; Jiang, X.J.; Scott, P.J. Freeform surface filtering using the lifting wavelet transform. *Precis. Eng.* **2013**, *37*, 187–202. [CrossRef]

28. Yesilli, M.C.; Chen, J.; Khasawneh, F.A.; Guo, Y. Automated surface texture analysis via Discrete Cosine Transform and Discrete Wavelet Transform. *Precis. Eng.* **2022**, *77*, 141–152. [CrossRef]
29. Gogolewski, D. Influence of the edge effect on the wavelet analysis process. *Meas. J. Int. Meas. Confed.* **2020**, *152*, 107314. [CrossRef]
30. Navarro, P.J.; Fernández-Isla, C.; Alcover, P.M.; Suardíaz, J. Defect detection in textures through the use of entropy as a means for automatically selecting the wavelet decomposition level. *Sensors* **2016**, *16*, 1178. [CrossRef]
31. Bruzzone, A.A.G.; Montanaro, J.S.; Ferrando, A.; Lonardo, P.M. Wavelet analysis for surface characterisation: An experimental assessment. *CIRP Ann. Manuf. Technol.* **2004**, *53*, 479–482. [CrossRef]
32. Gogolewski, D.; Makieła, W.; Nowakowski, Ł. An assessment of applicability of the two-dimensionalwavelet transform to assess the minimum chip thickness determination accuracy. *Metrol. Meas. Syst.* **2020**, *27*, 659–672.
33. Dutta, S.; Pal, S.K.; Sen, R. Progressive tool flank wear monitoring by applying discrete wavelet transform on turned surface images. *Meas. J. Int. Meas. Confed.* **2016**, *17*, 388–401. [CrossRef]
34. Nouhi, S.; Pour, M. Prediction of surface roughness of various machining processes by a hybrid algorithm including time series analysis, wavelet transform and multi view embedding. *Meas. J. Int. Meas. Confed.* **2021**, *184*, 109904. [CrossRef]
35. Shao, Y.; Du, S.; Tang, H. An extended bi-dimensional empirical wavelet transform based filtering approach for engineering surface separation using high definition metrology. *Meas. J. Int. Meas. Confed.* **2021**, *178*, 109259. [CrossRef]
36. Mahashar Ali, J.; Siddhi Jailani, H.; Murugan, M. Surface roughness evaluation of electrical discharge machined surfaces using wavelet transform of speckle line images. *Meas. J. Int. Meas. Confed.* **2020**, *149*, 107029. [CrossRef]
37. Ti64 M290 Material Data Sheet. Available online: https://3dformtech.fi/wp-content/uploads/2019/11/Ti-Ti64_9011-0014_9011-0039_M290_Material_data_sheet_11-17_en-1.pdf (accessed on 28 March 2023).
38. Todhunter, L.D.; Leach, R.K.; Lawes, S.D.A.; Blateyron, F. Industrial survey of ISO surface texture parameters. *CIRP J. Manuf. Sci. Technol.* **2017**, *19*, 84–92. [CrossRef]
39. Leach, R.; Thompson, A.; Senin, N.; Maskery, I. A metrology horror story: The additive surface. In Proceedings of the ASPEN/ASPE Spring Topical Meeting on Manufacture and Metrology of Structured and Freeform Surfaces for Functional Applications, Hong Kong, China, 14–17 March 2017.

Disclaimer/Publisher's Note: The statements, opinions and data contained in all publications are solely those of the individual author(s) and contributor(s) and not of MDPI and/or the editor(s). MDPI and/or the editor(s) disclaim responsibility for any injury to people or property resulting from any ideas, methods, instructions or products referred to in the content.

Article

Reduction in Errors in Roughness Evaluation with an Accurate Definition of the S-L Surface

Przemysław Podulka [1,*], Wojciech Macek [2], Ricardo Branco [3] and Reza Masoudi Nejad [4]

1. Faculty of Mechanical Engineering and Aeronautics, Rzeszow University of Technology, Powstanców Warszawy 8 Street, 35-959 Rzeszów, Poland
2. Faculty of Mechanical Engineering and Ship Technology, Gdańsk University of Technology, Narutowicza 11/12 Street, 80-233 Gdańsk, Poland
3. Department of Mechanical Engineering, Centre for Mechanical Engineering, Materials and Processes (CEMMPRE), University of Coimbra, 3030-788 Coimbra, Portugal
4. School of Mechanical and Electrical Engineering, University of Electronic Science and Technology of China, Chengdu 611731, China
* Correspondence: p.podulka@prz.edu.pl; Tel.: +48-17-743-2537

Citation: Podulka, P.; Macek, W.; Branco, R.; Nejad, R.M. Reduction in Errors in Roughness Evaluation with an Accurate Definition of the S-L Surface. *Materials* 2023, *16*, 1865. https://doi.org/10.3390/ma16051865

Academic Editor: Thomas Niendorf

Received: 26 January 2023
Revised: 19 February 2023
Accepted: 23 February 2023
Published: 24 February 2023

Copyright: © 2023 by the authors. Licensee MDPI, Basel, Switzerland. This article is an open access article distributed under the terms and conditions of the Creative Commons Attribution (CC BY) license (https://creativecommons.org/licenses/by/4.0/).

Abstract: Characterization of surface topography, roughly divided into measurement and data analysis, can be valuable in the process of validation of the tribological performance of machined parts. Surface topography, especially the roughness, can respond straightly to the machining process and, in some cases, is defined as a fingerprint of the manufacturing. When considering the high precision of surface topography studies, the definition of both S-surface and L-surface can drive many errors that influence the analysis of the accuracy of the manufacturing process. Even if precise measuring equipment (device and method) is provided but received data are processed erroneously, the precision is still lost. From that matter, the precise definition of the S-L surface can be valuable in the roughness evaluation allowing a reduction in the rejection of properly made parts. In this paper, it was proposed how to select an appropriate procedure for the removal of the L- and S-components from the raw measured data. Various types of surface topographies were considered, e.g., plateau-honed (some with burnished oil pockets), turned, milled, ground, laser-textured, ceramic, composite, and, generally, isotropic. They were measured with different (stylus and optical) methods, respectively, and parameters from the ISO 25178 standard were also taken into consideration. It was found that commonly used and available commercial software methods can be valuable and especially helpful in the precise definition of the S-L surface; respectively, its usage requires an appropriate response (knowledge) from the users.

Keywords: surface topography; surface texture; roughness; S-L surface; form removal; measurement noise

1. Introduction

Characterization of surface topography in the manufacturing process can be valuable in the analysis of the tribological performance of machined parts. Much valuable information can be received straightly from the analysis of surface roughness data, such as wear resistance [1], lubricant retention [2], friction [3], fatigue [4–6], sealing [7], analysis of energy consumption [8], eco-friendly strategies [9,10], or, generally, functional performance [11,12]. In many cases, surface topography is perceived as a fingerprint of the manufacturing process [13]. When considering the precision of surface topography studies, there must be validation of both the measurement and the data processes. Errors that occur when both operations are provided can cause an erroneous estimation of properties of properly manufactured parts leading to their classification as a lack and, unfortunately, their rejection [14]. Many types of errors can be found in surface topography studies. Roughly, they can be divided into those reflected in the measurement process [15–17] and those connected with the whole data analysis actions [18].

It was found in previous studies that even when a highly precise measurement technique is used, if the process of data calculation and evaluation is selected erroneously, the whole surface roughness analysis is not provided appropriately. The biggest errors in data processing can result in the biggest distortion in the whole surface topography analysis [19]. Considering errors in the field of data evaluation, especially the calculation of surface roughness parameters, errors in defining an appropriate reference plane are very often encountered. From the definition, a reference plane is a plane according to which the surface topography (roughness) parameters are calculated. From that matter, when defining this plane with unappropriated methods or, respectively, by inappropriate application of proper procedures, surface roughness parameters can be falsely estimated. Thus, errors can arise in the validation of the manufactured product [20].

Generally, the surface data, especially those associated with surface topography, can be roughly divided into form, waviness, and roughness [21]. Surface roughness parameters of machined parts are calculated after areal form removal, where form includes form and waviness components. These components of surface data are defined as L-components and are included in the L-surface [22]. Distortion in a proper definition of L-surface can be especially visible when analyzing surfaces containing deep or wide features, such as oil pockets or, generally, dimples [23–25]. It was found that texturing of the surface, especially when creating additional oil pockets by burnishing techniques, can significantly improve the properties of the machined surface [26–28]. From that point of view, precise characterization of multi-process surfaces [29,30] is of great importance.

The reduction in errors in the feature characterization is another encouraging task to be resolved [31]. In general, each of the actions provided on the surface topography data, including those with the feature-based characterization, is provided for a more direct relationship between characterization, manufacturing process, and surface function [32]. The effect of feature size, density, and distribution was found crucial in the validation of methods for both an areal form removal and high-frequency measurement noise reduction [33]. The measurement noise is a type of error, simplifying, added to the output signal, which occurs when the measuring instrument (e.g., roughness profilometer) is used [34]. The measurement noise can be analyzed in various domains; the bandwidth characterization was proposed previously [35]. One of the types of measurement noise is in the high-frequency domain [36,37].

The measurement noise, described in the high-frequency domain, can be derived from the instability of the mechanics received by influences from the environment. Nevertheless, in most cases, the high-frequency measurement noise outcome from the vibration [38]. Those components of the surface data are reflected in the S-components, and the surface including those errors can be defined as the S-surface or, respectively, as the noise surface [39]. Usually, the high-frequency spatial components are eliminated by the S-filter [40,41], which removes small-scale lateral basics from the surface [22]. Small-scale (S-) components can also be removed by the proposal of the nesting index [42] term; it is an extension of the concept of cut-off. It was proposed to be the S-nesting index value and should be proposed at a 3:1 ratio with the maximum sampling distance [43–45]. From that matter, the L-nesting index [46] can be defined for removing the large-scale components (such as form and waviness) and, correspondingly, the S-nesting index for removing the small-scale components (e.g., high-frequency measurement errors) from the surface data. However, the process of selection of the S-filter and L-filter nesting indexes must be studied for each type of surface separately with consideration of distortion of the surface topography features. For example, the surface texture of the ink-jet printed THV films was investigated after the application of an S-filter with a bandwidth (nesting index) equal to 2.5 µm and an L-filter with a cut-off (nesting index) of 250 µm to remove high-spatial frequency noise and, respectively, long-scale waviness/form from the raw measured data [47].

Generally, when calculating the roughness parameters, both components (L- and S-) must be removed from the raw measured data. In the result, the received roughness surface, after L-filtering and S-filtering, is derived as the S-L surface [48]. Reducing errors in the calculation of the S-L surface simultaneously influences the errors in roughness evaluation. There are many papers considering the selection of methods (S-filters and L-filters) for roughness evaluation; nevertheless, errors in the false estimation of the S-L surface were not comprehensively studied. Moreover, even if the selection of the S-L surface depends on the type of detail considered (machining process and its parameters), the requirements from the analysis provided in this paper increase. In this paper, a selection of proper procedures for the definition of the S-L surface of different topographies is presented, especially with an indication of the distortion of selected features, such as dimples, scratches, and valleys. It is also considered that the reduction in errors in roughness evaluation has a straight impact on the values of ISO 25178 parameters.

2. Materials and Methods

2.1. Analysed Details

Various types of surface topographies were considered, as follows: deterministic one-process turned piston skirts, ground (speed of 28 m/s, in-feed of 10 m/min, cross-feed of 1 mm/pitch, and depth of 0.02 mm), milled (depth of machining of 0.4 mm, speed of multi-cut head of 140 rev/min, and feeds of 0.3), laser-textured (with different angles of texturing, 30°, 60°, 90°, and 120°), composite, ceramic and, respectively, generally isotropic. More than 10 surfaces from each type of topography were measured and studied. Further, some of them were examined and presented in more detail. Additionally, the analyses were improved by incorporating modeled data. Then, the data were compared to the measured results in order to make general recommendations. Figure 1 shows examples of each type of surface (turned (**a**), laser-textured (**b**), ground (**c**), ceramic (**d**), and composite (**e**)) with contour map plots (left column), areal autocorrelation functions (middle column) and selected ISO 25178 surface roughness parameters (right column).

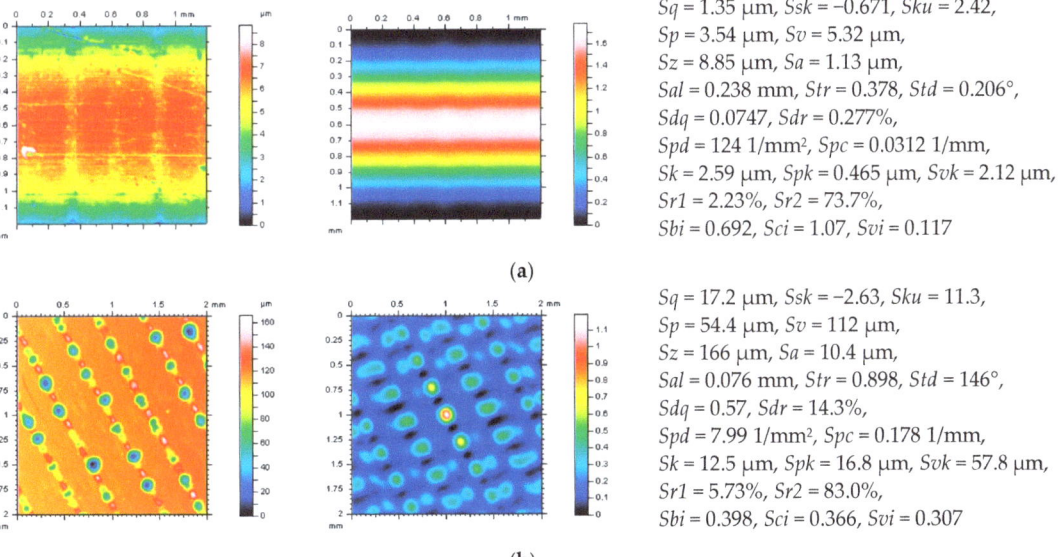

Figure 1. Cont.

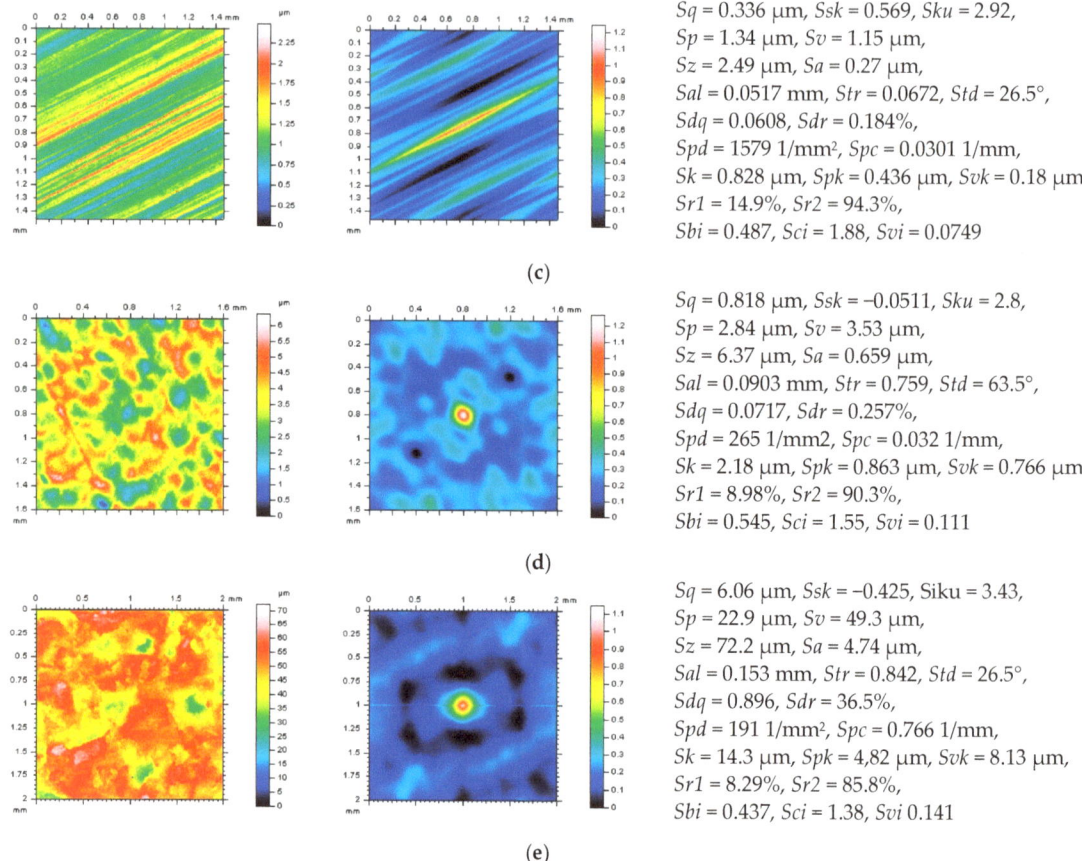

Figure 1. Contour map plots (left column), the 3D ACF (middle column) and ISO 25178 surface topography parameters (right column) of turned (**a**), laser−textured (**b**), ground (**c**), ceramic (**d**), and composite (**e**) surface.

The following ISO 25178 roughness parameters from various groups were measured and studied: root-mean-square height Sq, skewness Ssk, kurtosis Sku, maximum peak height Sp, maximum valley depth Sv, the maximum height of surface Sz, arithmetic mean height Sa from amplitude parameters; areal material ratio Smr, inverse areal material ratio Smc, extreme peak height Sxp from functional parameters; auto-correlation length Sal, texture parameter Str, texture direction Std from spatial parameters; root-mean-square gradient Sdq and developed interfacial areal ratio Sdr from hybrid parameters; peak density Spd and arithmetic mean peak curvature Spc from feature parameters; surface bearing index Sbi, core fluid retention index Sci and valley fluid retention index Svi from functional indices; and core roughness depth Sk, reduced summit height Spk, reduced valley depth Svk, upper bearing area $Sr1$, and lower bearing area $Sr2$ from the Sk family parameters.

2.2. Measurement Process

All of the analyzed surfaces were measured with contact (stylus) and non-contact (optical) methods to improve proposals for different measurement techniques.

The contact technique was based on a Talyscan 150 stylus instrument (Taylor Hobson, Warrenville, IL, USA), equipped with a nominal tip radius of 2 μm, approximately, a height resolution equal to 10 nm, a measured area of 5 by 5 mm (1000 × 1000 measured points), the sampling interval 5 μm, and the measurement speed 0.75 mm/s.

The non-contact measurement device was the white light interferometer Talysurf CCI Lite (produced by Taylor Hobson Ltd., Leicester, U.K., version 2.8.2.95), employed with a height resolution of 0.01 nm, a measured area of 3.35 by 3.35 mm (1024 × 1024 measured points), and a spacing of 3.27 μm. A Nikon 5×/0.13 TI objective was utilized.

For both analyses, areal digital filters from the TalyMap Gold (Digital Surf) software were employed to receive the ISO 25178 roughness parameters. Moreover, all of the functions proposed and validated in this paper were used from this source as well.

2.3. Applied Methods

For the characterization of surface topography with evaluation (calculation) of the roughness, both data analysis methods (definition of S-surface and L-surface) must be provided with error minimization. In this proposal, selected functions available in commonly used commercial software were utilized.

Often applied for the characterization of surface topography is an Autocorrelation Function (ACF). This function is described by the ISO standards and many research items [49]. In many primary studies, the ACF was proposed for the analysis of roughness isotropy [50] or anisotropy [51], description and realizations of homogeneous and isotropic two-dimensional Gaussian random processes [52], isotropic exponential and transformed exponential multiscale correlations [53], measurements of the variance of surface height obtained in several scattering geometries and also for stylus measurements [54], statistical computations of the root mean square (RMS) height, skewness (Ssk), and kurtosis (Sku) of the roughness height distribution [55], direction parallel and perpendicular to grooves, classification of data (signal) to the individual groups [56], identification of the periodicity and randomness [57], angular distribution characterization [58], the relationship between the height of a one-dimensionally rough surface and the intensity distribution of the light scattered by surface [59], vertical and lateral information about surface roughness [60], statistical irregularities of the waveguide substrate [61], characterization of the random component of the surface profile [62], determination of the domination in the frequency spectrum [63], and, frequently, characterization of the roughness measurement of the machined surfaces [64].

The ACF can also be valuable in the modeling of surface data [65], such as generating non-Gaussian surfaces with specified standard deviation [66], modeling bidirectional soil surfaces [67], influencing surface-induced resistivity of gold films [68], scale-dependent roughness modeling [69] or horizontal components [70], random generation of rough surface [71], simulations in ultrasonic assisted magnetic abrasive finishing [72], or in electro-discharge machining [73] processes when interrelationship between surface texture parameters and process parameters are emphasized. The ACF can also be valuable in the analysis of similarity and fractality in the modeling of roughness [74]. In summary, the ACF describes the dependence of the values of the data at one position on the values at another position [75].

When measuring the surface roughness, the scan resolution of scanning probe microscopy must be considered [76]. It was found that the shape of the ACF is sensitive to the measurement resolution [77]. It was also found that for very smooth surfaces, e.g., rolled or harrowed fields, the fractal process can determine mainly the overall shape of the ACF. Continuing, when considering very rough surfaces, the shape of the ACF can be determined by the single-scale process as well [78]. The ACF shape at various spatial scales, with RMS height and correlation length statistics, can be crucial in the analysis of roughness properties of the different tillage classes [79]. The ACF shape has a strong influence on the backscatter simulation results [80] as well. The influence of profile length on both the roughness parameters and the ACF shape was studied. It was assumed that the values of roughness parameters increase asymptotically with the increasing profile length [81]. The profile line shapes of the ACF intensities are obtained at different heights, and it is shown that the shapes are affected by noise [82], depending on the frequency type of noise [83]. By studying the shape of the ACF, the detection and reduction in selected types (frequencies) of measurement errors (noise) can be significantly improved [84].

Similarly to the ACF, also often applied in surface topography characterization, is the spectral analysis that can be a reliable indicator of roughness [85]. A typical example of this type of study is the application of the power spectral density (PSD) [86]. It was presented in the surface analysis that the RMS roughness depends on the length scale used for the measurement so, correspondingly, the RMS value of surface roughness is not a scale-invariant quantity [87]. From that matter, a precise description of surface morphology requires more sophisticated tools, and the PSD is classified as such a method [88]. Moreover, the PSD can be a preferred quantity when specifying surface roughness, especially considering the draft international drawing standard for surface texture [89].

Compared to the ACF, the PSD can support the modeling of surface roughness [90] as well. Profile generation [91] is received with an application of Fourier transformation. In the modeling of the surface roughness of thin films, the PSD was proposed through selected correlation [92]. Improvements in Fourier techniques to characterize the wavefront of optical components can also be received through the usage of PSD [93], especially when considering morphological parameters [94]. The vertical and lateral information of the surface profile can also be obtained with PSD applications [95]. Combining PSD with other methods, the profile roughness can be characterized by the PSD curve first and then by formation mechanisms of different frequency regions analyzed in more detail [96]. The PSD distribution can be used to explain the influence of tool feed, spindle speed, and, respectively, material-induced vibrations on surface roughness [97]. Generally, the PSD characterization can give a full description of the spatial frequency spectrum present on the surface, which is a result of interactions between all the machining parameters [98].

Both methods, ACF and PSD, were proposed and can be applied. In some cases, the applications must be provided simultaneously [99], especially when defining an appropriate L- or S-surface.

2.3.1. Supporting an Areal Form Removal with Thresholding Method

The definition of an L-surface, removing from the raw measured data long-components, described as form and waviness, can be supported by the thresholding methods. Generally, the typical thresholding of the surface topography is on its height, considering the segmentation of the analyzed data. However, a simple thresholding method cannot be classified as stable when surfaces have stochastic content. In this case, it can produce many insignificant features, so it can cause problems for many parameters, e.g., the number of defects and the density of features. Consequently, the thresholding method was proposed in many previous studies considering an analysis of the surface topography. Usage of its with wavelet decompositions [100,101] or Wiener filtration [102] was proposed for denoising the roughness measured signal.

The thresholding method was found especially valuable in reducing data processing errors of surfaces containing deep or wide features, such as dimples [103,104]. However, the selection of the thresholded value is another task to be comprehensively studied and adequately resolved. Generally, the application of the thresholding method when analyzing and defining the L-surface is to reduce the influence of the deep/wide features on the position (calculation) of the received plane. It was found in previous studies that size (depth, width) [105,106], density (number) [107], and location [108] (especially edge distribution [109–111]) have a considerable influence on the areal form removal process.

Figure 2 shows proposals for the selection of thresholding value graphically justified for the laser-textured surface (measured data in Figure 2a). The presented A1 and A2 values (Figure 2d) were located in the areas where the amplitude of changes was the greatest. It can be easily transferred on the Abbott–Firestone (Figure 2c) and material ratio (Figure 2b) curves. The plane corresponding to the L-surface (Figure 2f) can be more easily calculated (with reducing data processing errors) and positioned when the analyzed surface data do not contain deep and wide features (Figure 2e), such as laser-texturing traces. For the received surface after an areal form removal (Figure 2g), the values of the Sk parameter, core roughness depth, calculated as the distance between A1 and A2 values, decreased from 12.5 μm to 3.25 μm (see Figure 2h).

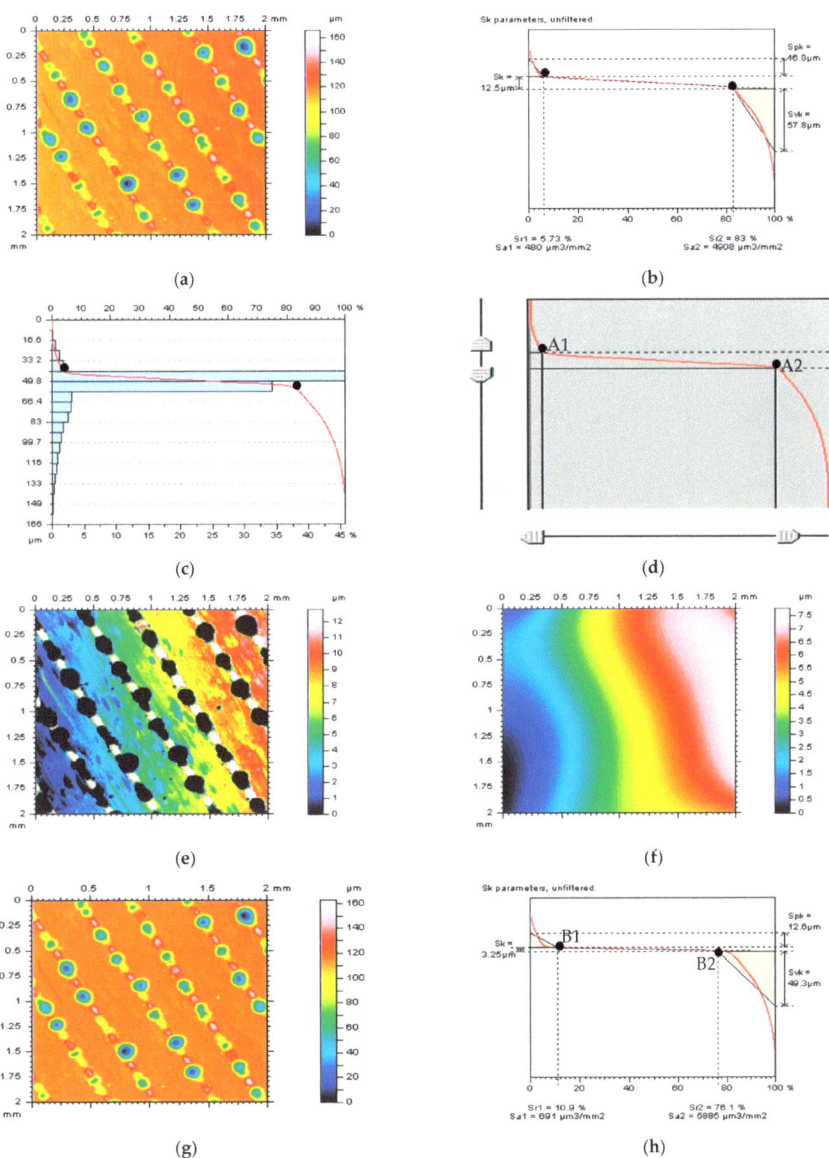

Figure 2. Contour map plots of measured laser-textured surface (**a**), its material ratio (**b**) and Abbott–Firestone curves (**c**), the method of selection of thresholding values A1 and A2 (**d**), the thresholded surface data (**e**), the Poly4 L-surface received from the thresholded surface (**f**), surface received by removal of the Poly4 L-surface from the raw measured data (**g**), and its material ratio curve (**h**).

The thresholding method can also be extremely valuable in validating the areal form removal algorithms. In Figure 3, the thresholding method was applied after this process. Removal of the deep features caused a better recognition of the distortion in the theoretically flat surface. The surfaces after an areal form removal (left column) were thresholded to remove the valleys (middle column), receiving non-dimple data (right column). The more the received L-surface was flat, the better the algorithm for form removal was rated.

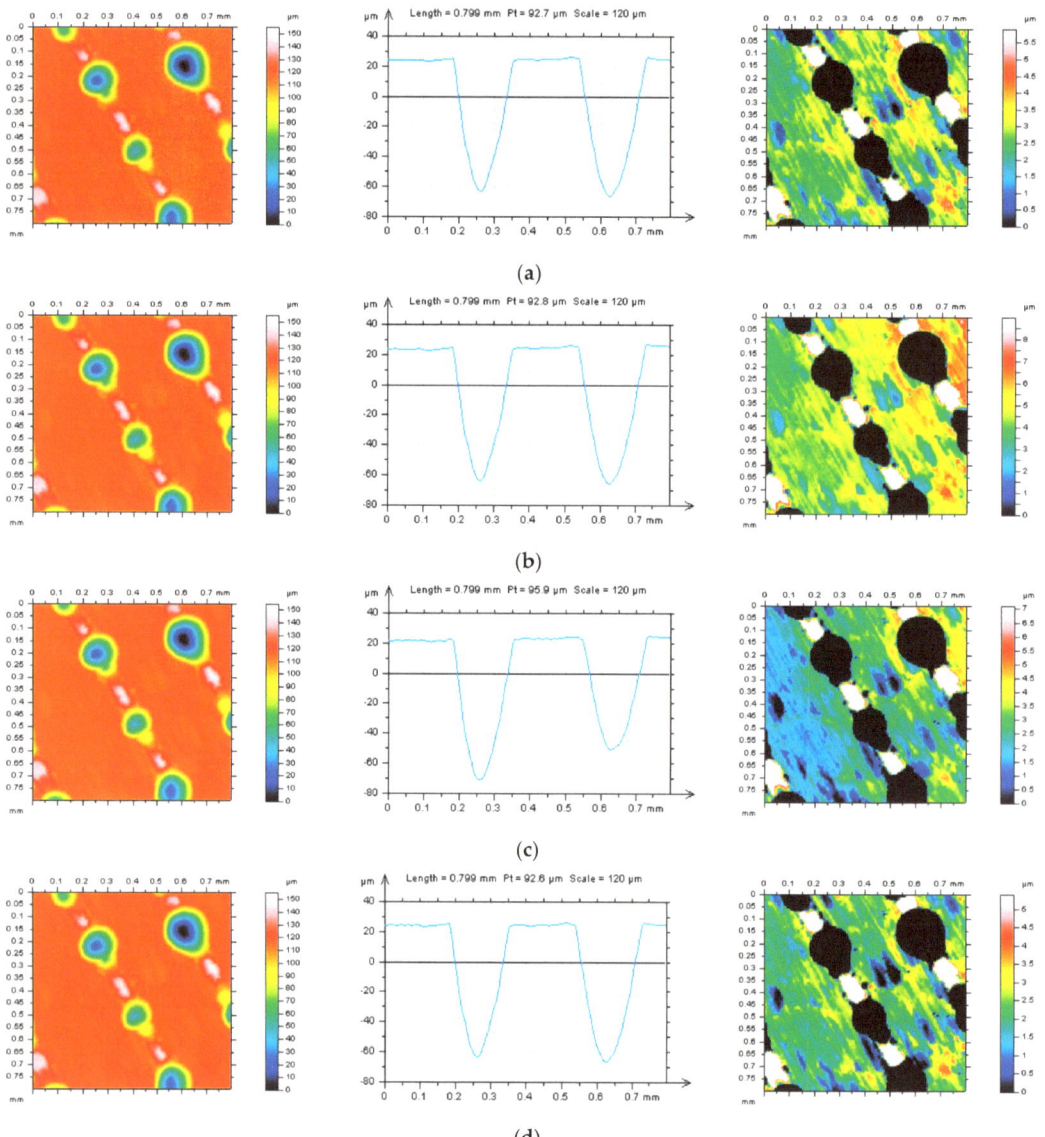

Figure 3. Cont.

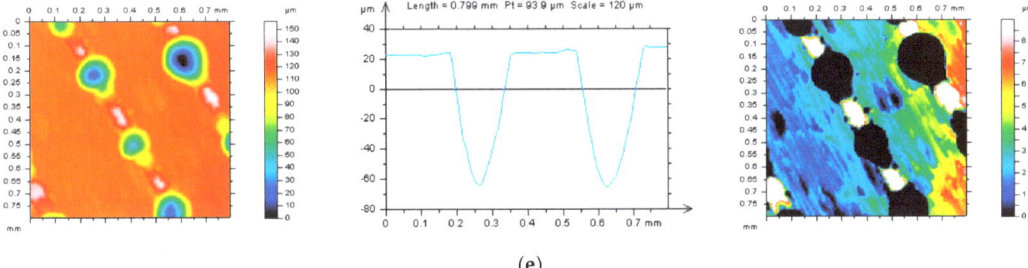

(e)

Figure 3. Contour map plots (left column), selected profiles (middle column), and A1−A2 thresholded surface (right column) received from the edge area of laser-textured surface topography after the definition of L−surface by application of Poly2 (**a**), Poly4 (**b**), GRF (**c**), RGRF (**d**), and SF (**e**), cut−off = 0.8 mm.

2.3.2. Improvement and Validation of Procedures for the Definition and Reduction in High-Frequency Measurement Noise

Considering the improvement in the detection and reduction in high-frequency measurement noise, it was proposed in previous studies to provide multithread analysis. This approach would help in reducing the errors in both processes. Except for the application of ACF and PSD functions, both studies, considering an areal (3D) and profile (2D) [112] characterization, can be valuable. Calculating the ACF and PSD for profiles and surfaces can indicate the occurrence of high-frequency measurement errors.

It was found in previous studies that, in some cases, the profile definition of noise can be more reliable than an areal. In practice, an even surface is characterized by an areal performance, which is crucial in the tribological characteristics of the details properties; extraction of profiles can be essential. Consequently, the identification of high-frequency noise for profile PSD and ACF analysis gave more direct results.

Moreover, the influence of the direction of profile characterization has a considerable impact on the validation of noise removal procedures [19]. The direction has an impact on the results in such a measurement process. For instance, in atomic force microscopy (AFM), the direction parallel to the scanning axis is sampled in less than topography in the perpendicular direction that may take several minutes to measure: the latter is, therefore, much more prone to artifacts from drift. A proposed solution was repeating the measurement on the same surface in three different directions: horizontal, vertical, and oblique. The influence of surface orientation on the variability of measurement results has already been comprehensively studied [113].

Figure 4 (left column) presents profiles received by extraction (from the ground surface) in different directions. Except for the traditional horizontal (a), vertical (b), and random (oblique) (c) directions, the treatment trace (d,e) was utilized. This technique can depend on the peak or valley location. The treatment trace peak direction is consistent with the direction of the peak trace on the machined surface, and the treatment trace valley is in line with the direction of the valley trace. It was found that the validation of the treatment trace technique depends on the peak and valley details of the type (plateauhoned, turned, ground, laser-textured, or, generally, textured) of the analyzed surface topography. From the results obtained for the PSD (middle column) and ACF (right column), the horizontal, vertical, and oblique directions did not allow for the detection of high-frequency measurement noise from the results of surface roughness. For the peak characterization (Figure 4d), the PSD did not justify if the noise existed, but respective differences in the shape of ACF could indicate some noise occurrence. Both methods (PSD and ACF) indicated that high-frequency errors can occur in the results of surface roughness measurements when the treatment trace valley direction was selected (Figure 4e). Response from that matter is that, if the amplitude of the surface data is relatively high, the detection of high-frequency measurement errors from the roughness measurement is difficult, even when the multithreaded analysis is provided. It is suggested to define the profile with the lowest height amplitude when detecting the high-frequency measurement noise with the directional extraction method.

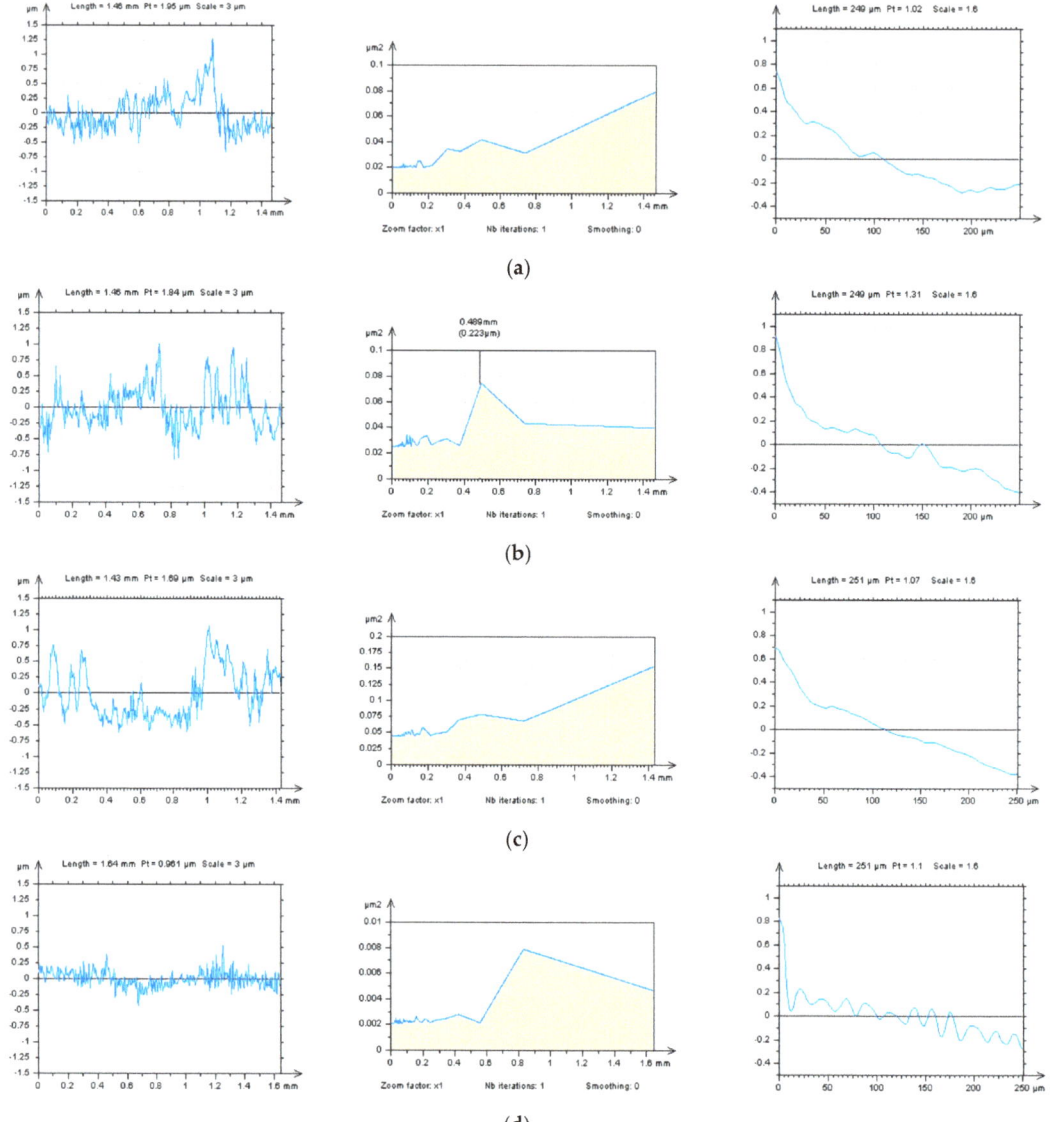

Figure 4. *Cont.*

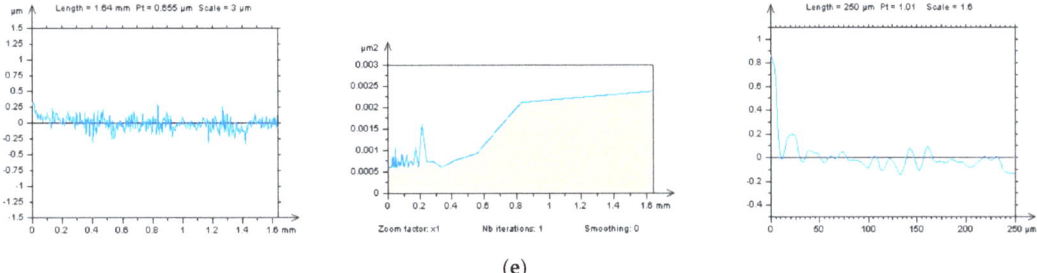

(e)

Figure 4. Profiles (left column), their PSD (middle column) and ACF (right column) received from the ground surface by extraction in: horizontal (**a**), vertical (**b**), random (**c**), treatment−trace peak (**d**), and treatment trace valley (**e**) directions.

The accuracy in the reduction in high-frequency measurement errors can be lost, even if the detection is provided appropriately, i.e., the noise removal procedure (e.g., digital filtering) can increase the distortion of results. From that issue, an analysis of the noise surface (NS) was suggested. The NS is a surface that consists of high-frequency measurement noise. In practice, it is a surface received by S-filtering with an application of the S-operator [44] or, simplifying, S-filter. It was proposed that precisely defined NS should consist of only the high-frequency components and, respectively, should be isotropic. In Figure 5, various NS plots were presented. They were received by various digital filtering methods, especially those commonly used (available in commercial software). Except for the analysis of contour map plots (Figure 5b) of the NS, and similarity in the PSD functions (Figure 5c), the differences indicating the algorithm precision were visible in the texture direction (TD) graphs calculated for the NS (Figure 5d).

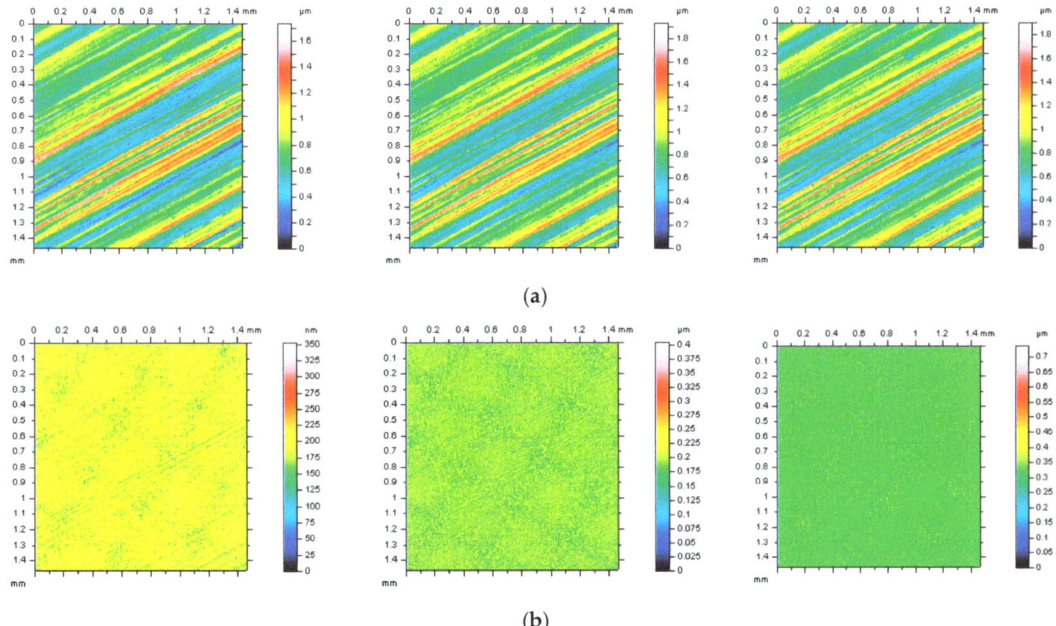

Figure 5. *Cont.*

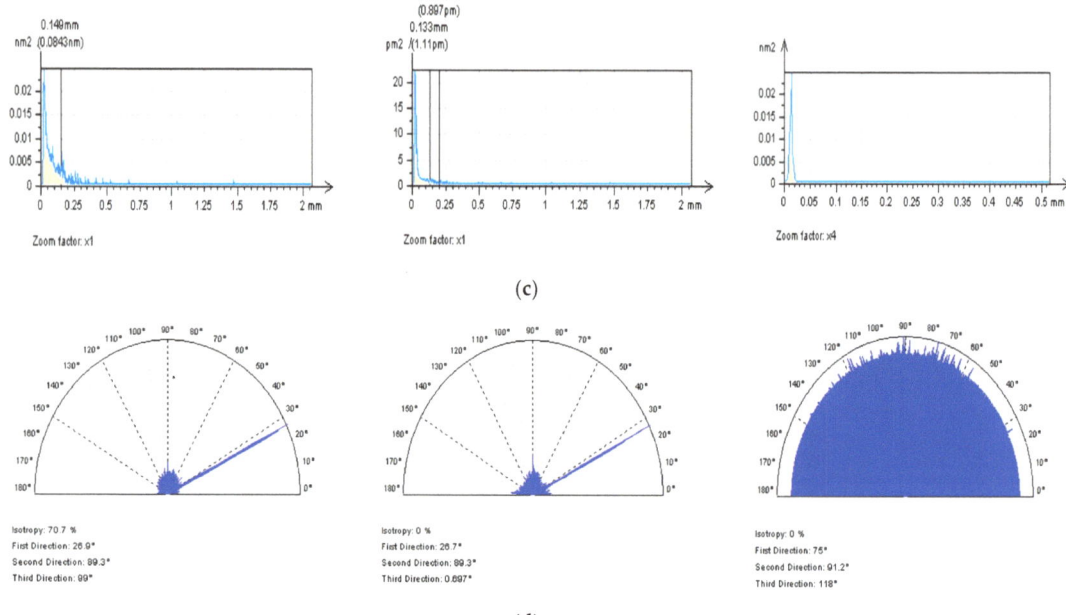

(c)

(d)

Figure 5. Contour map plots of the ground surface after S−surface removal (**a**), defined S−surface (**b**), PSDs (**c**), and TDs (**d**) of S−surfaces received by application of GRF (left column), SF (middle column), and FFTF (right column), cut−off = 0.10 mm.

3. Results

Studies of ST were divided into two subsections. Firstly, in Section 3.1, the errors in the definition of L-surface were presented, and their reduction was proposed. Secondly, in Section 3.2, proposals of procedure for the minimization of errors in the S-surface definition were presented. For both methods, results were validated and presented.

The main course of the studies was to, firstly, select an appropriate method (e.g., degree of least-square fitted polynomial plane) for an areal form removal (definition of the L-surface) and then to identify the digital filter (with cut-off value) causing the smallest errors in processed data (and ISO 25178 parameters). Both operations (definition of L-surface and S-surface) were improved with an application of commonly used (available in commercial software) functions, such as ACF, PSD, and TD graphs.

3.1. Reduction in Errors in the Definition of the L-Surface

The definition of an L-surface usually depends on the precision in the minimization of surface topography feature distortion. The exaggregation of features, such as dimples, oil pockets, scratches, or, generally, dimples, increased when they were located near the edge of the analyzed detail. Moreover, when the surface contained deep and wide dimples, distortion increased enormously as well. Some proposals can be found with valley extraction [110]. Not only the application of the too-large degree of the least-square fitted polynomial (LSFP) plane [33] can cause a dimple distortion; digital filtration, such as regular Gaussian regression (GRF), robust Gaussian regression (RGRF), or spline (SF) filters, can grossly distort selected surface topography features. The bi-square modification of polynomials of the nth degree can reduce those errors; nevertheless, their application requires mindful users [114]. Considering the distribution of features, the areas located between deep and wide features and the edge of the analyzed details were also vulnerable to greater distortion, contrary to the areas where such features were not located. This disadvantage was especially visible for digital filtering, even though the bandwidth was enlarged. For some solutions, the cut-off was proposed to be enlarged; nevertheless, on the other hand, it may have caused the form was not removed entirely. An exemplary solution can be found when the features are not distorted, but the out-of-feature [13] surface is completely flat. To receive these data, features must be excluded from the surface. One of the methods for extraction (removal) of features from the raw measured data is an application of the thresholding method, widely presented and proposed in this study.

In Figure 6, selected profiles and their hole/peak area diagrams were presented. They were received from the surface after an areal form removal by various methods. Increasing the degree of LSFP resulted in distorted edge-located areas where the dimples occurred. It was also found that exaggeration was enlarged when the size (depth and width) of the feature increased. The greater the features, the larger the distortion (Figure 6a–c). Application of GRF and SF seems to be the most encouraging solution; nevertheless, not always the entire form, especially waviness, was eliminated. From that matter, the application of the 2nd degree of LSFP (Poly2) seems to be the most suitable for the definition of the L-surface. However, it must be considered that a low (e.g., second) degree of a polynomial would not entirely remove the form from the measured raw data. Some proposals can be found in the application, firstly the polynomial of the second degree (removal of shape) and then the usage of digital filtering (e.g., GRF or RGRF) to eliminate the waviness. However, increasing the number of methods applied can significantly extend the time of data processing and, unfortunately, enlarge the number of errors in data analysis. It is best to remove the form (shape and waviness) entirely.

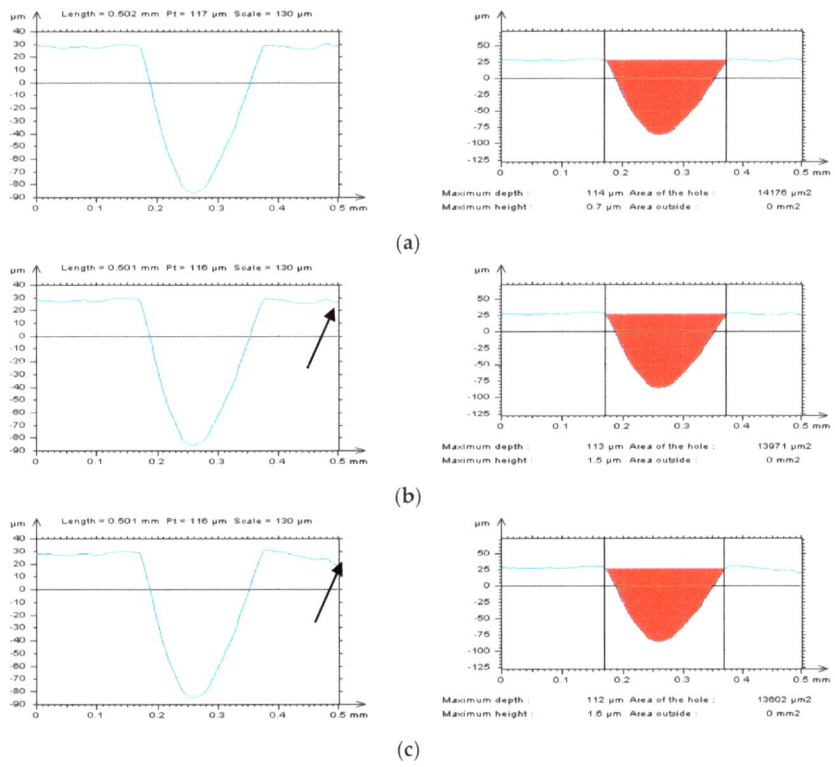

Figure 6. *Cont.*

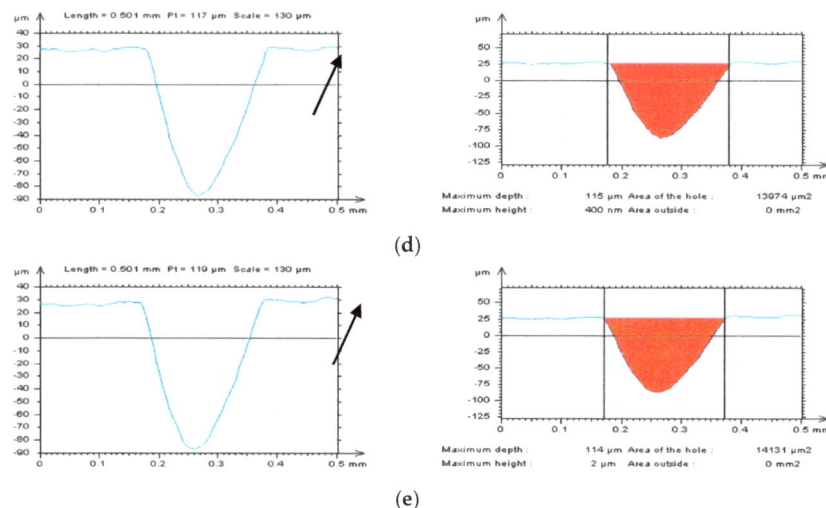

(d)

(e)

Figure 6. The dimple profiles (left column) and their hole/peak area diagrams (right column) received from the edge area of laser−textured surface topography after the definition of L−surface by Poly2 (**a**), Poly6 (**b**), Poly10 (**c**), GRF (**d**), and SF (**e**), cut−off = 0.8 mm.

Distortion in the features can be particularly noted for the profiles presented in Figure 7 (left column). Except for the exaggregation of features and edge-located areas of details, tilt can also be found, even if the surface was previously leveled [115] (according to the guidance provided by the software). It is another interference with the data and, correspondingly, can enlarge the possibility of data distortion. From all of the methods presented in Figure 7, Poly4 or RGRF seem to give the most encouraging results. However, additional leveling is required. When defining an appropriate L-surface, firstly, the distortion of features is not allowed; secondly, the out-of-feature part of the surface should be flat. The thresholding method can be proposed for both improving the form removal methods and validation of the approach already applied.

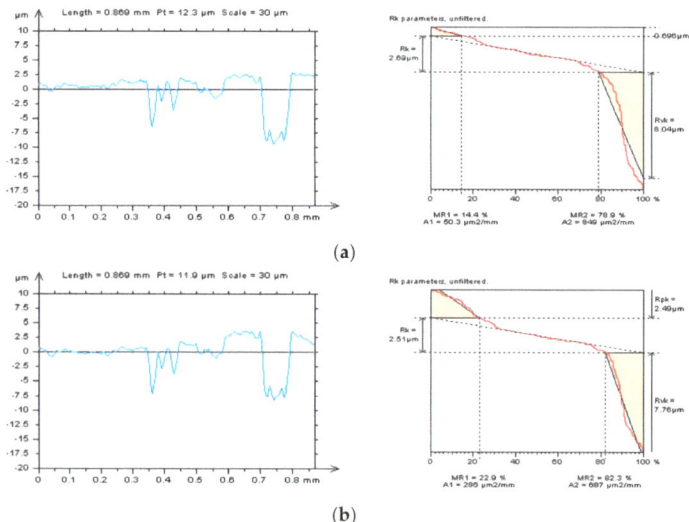

(a)

(b)

Figure 7. *Cont.*

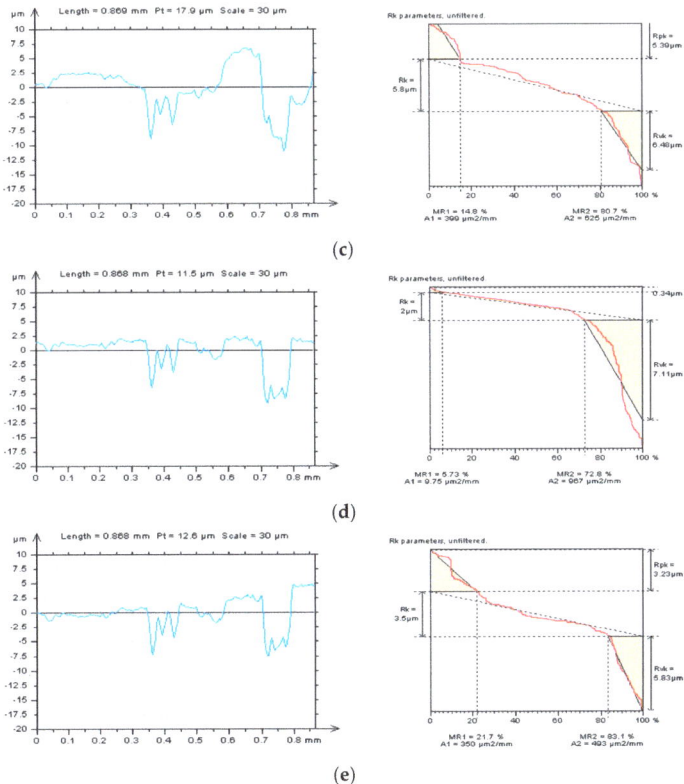

Figure 7. Selected profiles (left column) and their graphical studies of Rk—group parameters (right column) received from the laser—textured surface topography after the definition of L—surface by Poly4 (**a**), Poly8 (**b**), Poly12 (**c**), RGRF (**d**), and SF (**e**), cut—off = 0.8 mm.

3.2. Selection of a Method for S-Surface Definition with a Suppression of the High-Frequency Noise

The process of selection of the method for the reduction in a high-frequency measurement noise was proposed with areal (3D), profile (2D), ACF (areal and profile), PSD (2D and 3D), and TD analyses. It was suggested that all of the required properties should be studied with a multithreaded characterization. The measurement noise, similar to the uncertainty [41,116,117], can be reduced by repeating the measurement process of the same probe (detail). However, noise, especially in the high-frequency domain, can be characterized as separated data from those measured raw. The results received after S-filtration were defined as noise surface (NS) [39]. Some significant properties of the NS were defined and analyzed, considering the validation of noise removal methods, such as Gaussian (GRF or RGRF), spline (SF), median denoising (MDF), and fast Fourier transform (FFTF) filters, all available in commercial software.

According to the first NS property, it should contain only the required noise frequencies. In the considered case, only high-frequency components must be defined in the NS. Some non-noise features can be received when analyzing the isometric view of the surface. In Figure 8a, three various NS were presented and obtained after the application of the GRF, SF, and MDF methods (cut-off = 0.10 mm), respectively, from left to right. From that analysis, the NS received by GRF and SF filtration included some unwanted elements, indicated by the arrows, and it seems that the MDF is the most encouraging method.

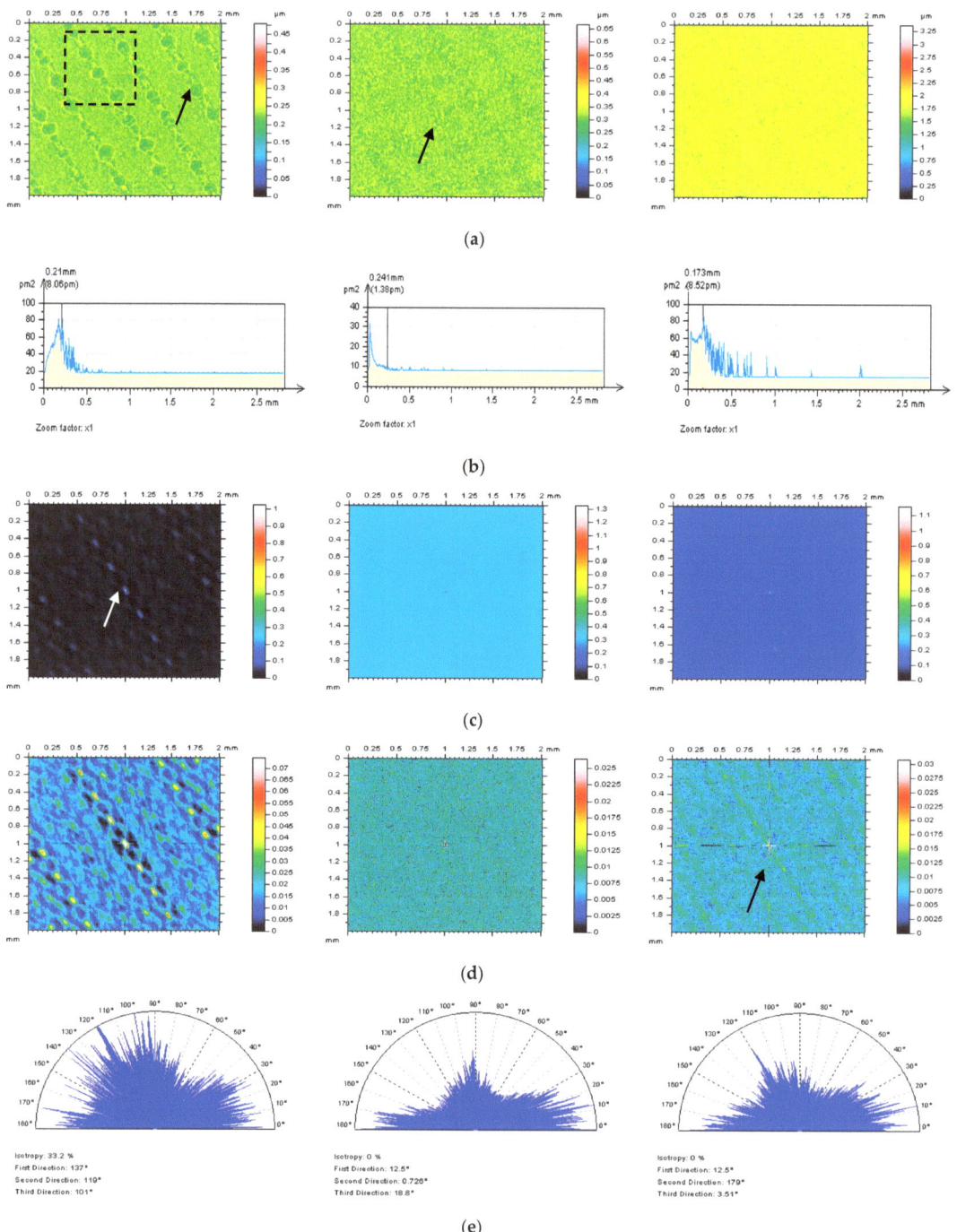

Figure 8. Analysis of NS: contour map plots (**a**), its PSDs (**b**) and ACFs (**c**), thresholded ACFs (**d**) and TDs (**e**) received after application of GRF (left column), SF (middle column), and MDF (right column), cut−off = 0.010 mm; studies provided for the laser-textured detail.

Considering the second NS property, it should be in the domain of noise. In the analyzed example, the NS should be in the high-frequency domain. For validation of this issue, the PSDs (Figure 8b) were considered. From that matter, all of those three filters gave suitable results; nevertheless, NS created by the SF accumulated the most high-frequency components.

As the third issue, the ACF of NS should be isotropic, as the NS itself. In Figure 8c, the 3D ACFs were presented for each of the NS. The GRF method created NS with ACF containing non-noise components. Moreover, the ACF consisted of some non-noise features (identified by the arrows). The ACFs received after SF and MDF filtering seemed not to contain any non-high-frequency-noise components. For the validation of this property, the thresholding method for ACF characterization was proposed (Figure 8d). Application of the thresholding technique, with considerable A1 and A2 values, confirmed non-noise components on the GRF NS, but it also indicated that NS obtained by MDF filtration contained the unwanted elements (presented by the arrows). According to those results, the MDF seems to be unsuitable for the extraction of high-frequency errors from the results of surface topography measurement of laser-textured surfaces.

The isotropic property of NS can be additionally studied with an analysis of TD graphs. Figure 8e presents TD graphs of all the three NSs studied in this work. From that issue, all three compared algorithms (GRF, SF, and MDF) gave no reliable responses. The isotropic property was not received for each of the filters matched. However, if selection must be proposed from those methods, the SF seems to give the most appropriate response (from those analyzed) for the suppression of high-frequency noise from the results of laser-textured surface topography measurements.

Some improvements in the validation of the method can be obtained when NS enlarged details are considered. An example of an enlarged detail (0.8 mm × 0.8 mm) was presented in Figure 9. In this case, the isometric views (Figure 9a) indicated that NS received by MDF contained non-noise components (it was indicated by the arrow). Contrary to the analysis of the whole detail (Figure 8), the SF NS did not contain non-noise components; it could be falsely estimated by the eye-view analysis. From this example, the multithreaded studies seem to be more justified. In terms of other properties, the PSD validation of the noise frequency dominance, the exclusion of occurrence of the non-noise feature with ACF and thresholded ACF studies, and the TD graphs analysis gave similar responses to those presented for larger detail. This (enlargement) method can validate the analysis of larger details and help in reducing the errors of the S-surface definition. The selection of an appropriate method for high-frequency noise reduction can be improved.

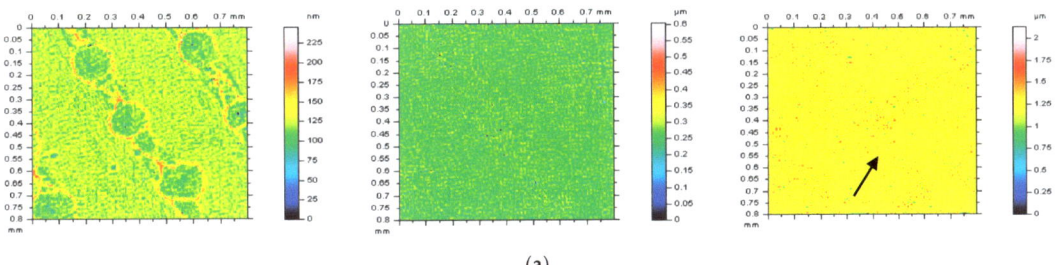

(a)

Figure 9. *Cont.*

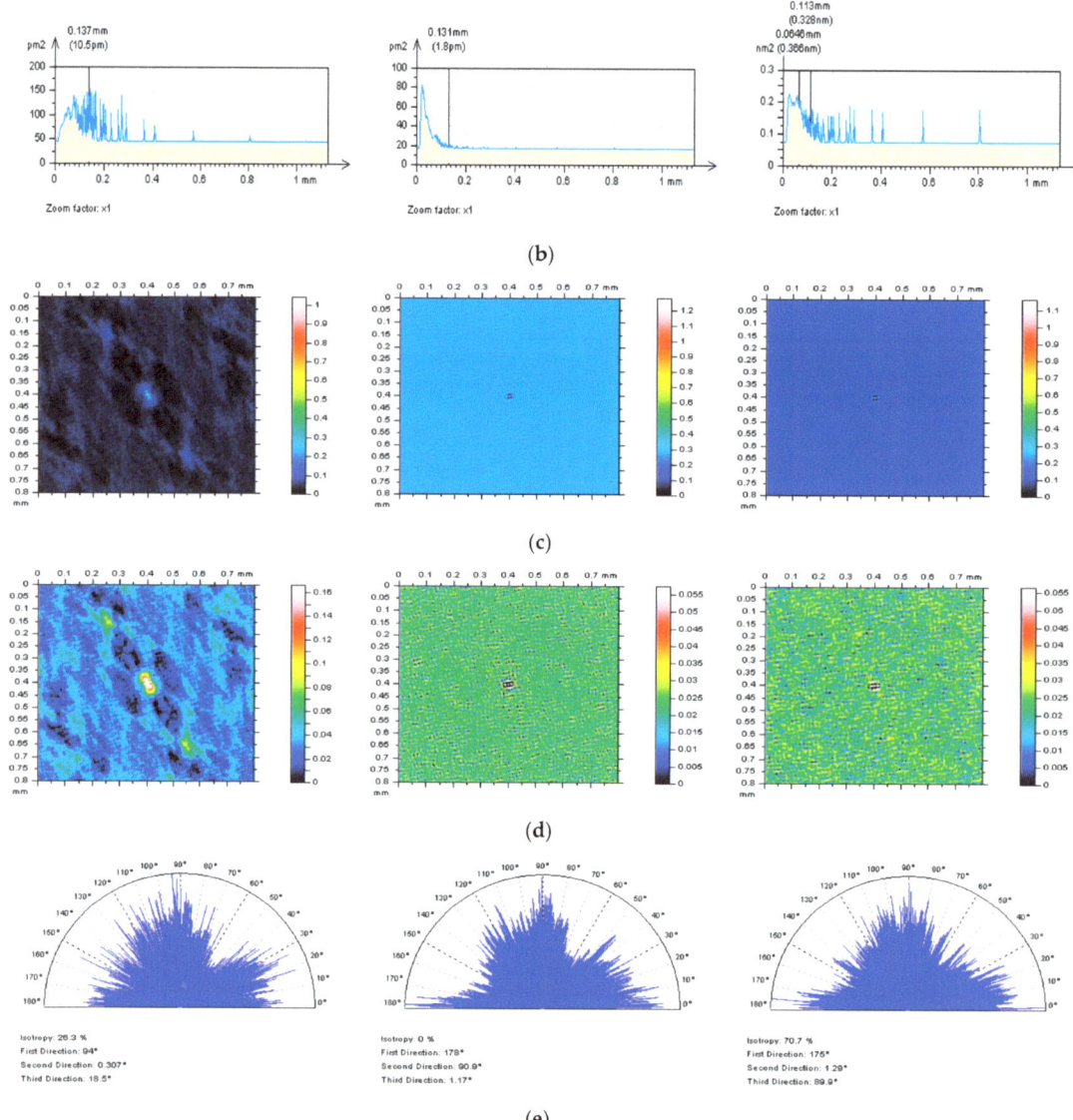

Figure 9. Analysis of extracted details from the NS: contour map plots (**a**), its PSDs (**b**) and ACFs (**c**), thresholded ACFs (**d**) and TDs (**e**) received after application of GRF (left column), SF (middle column) and MDF (right column), cut−off = 0.010 mm; description of the enlarged detail location was presented in Figure 8a (first left).

4. The Outlook

Despite the many studies presented, there are still many issues that can be addressed. There are some examples below:

1. The proposal of selection of cut-off (as the 3× sampling interval for a stylus, or 3× spacing for optical methods) must be studied and validated for isotropic surfaces. The validation of this type of topographies can be difficult with the methods proposed;

2. The analysis of some isotropic surfaces not containing some treatment traces and directional studies was not comprehensively analyzed. The treatment trace profile characterization may not respond adequately according to the proposals raised;
3. The definition and selection of the thresholding value must be precise for each of the surface types separately. Surfaces after different types of machining can receive various ranges of thresholding values;
4. Errors received by false estimation of thresholding value were not comprehensively studied in this paper. Distortions for each type of machined surface should be described separately as well.

5. Conclusions

From all of the studies presented, the following conclusions can be drawn:

1. The false estimation of L-surface when removing the form (shape and waviness) from the measured data can cause huge errors in the calculation of ISO 25178 surface topography parameters and can be the source of classification of properly made parts as lacks, leading to their rejection;
2. The distortion of L-surface positioning increases when the surface contains some deep or wide features, such as dimples, oil pockets, scratches, and valleys. The exaggeration can increase with the enlargement of the feature sizes, density, and distance from the edge of the analyzed detail. Special care must be taken when such features are edge located;
3. To reduce the errors in the definition of L-surface, the thresholding method is proposed. Contrary to the valley-excluding method, the present analysis is faster and does not require additional digital actions allowing to exclude some errors that can arise when the user does not entirely select the feature detail;
4. When selecting the thresholding value, reference to the material ratio and the Abbott–Firestone curves can be advantageous. The thresholding values received for all three functions should be similar or, correspondingly, the difference must be negligible;
5. When selecting the procedure for the definition of S-surface, functions of power spectral density, autocorrelation, and texture direction seem to be required. They must be supported with a mindful analysis of the isometric view of the noise surface;
6. For the characterization of the noise surface properties, thresholding techniques can be beneficial. Supporting this method with a selection of enlarged details can improve the validation of the approach for high-frequency measurement noise detection and reduction;
7. The thresholding method can be advantageous in the process of selection of cut-off values for both the L-surface and the S-surface definitions. Excluding deep or wide features, e.g., thresholding technique, it can reduce the errors in positioning of the reference plane (L-surface) and improve the processes of detection (definition) and reduction (removal) of the high-frequency noise. Moreover, this approach can be found even more crucial in varying the bandwidth value on the type of analyzed surface (e.g., laser-textured).

Author Contributions: Conceptualization, P.P.; methodology, P.P.; writing—original draft preparation, P.P. and W.M.; Investigation, P.P. and W.M.; validation P.P., W.M. and R.B.; writing—review and editing, W.M., R.B. and R.M.N.; Supervision, P.P.; funding acquisition, P.P., W.M., R.B. and R.M.N. All authors have read and agreed to the published version of the manuscript.

Funding: This research is sponsored by FEDER funds through the program COMPETE—Programa Operacional Factores de Competitividade—and by national funds through FCT—Fundação para a Ciência e a Tecnologia—under the project UIDB/00285/2020.

Institutional Review Board Statement: Not applicable.

Informed Consent Statement: Not applicable.

Data Availability Statement: Data sharing is not applicable to this article.

Conflicts of Interest: The authors declare no conflict of interest.

Parameters and Abbreviations

The following abbreviations (left column) and parameters (right column) are used in the manuscript:

ACF autocorrelation function	Sa arithmetic mean height, μm
AFM Atomic Force Microscopy	Sal auto-correlation length, mm
FFTF Fast Fourier Transform Filter	Sbi surface bearing index
GRF Gaussian regression filter	Sci core fluid retention index
L-filter filter used for definition of the L-surface	Sdq root mean square gradient
L-surface long-wavelength surface	Sdr developed interfacial areal ratio, %
MDF median denoising filter	Sk core roughness depth, μm
NS noise surface	Sku kurtosis
POLY2 least-square polynomial of the 2nd degree	Smc inverse areal material ratio, μm
POLY4 least-square polynomial of the 4th degree	Smr areal material ratio, %
POLY6 least-square polynomial of the 6th degree	Sp maximum peak height, μm
POLY8 least-square polynomial of the 8th degree	Spc arithmetic mean peak curvature, 1/mm
POLY10 least-square polynomial of the 10th degree	Spd peak density, $1/\text{mm}^2$
POLY12 least-square polynomial of the 12th degree	Spk reduced summit height, μm
PSD power spectral density	Sq root mean square height, μm
RGRF robust Gaussian regression filter	$Sr1$ upper bearing area, %
RMS root mean square height	$Sr2$ lower bearing area, %
S-filter removes small-scale lateral components	Ssk skewness
S-L surface a surface received after S- and L- filtering	Std texture direction, °
S-surface small-wavelength surface	Str texture parameter
SF spline filter	Sxp extreme peak height, μm
TD texture direction (graph)	Sv maximum valley depth, μm
	Svi valley fluid retention index
	Svk reduced valley depth, μm
	Sz the maximum height of the surface, μm

References

1. Padhan, M.; Marathe, U.; Bijwe, J. Surface topography modification, Film transfer and Wear mechanism for fibre reinforced polymer composites—An Overview. *Surf. Topogr. Metrol. Prop.* **2020**, *8*, 043002. [CrossRef]
2. Liu, Q.; Yang, Y.; Huang, M.; Zhou, Y.; Liu, Y.; Liang, X. Durability of a lubricant-infused Electrospray Silicon Rubber surface as an anti-icing coating. *Appl. Surf. Sci.* **2015**, *346*, 68–76. [CrossRef]
3. Shi, R.; Wang, B.; Yan, Z.; Wang, Z.; Dong, L. Effect of Surface Topography Parameters on Friction and Wear of Random Rough Surface. *Materials* **2019**, *12*, 2762. [CrossRef] [PubMed]
4. Macek, W. Fracture Areas Quantitative Investigating of Bending-Torsion Fatigued Low-Alloy High-Strength Steel. *Metals* **2021**, *11*, 1620. [CrossRef]
5. Macek, W. Correlation between Fractal Dimension and Areal Surface Parameters for Fracture Analysis after Bending-Torsion Fatigue. *Metals* **2021**, *11*, 1790. [CrossRef]
6. Macek, W.; Branco, R.; Costa, J.D.; Trembacz, J. Fracture Surface Behavior of 34CrNiMo6 High-Strength Steel Bars with Blind Holes under Bending-Torsion Fatigue. *Materials* **2021**, *15*, 80. [CrossRef]
7. Wizner, M.; Jakubiec, W.; Starczak, M. Description of surface topography of sealing rings. *Wear* **2011**, *271*, 571–575. [CrossRef]
8. Anderberg, C.; Dimkovski, Z.; Rosén, B.-G.; Thomas, T.R. Low friction and emission cylinder liner surfaces and the influence of surface topography and scale. *Tribol. Int.* **2019**, *133*, 224–229. [CrossRef]
9. Yıldırım, C.V. Investigation of hard turning performance of eco-friendly cooling strategies: Cryogenic cooling and nanofluid based MQL. *Tribol. Int.* **2020**, *144*, 106127. [CrossRef]
10. Awale, A.S.; Vashista, M.; Yusufzai, M.Z.K. Application of eco-friendly lubricants in sustainable grinding of die steel. *Mater. Manuf. Process.* **2021**, *36*, 702–712. [CrossRef]
11. Grzesik, W. Prediction of the Functional Performance of Machined Components Based on Surface Topography: State of the Art. *J. Mater. Eng. Perform.* **2016**, *25*, 4460–4468. [CrossRef]
12. Bruzzone, A.A.G.; Costa, H.L.; Lonardo, P.M.; Lucca, D.A. Advances in engineered surfaces for functional performance. *CIRP Annals* **2008**, *57*, 750–769. [CrossRef]
13. Podulka, P. Reduction of Influence of the High-Frequency Noise on the Results of Surface Topography Measurements. *Materials* **2021**, *14*, 333. [CrossRef] [PubMed]
14. Blunt, L.; Jiang, X. (Eds.) *Advanced Techniques for Assessment of Surface Topography*; Kogan Pages: London, UK, 2003.

15. Pawlus, P.; Wieczorowski, M.; Mathia, T. *The Errors of Stylus Methods in Surface Topography Measurements*; Zapol: Szczecin, Poland, 2014.
16. Thompson, A.; Senin, N.; Giusca, C.; Leach, R. Topography of selectively laser melted surfaces: A comparison of different measurement methods. *CIRP Ann.* **2017**, *66*, 543–546. [CrossRef]
17. Cheng, F.; Zou, J.; Su, H.; Wang, Y.; Yu, Q. A Differential Measurement System for Surface Topography Based on a Modular Design. *Appl. Sci.* **2020**, *10*, 1536. [CrossRef]
18. Podulka, P. Selection of Methods of Surface Texture Characterisation for Reduction of the Frequency-Based Errors in the Measurement and Data Analysis Processes. *Sensors* **2022**, *22*, 791. [CrossRef]
19. Podulka, P. Proposals of Frequency-Based and Direction Methods to Reduce the Influence of Surface Topography Measurement Errors. *Coatings* **2022**, *12*, 726. [CrossRef]
20. Leach, R.; Evans, C.; He, L.; Davies, A.; Duparré, A.; Henning, A.; Jones, C.W.; O'Connor, D. Open questions in surface topography measurement: A roadmap. *Surf. Topogr. Metrol. Prop.* **2015**, *3*, 13001. [CrossRef]
21. Raja, J.; Muralikrishnan, B.; Fu, S. Recent advances in separation of roughness, waviness and form. *Precis. Eng.* **2002**, *26*, 222–235. [CrossRef]
22. ISO 25178-2:2021; Geometrical Product Specifications (GPS)—Surface Texture: Areal—Part 2: Terms, Definitions and Surface Texture Parameters. International Organization for Standardization: Geneva, Switzerland, 2021.
23. Podulka, P. The effect of valley depth on areal form removal in surface topography measurements. *Bull. Pol. Acad. Sci. Tech. Sci.* **2019**, *67*, 391–400. [CrossRef]
24. Grabon, W.; Pawlus, P.; Galda, L.; Dzierwa, A.; Podulka, P.J. Problems of surface topography with oil pockets analysis. *Phys. Conf. Ser.* **2011**, *311*, 012023. [CrossRef]
25. Podulka, P. Bisquare robust polynomial fitting method for dimple distortion minimisation in surface quality analysis. *Surf. Interface Anal.* **2020**, *52*, 875–881. [CrossRef]
26. Koszela, W.; Pawlus, P.; Galda, L. The effect of oil pockets size and distribution on wear in lubricated sliding. *Wear* **2007**, *263*, 1585–1592. [CrossRef]
27. He, Y.; Yang, J.; Wang, H.; Gu, Z.; Fu, Y. Micro-dimple and micro-bulge textures:Influence of surface topography types on stick–slip behavior under starved lubrication. *Appl. Surf. Sci.* **2022**, *585*, 152501. [CrossRef]
28. Podgornik, B.; Jerina, J. Surface topography effect on galling resistance of coated and uncoated tool steel. *Surf. Coat. Technol.* **2012**, *206*, 2792–2800. [CrossRef]
29. Hu, S.; Huang, W.; Shi, X.; Peng, Z.; Liu, X.; Wang, Y. Multi-Gaussian Stratified Modeling and Characterization of Multi-process Surfaces. *Tribol. Lett.* **2018**, *66*, 117. [CrossRef]
30. Cogdell, J.D. A convolved multi-Gaussian probability distribution for surface topography applications. *Precis. Eng.* **2008**, *32*, 34–46. [CrossRef]
31. Podulka, P. Feature-Based Characterisation of Turned Surface Topography with Suppression of High-Frequency Measurement Errors. *Sensors* **2022**, *22*, 9622. [CrossRef] [PubMed]
32. Jiang, X.; Senin, N.; Scott, P.J.; Blateyron, F. Feature-based characterisation of surface topography and its application. *CIRP Ann.-Manuf. Technol.* **2021**, *70*, 681–702. [CrossRef]
33. Podulka, P. The Effect of Surface Topography Feature Size Density and Distribution on the Results of a Data Processing and Parameters Calculation with a Comparison of Regular Methods. *Materials* **2021**, *14*, 4077. [CrossRef]
34. ISO 2016 25178-600; Geometrical Product Specification (GPS)—Surface Texture: Areal Part 600: Metrological Characteristics for Areal-Topography Measuring Methods. International Organization for Standardization: Geneva, Switzerland, 2016.
35. Leach, R.K.; Haitjema, H. Bandwidth characteristics and comparisons of surface texture measuring instruments. *Meas. Sci. Technol.* **2010**, *21*, 032001. [CrossRef]
36. Zuo, X.; Peng, M.; Zhou, Y. Influence of noise on the fractal dimension of measured surface topography. *Measurement* **2020**, *152*, 107311. [CrossRef]
37. Sun, J.; Song, Z.; Heb, G.; Sang, Y. An improved signal determination method on machined surface topography. *Precis. Eng.* **2018**, *51*, 338–347. [CrossRef]
38. Podulka, P. Suppression of the High-Frequency Errors in Surface Topography Measurements Based on Comparison of Various Spline Filtering Methods. *Materials* **2021**, *14*, 5096. [CrossRef]
39. Podulka, P. Improved Procedures for Feature-Based Suppression of Surface Texture High-Frequency Measurement Errors in the Wear Analysis of Cylinder Liner Topographies. *Metals* **2021**, *11*, 143. [CrossRef]
40. Giusca, C.L.; Claverley, J.D.; Sun, W.; Leach, R.K.; Helmli, F.; Chavigner, M.P.J. Practical estimation of measurement noise and flatness deviation on focus variation microscopes. *CIRP Ann.* **2014**, *63*, 545–548. [CrossRef]
41. Haitjema, H. Uncertainty in measurement of surface topography. *Surf. Topogr. Metrol. Prop.* **2015**, *3*, 035004. [CrossRef]
42. ISO 16610-1:2015; Geometrical Product Specifications (GPS)–Filtration—Part 3: Terms and Definitions. International Organization for Standardization: Geneva, Switzerland, 2015.
43. Maculotti, G.; Feng, X.; Su, R.; Galetto, M.; Leach, R. Residual flatness and scale calibration for a point autofocus surface topography measuring instrument. *Meas. Sci. Technol.* **2019**, *30*, 075005. [CrossRef]
44. ISO 25178-3:2012; Geometrical Product Specifications (GPS)—Surface Texture: Areal—Part 3: Specification Operators. International Organization for Standardization: Geneva, Switzerland, 2012.

45. Li, Z.; Gröger, S. Investigation of noise in surface topography measurement using structured illumination microscopy. *Metrol. Meas. Syst.* **2021**, *28*, 4. [CrossRef]
46. Pomberger, S.; Stoschka, M.; Leitner, M. Cast surface texture characterisation via areal roughness. *Precis. Eng.* **2019**, *60*, 465–481. [CrossRef]
47. Gomez, C.; Campanelli, C.; Su, R.; Leach, R. Surface-process correlation for an ink-jet printed transparent fluoroplastic. *Surf. Topogr. Metrol. Prop.* **2020**, *8*, 034002. [CrossRef]
48. Podulka, P. Roughness Evaluation of Burnished Topography with a Precise Definition of the S-L Surface. *Appl. Sci.* **2022**, *12*, 12788. [CrossRef]
49. Sutowska, M.; Łukianowicz, C.; Szada-Borzyszkowska, M. Sequential Smoothing Treatment of Glass Workpieces Cut by Abrasive Water Jet. *Materials* **2022**, *15*, 6894. [CrossRef] [PubMed]
50. Nayak, P.R. Some aspects of surface roughness measurement. *Wear* **1973**, *26*, 165–174. [CrossRef]
51. Sayles, R.S.; Thomas, T.R. The spatial representation of surface roughness by means of the structure function: A practical alternative to correlation. *Wear* **1977**, *42*, 263–276. [CrossRef]
52. Dodds, C.J.; Robson, J.D. The description of road surface roughness. *J. Sound Vib.* **1973**, *31*, 175–183. [CrossRef]
53. Manninen, A.T. Multiscale surface roughness and backscattering Summary. *J. Electromagn. Waves Appl.* **1997**, *11*, 471–475. [CrossRef]
54. Chandley, P.J. Surface roughness measurements from coherent light scattering. *Opt. Quant. Electron.* **1976**, *8*, 323–327. [CrossRef]
55. Taylor, R.P. Surface Roughness Measurements on Gas Turbine Blades. *ASME. J. Turbomach.* **1990**, *112*, 175–180. [CrossRef]
56. Le Bosse, J.C.; Hansali, G.; Lopez, J.; Dumas, J.C. Characterisation of surface roughness by laser light scattering: Diffusely scattered intensity measurement. *Wear* **1999**, *224*, 236–244. [CrossRef]
57. Roy, S.; Bhattacharyya, A.; Banerjee, S. Analysis of effect of voltage on surface texture in electrochemical grinding by autocorrelation function. *Tribol. Int.* **2007**, *40*, 1387–1393. [CrossRef]
58. Vorburger, T.V.; Marx, E.; Lettieri, T.R. Regimes of surface roughness measurable with light scattering. *Appl. Opt.* **1993**, *32*, 3401–3408. [CrossRef] [PubMed]
59. Marx, E.; Leridon, B.; Lettieri, T.R.; Song, J.-F.; Vorburger, T.V. Autocorrelation functions from optical scattering for one-dimensionally rough surfaces. *Appl. Opt.* **1993**, *32*, 67–76. [CrossRef] [PubMed]
60. Schröder, S.; Duparré, A.; Coriand, L.; Tünnermann, A.; Penalver, D.H.; Harvey, J.E. Modeling of light scattering in different regimes of surface roughness. *Opt. Express* **2011**, *19*, 9820–9835. [CrossRef] [PubMed]
61. Egorov, A.A. Using waveguide scattering of laser radiation for determining the autocorrelation function of statistical surface roughness within a wide range of changes of the roughness correlation interval. *Quantum Electron.* **2002**, *32*, 357. [CrossRef]
62. Stout, K.J. Surface roughness ~ measurement, interpretation and significance of data. *Mater. Des.* **1981**, *2*, 260–265. [CrossRef]
63. Manninen, A.T. Multiscale surface roughness description for scattering modelling of bare soil. *Phys. A* **2003**, *319*, 535–551. [CrossRef]
64. Dhanasekar, B.; Mohan, N.K.; Bhaduri, B. Ramamoorthy, B. Evaluation of surface roughness based on monochromatic speckle correlation using image processing. *Precis. Eng.* **2008**, *32*, 196–206. [CrossRef]
65. Patrikar, R.M. Modeling and simulation of surface roughness. *Appl. Surf. Sci.* **2004**, *228*, 213–220. [CrossRef]
66. Wang, W.-Z.; Chen, H.; Hu, Y.-Z.; Wang, H. Effect of surface roughness parameters on mixed lubrication characteristics. *Tribol. Int.* **2006**, *39*, 522–527. [CrossRef]
67. Dusséaux, R.; Vannier, E. Soil surface roughness modelling with the bidirectional autocorrelation function. *Biosyst. Eng.* **2022**, *220*, 87–102. [CrossRef]
68. Munoz, R.C.; Vidal, G.; Kremer, G.; Moraga, L.; Arenas, C.; Concha, A. Surface roughness and surface-induced resistivity of gold films on mica: Influence of roughness modelling. *J. Phys. Condens. Matter* **2000**, *12*, 2903. [CrossRef]
69. Rees, W.G.; Arnold, N.S. Scale-dependent roughness of a glacier surface: Implications for radar backscatter and aerodynamic roughness modelling. *J. Glaciol.* **2006**, *52*, 214–222. [CrossRef]
70. Gharechelou, S.; Tateishi, R.; Johnson, B.A. A Simple Method for the Parameterization of Surface Roughness from Microwave Remote Sensing. *Remote Sens.* **2018**, *10*, 1711. [CrossRef]
71. Patir, N. A numerical procedure for random generation of rough surfaces. *Wear* **1978**, *47*, 263–277. [CrossRef]
72. Misra, A.; Pandey, P.M.; Dixit, U.S. Modeling and simulation of surface roughness in ultrasonic assisted magnetic abrasive finishing process. *Int. J. Mech. Sci.* **2017**, *133*, 344–356. [CrossRef]
73. Petropoulos, G.; Vaxevanidis, N.M.; Pandazaras, C. Modeling of surface finish in electro-discharge machining based upon statistical multi-parameter analysis. *J. Mater. Process. Tech.* **2004**, *155–156*, 1247–1251. [CrossRef]
74. Borodich, F.M.; Onishchenko, D.A. Similarity and fractality in the modelling of roughness by a multilevel profile with hierarchical structure. *Int. J. Solids Struct.* **1999**, *36*, 2585–2612. [CrossRef]
75. Krolczyk, G.M.; Raos, P.; Legutko, S. Experimental analysis of surface roughness and surface texture of machined and fused deposition modelled parts. *Teh. Vjes.* **2014**, *21*, 217–221.
76. Fubel, A.; Zech, M.; Leiderer, P.; Klier, J.; Shikin, V. Analysis of roughness of Cs surfaces via evaluation of the autocorrelation function. *Surf. Sci.* **2007**, *601*, 1684–1692. [CrossRef]
77. Ogilvy, J.A.; Foster, J.R. Rough surfaces: Gaussian or exponential statistics? *J. Phys. D Appl. Phys.* **1989**, *22*, 1243–1251. [CrossRef]

78. Zhixiong, L.; Nan, C.; Perdok, U.D.; Hoogmoed, W.B. Characterisation of Soil Profile Roughness. *Biosyst. Eng.* **2005**, *91*, 369–377. [CrossRef]
79. Davidson, M.W.J.; Le Toan, T.; Mattia, F.; Satalino, G.; Manninen, T.; Borgeaud, M. On the characterization of agricultural soil roughness for radar remote sensing studies. *IEEE Trans. Geosci. Remote* **2000**, *38*, 630–640. [CrossRef]
80. Loew, A.; Mauser, W. A semiempirical surface backscattering model for bare soil surfaces based on a generalized power law spectrum approach. *IEEE Trans. Geosci. Remote* **2006**, *44*, 1022–1035. [CrossRef]
81. Callens, M.; Verhoest, N.E.C.; Davidson, M.W.J. Parameterization of tillage-induced single-scale soil roughness from 4-m profiles. *IEEE Trans. Geosci. Remote* **2006**, *44*, 878–888. [CrossRef]
82. Hamed, A.M.; Saudy, M. Computation of surface roughness using optical correlation. *Pramana—J. Phys.* **2007**, *68*, 831–842. [CrossRef]
83. Podulka, P. Fast Fourier Transform detection and reduction of high-frequency errors from the results of surface topography profile measurements of honed textures. *Eksploat. Niezawodn.* **2021**, *23*, 84–89. [CrossRef]
84. Podulka, P. Resolving Selected Problems in Surface Topography Analysis by Application of the Autocorrelation Function. *Coatings* **2023**, *13*, 74. [CrossRef]
85. Sagy, A.; Brodsky, E.E.; Axen, G.J. Evolution of fault-surface roughness with slip. *Geology* **2007**, *35*, 283–286. [CrossRef]
86. Jacobs, T.D.B.; Junge, T.; Pastewka, L. Quantitative characterisation of surface topography using spectral analysis. *Surf. Topogr. Metrol. Prop.* **2017**, *5*, 013001. [CrossRef]
87. González Martínez, J.F.; Nieto-Carvajal, I.; Abad, J.; Colchero, J. Nanoscale measurement of the power spectral density of surface roughness: How to solve a difficult experimental challenge. *Nanoscale Res. Lett.* **2012**, *7*, 174. [CrossRef] [PubMed]
88. Fang, S.; Haplepete, S.; Chen, W.; Helms, C.; Edwards, H. Analyzing atomic force microscopy images using spectral methods. *J. Appl. Phys.* **1997**, *82*, 5891–5898. [CrossRef]
89. Elson, J.M.; Bennett, J.M. Calculation of the power spectral density from surface profile data. *Appl. Opt.* **1995**, *34*, 201–208. [CrossRef] [PubMed]
90. Sun, L. Simulation of pavement roughness and IRI based on power spectral density. *Math. Comput. Simulat.* **2003**, *61*, 77–88. [CrossRef]
91. Andren, P. Power spectral density approximations of longitudinal road profiles. *Int. J. Veh. Des.* **2005**, *40*, 2–14. [CrossRef]
92. Mwema, F.M.; Akinlabi, E.T.; Oladijo, O.P.; Oladijo, O.P. The Use of Power Spectral Density for Surface Characterization of Thin Films. In *Photoenergy and Thin Film Materials*; Scrivener Publishing LLC: Beverly, MA, USA, 2019; pp. 379–411. [CrossRef]
93. Lawson, J.K.; Wolfe, C.R.; Manes, K.R.; Trenholme, J.B.; Aikens, D.M.; English, R.E. Specification of Optical Components Using the Power Spectral Density Function. *Proc. Soc. Photo-Opt. Ins.* **1995**, *2536*, 38–50. [CrossRef]
94. Senthilkumar, M.; Sahoo, N.K.; Thakur, S.; Tokas, R.B. Characterization of microroughness parameters in gadolinium oxide thin films: A study based on extended power spectral density analyses. *Appl. Surf. Sci.* **2005**, *252*, 1608–1619. [CrossRef]
95. Vepsalainen, L.; Paakkonen, P.; Suvanto, M.; Pakkanen, T.A. Frequency analysis of micropillar structured surfaces: A characterization and design tool for surface texturing. *Appl. Surf. Sci.* **2012**, *263*, 523–531. [CrossRef]
96. Li, Q.; Deng, Y.; Li, J.; Shi, W. Roughness characterization and formation mechanism of abrasive air jet micromachining surface studied by power spectral density. *J. Manuf. Process.* **2020**, *57*, 737–747. [CrossRef]
97. Tanaka, H.; Okui, K.; Oku, Y.; Takezawa, H.; Shibutani, Y. Corrected power spectral density of the surface roughness of tire rubber sliding on abrasive material. *Tribol. Int.* **2021**, *153*, 106632. [CrossRef]
98. Mishra, V.; Khan, G.S.; Chattopadhyay, K.D.; Nand, K.; Sarepaka, R.V. Effects of tool overhang on selection of machining parameters and surface finish during diamond turning. *Measurement* **2014**, *55*, 353–361. [CrossRef]
99. Podulka, P. Proposal of frequency-based decomposition approach for minimization of errors in surface texture parameter calculation. *Surf. Interface Anal.* **2020**, *52*, 882–889. [CrossRef]
100. Khoshelham, K.; Altundag, D. Wavelet de-noising of terrestrial laser scanner data for the characterization of rock surface roughness. *Opt. Laser Remote Sens.* **2010**, *38*, 373–378.
101. Khoshelham, K.; Altundag, D.; Ngan-Tillard, D.; Menenti, M. Influence of range measurement noise on roughness characterization of rock surfaces using terrestrial laser scanning. *Int. J. Rock Mech. Min.* **2011**, *48*, 1215–1223. [CrossRef]
102. Simunovic, G.; Svalina, I.; Simunovic, K.; Saric, T.; Havrlisan, S.; Vukelic, D. Surface roughness assessing based on digital image features. *Adv. Prod. Eng. Manag.* **2016**, *11*, 93–104. [CrossRef]
103. Podulka, P. Thresholding Methods for Reduction in Data Processing Errors in the Laser-Textured Surface Topography Measurements. *Materials* **2022**, *15*, 5137. [CrossRef]
104. Galda, L.; Sep, J.; Prucnal, S. The effect of dimples geometry in the sliding surface on the tribological properties under starved lubrication conditions. *Tribol. Int.* **2016**, *99*, 77–84. [CrossRef]
105. Podulka, P. Selection of reference plane by the least squares fitting methods. *Adv. Sci. Technol. Res. J.* **2016**, *10*, 164–175. [CrossRef]
106. Galda, L.; Pawlus, P.; Sep, J. Dimples shape and distribution effect on characteristics of Stribeck curve. *Tribol. Int.* **2009**, *42*, 1505–1512. [CrossRef]
107. Podulka, P. The effect of valley location in two-process surface topography analysis. *Adv. Sci. Technol. Res. J.* **2018**, *12*, 97–102. [CrossRef]
108. Dadouche, A.; Conlon, M.J. Operational performance of textured journal bearings lubricated with a contaminated fluid. *Tribol. Int.* **2016**, *93*, 377–389. [CrossRef]

109. Janecki, D. Edge effect elimination in the recursive implementation of Gaussian filters. *Precis. Eng.* **2012**, *36*, 128–136. [CrossRef]
110. Podulka, P. Edge-area form removal of two-process surfaces with valley excluding method approach. *Matec. Web. Conf.* **2019**, *252*, 05020. [CrossRef]
111. Hanada, H.; Saito, T.; Hasegawa, M.; Yanagi, K. Sophisticated filtration technique for 3D surface topography data of rectangular area. *Wear* **2008**, *264*, 422–427. [CrossRef]
112. Macek, W.; Branco, R.; Szala, M.; Marciniak, Z.; Ulewicz, R.; Sczygiol, N.; Kardasz, P. Profile and Areal Surface Parameters for Fatigue Fracture Characterisation. *Materials* **2020**, *13*, 3691. [CrossRef]
113. Dzierwa, A.; Reizer, R.; Pawlus, P.; Grabon, W. Variability of areal surface topography parameters due to the change in surface orientation to measurement direction. *Scanning* **2014**, *36*, 170–183. [CrossRef]
114. Peta, K.; Mendak, M.; Bartkowiak, T. Discharge Energy as a Key Contributing Factor Determining Microgeometry of Aluminum Samples Created by Electrical Discharge Machining. *Crystals* **2021**, *11*, 1371. [CrossRef]
115. Senin, N.; Blunt, L.; Tolley, M. The use of areal surface topography analysis for the inspection of micro-fabricated thin foil laser targets for ion acceleration. *Meas. Sci. Technol.* **2012**, *23*, 105004. [CrossRef]
116. Leach, R.; Haitjema, H.; Su, R.; Thompson, A. Metrological characteristics for the calibration of surface topography measuring instruments: A review. *Meas. Sci. Technol.* **2020**, *32*, 032001. [CrossRef]
117. Leach, R.; Giusca, C.; Rickens, K.; Riemer, O.; Rubert, P. Development of material measures for performance verifying surface topography measuring instruments. *Surf. Topogr. Metrol. Prop.* **2014**, *2*, 025002. [CrossRef]

Disclaimer/Publisher's Note: The statements, opinions and data contained in all publications are solely those of the individual author(s) and contributor(s) and not of MDPI and/or the editor(s). MDPI and/or the editor(s) disclaim responsibility for any injury to people or property resulting from any ideas, methods, instructions or products referred to in the content.

Article

Analysis on the Morphology and Interface of the Phosphate Coating Prepared on X39Cr13 and S355J2 Steels

Monika Gwoździk [1,*], Mirosław Bramowicz [2] and Sławomir Kulesza [2]

1. Faculty of Production Engineering and Materials Technology, Czestochowa University of Technology, Armii Krajowej Street 19, 42-201 Czestochowa, Poland
2. Faculty of Technical Sciences, University of Warmia and Mazury in Olsztyn, Oczapowskiego 11, 10-719 Olsztyn, Poland; mbramowicz@moskit.uwm.edu.pl (M.B.); slawomir.kulesza@uwm.edu.pl (S.K.)
* Correspondence: monika.gwozdzik@pcz.pl

Abstract: The article presents the results of the characterization of the geometric structure of the surface of unalloyed structural steel and alloyed (martensitic) steel subjected to chemical processing. Prior to phosphating, the samples were heat-treated. Both the surfaces and the cross-sections of the samples were investigated. Detailed studies were made using scanning electron microscopy (SEM), XRD, metallographic microscopy, chemical composition analysis and fractal analysis. The characteristics of the surface geometry involved such parameters as circularity, roundness, solidity, Feret's diameter, watershed diameter, fractal dimensions and corner frequencies, which were calculated by numerical processing of SEM images.

Keywords: SEM; image processing; watershed; shape factor; Feret's diameter

Citation: Gwoździk, M.; Bramowicz, M.; Kulesza, S. Analysis on the Morphology and Interface of the Phosphate Coating Prepared on X39Cr13 and S355J2 Steels. *Materials* **2024**, *17*, 2805. https://doi.org/10.3390/ma17122805

Academic Editor: Alexander Yu Churyumov

Received: 20 May 2024
Revised: 29 May 2024
Accepted: 5 June 2024
Published: 8 June 2024

Copyright: © 2024 by the authors. Licensee MDPI, Basel, Switzerland. This article is an open access article distributed under the terms and conditions of the Creative Commons Attribution (CC BY) license (https:// creativecommons.org/licenses/by/ 4.0/).

1. Introduction

Phosphating is the most often used surface treatment and finishing process for ferrous and non-ferrous metals. This is a low-cost and fast procedure that develops corrosion and wear resistance on the surface [1–5] while also improving the adhesive and lubricating properties of the material. For that reason, phosphating plays a very important role in the automobile, processing and domestic appliances industries [6] to protect steel [7] and its alloys against corrosion. The use of phosphating on steel has been shown in Table 1.

Other than that, phosphate coatings are also prepared to cover alloys of such metals as zinc, cadmium, aluminum and magnesium. Phosphating creates a protective layer on the surface once the metal matrix is immersed in a phosphate solution as a result of chemical reactions between a dilute solution of phosphoric acid and, for example, zinc, iron and manganese. The phosphating mechanism is described by Yan et al. in paper [8]:

$$Fe \rightarrow Fe^{2+} + 2e^- \qquad (1)$$

$$H^+ + 2e^- \rightarrow H_2 \uparrow \qquad (2)$$

$$HPO_4^{2-} + Me^{2+} + yH_2O \rightarrow MeHPO_4 \cdot yH_2O \qquad (3)$$

$$2PO_4^{3-} + xMe^{2+} + yH_2O \rightarrow Me_x(PO_4)_2 \cdot yH_2O \qquad (4)$$

where Me^{2+} represents metal cations Zn^{2+}, Fe^{2+}, Mn^{2+} and Ni^{2+}.

The composition of the phosphate bath affects the properties of the resulting coatings. According to the authors of [9], the addition of sodium molybdate to the bath increased corrosion resistance of the coatings. They found that the corrosion current decreased with increased content of Na_2MoO_4. In this case, the coating was used as an intermediate protective layer to improve the adhesion of the final paint layer to automobile iron castings.

The researchers conducted two tests using (a) salt spray and (b) atmospheric species. The anti-corrosion effect on car castings was demonstrated in both cases. Unalloyed steels are more often subjected to the phosphating process than alloyed ones, because the presence of a passive oxide layer on the surface of corrosion-resistant steels makes them less susceptible to phosphating. However, some research on phosphating alloyed steels has been conducted [10]. Oskuie et al. [10] reported that tri-cation phosphate coating of Zn, Ca and Fe was grown electrochemically on 316 steel. A cathodic current was used as an accelerator for the phosphating process. The higher electrophosphating current density was shown to cause finer coating crystals that deteriorated the quality of the layer. Manna [11] tested phosphate coatings on steel with a ferritic–pearlitic structure, tempered martensite and tempered martensite with an oxide layer. In order to form the coating, a bath free from nitric acid was used. The test results showed that the structure of the substrate affected the thickness of the deposited phosphate coating. In turn, Ivanova [12] tested phosphate coatings based on pure Zn and Zn + Mn mixtures grown on carbon steels to determine the thickness of the coatings and the extent to which the core of the material (substrate) dissolved. It turned out that manganese phosphate greatly affected the obtained coating, reducing its thickness regardless of solution concentration and temperature that ended up in an increase of the mass of dissolved substrate metal. The coatings deposited in Zn-Mn baths consisted of the following phases: hopeite, phosphophyllite, quasihopeite, strunzite and their mixtures. Borko et al. [13] described how Domex 700 steel behaved in a 0.1 M NaCl environment. Prior to corrosion resistance measurements, the surface of the steel was sequentially treated by means of (1) grinding, (2) phosphating and (3) shot-peening. It turned out that the obtained MnP coating evenly and continuously covered the entire substrate both after grinding and shot-blasting. However, a more uniform layer with fewer defects was obtained after grinding. In turn, the shot-blasting contributed to the deterioration of thermodynamic and kinetic stability (corrosion resistance) of the coatings, with the opposite effect achieved after grinding and manganese phosphating. Taking into account the morphology and weight of phosphate coatings, three stages of the process can be distinguished: (1) corrosion of the substrate, (2) nucleation of the isolated phosphate crystals and (3) growth of the continuous phosphate coating [14]. Deposited coating is composed of crystals of disubstituted and trisubstituted metal phosphates [7]. On the other hand, Fang et al. [14] demonstrated that the deposited coating contained many close-packed lump crystallites mainly composed of $(Mn,Fe)_5H_2(PO_4)_4 \cdot 4H_2O$ complexes.

Table 1. Application of phosphate coatings for steel.

	Application of Phosphate Coatings
Steel	A layer facilitating cold-forming of steel. The phosphate coating in this case acts as a layer that prevents contact between the processed steel and the material from which the tool is made.
	Temporary protection of products during transport, storage and operation
	Anti-friction layer—reduces the coefficient of friction and also reduces the wear of interacting parts. This coating prevents welding of mating metals and quiets the element's operation system. Moreover, it reduces surface irregularities after mechanical processing and shortens the running-in period.
	Primer layer—increases the corrosion properties and adhesion of paint coatings. Corrosion resistance increases significantly after covering it with a layer of oil, paint or varnish.
	Insulating layer.

Fractal analysis is becoming increasingly popular in research of the outer surface layers [15], because it gives insight into various aspects of the geometric structure extending over several orders of magnitude. It is also possible to learn the relationships between the fractal and stereometric characteristics of the technological surface layers. In that framework, each surface can be characterized in terms of a single parameter—fractal dimension. Estimation of fractal dimensions is relatively simple and enables the analysis of surface variability from various images obtained, among others, using SEM microscopy [16]. Rovani et al. in paper [17] published results on phosphating of AISI steels previously subjected to heat treatment: hardening + tempering followed by shot-blasting. In the next steps, several different layers were applied: a zinc phosphate, a phenolic resin (base varnish) and a topcoat based on MoS$_2$. It was shown that phosphating itself significantly influenced the surface texture of the resin-bonded coating, taking into account changes in the surface texture ratio (S_{tr} parameter). Kurella et. al. in paper [18] showed that fractal dimensional analysis helps to interpret multi-scale surfaces. Moreover, fractal dimension analysis allows for the study of the surface of the materials in terms of topographic and chemical changes [19].

The aim of this paper was application of fractal analysis to SEM images in order to characterize the geometric structure of the phosphate layer, that is, to show that the quality of the obtained coatings can be determined through fractal analysis To this end, two steel samples that differ significantly in their chemical composition were selected and their spatial structures of the surface layers were determined. Such characterization might be also useful to optimize manufacturing processes or to identify the degradation effects.

2. Materials and Methods

Two steel samples were investigated in this study: X39Cr13 (corrosion-resistant martensitic steel [20]) and S355J2 (non-alloyed structural steel [21]). The chemical composition of both materials is presented in Table 2.

Table 2. Chemical composition of the steels under investigation.

Type of Steel	Acc. to	Chemical Composition, % Mass.					
		C	Si	Mn	P	S	Cr
X39Cr13	analysis	0.42	0.39	0.55	0.020	0.004	13.73
	EN 10088-2 [22]	0.36 ÷ 0.42	max. 1.00	max. 1.00	max. 0.040	max. 0.015	12.50 ÷ 14.50
S355J2	analysis	0.16	0.42	1.43	0.021	0.024	-
	EN 10025-2 [23]	max. 0.20	max. 0.55	max. 1.60	max. 0.025	max. 0.025	-

In the beginning, both steel samples were heat-treated in accordance with the specific processing guidelines. On one hand, the sample of X39Cr13 steel was hardened for 20 min at the austenitization temperature of 1050 °C followed by tempering for two hours at 300 °C. On the other hand, the sample of S355J2 steel was annealed for two hours at a temperature of 890 °C. Then, the surfaces were cleaned to remove residues and contaminants and degreased to allow the phosphate coating to be deposited. Prior to phosphating, prepared surfaces were activated in an aqueous solution (10%) of hydrochloric and sulfuric acids. Activation took 5 min at room temperature (22 °C). Finally, the phosphating was performed in a bath containing $MnHPO_4$ (3 g/L), $Mn(NO_3)_2$ (10 g/L), ZnO (5 g/L), H_3PO_4 (20 g/L), NaF (1 g/L). The process parameters were as follows: T = 52 °C, t = 1 h. The tests were performed both on the processed surfaces and in the cross-sections of the samples. For this purpose, square pieces measuring 10 × 10 mm^2 were cut out from the samples. The preparation of the metallographic specimens involved grinding and polishing. Selected samples were also etched in 5% nitric acid (for non-alloyed steel) and iron chloride (for alloyed steel). Detailed studies included observations in light microscope (LM Olympus GX41) and scanning electron microscope (SEM Jeol JSM-6610L) for metallographic examinations, chemical composition analysis and fractal analysis. The XRD experiments were carried out on a Seiffert 3003T/T diffractometer. A CoKα radiation cobalt lamp was used (λ = 1.79026 Å). The X-ray tube was operated at 40 kV and 30 mA. The XRD patterns were collected in 2 ranges between 5° and 90°. SEM images are composed of grayscale pixels corresponding to pseudo-heights that can be processed in order to derive various characteristics of the surface geometry. In the present paper, the two following approaches were used: (1) fractal analysis that makes use of scaling invariance between samples of averaged surface profiles [24–30] and (2) statistical approach working on the separated segments of original images. In the first method, original SEM images (example shown in Figure 1A) were averaged along the rows of the slow scan axis in order to obtain roughness profiles (Figure 1B), which were then processed into discrete structure functions according to the formula [31]:

$$S(\tau) = \frac{1}{N-m} \sum_{n=1}^{N-m} (z_{n+m} - z_n)^2 \qquad (5)$$

where τ is the discrete shift between original profile and its copy, $m = \tau/\Delta$ is the integer number, Δ—the scan step, z_k—the k-th sample of the mean profile and N—the number of samples in each profile. Figure 1C shows the plot of the structure function for S355J2 steel under 300× magnification. Thomas and Thomas [32] showed that for sufficiently small shifts τ, the one-dimensional structure function obeys the power–law dependence in the form:

$$S(\tau) = K\tau^{2(2-D)} \qquad (6)$$

where D is the unitless quantity referred to as fractal dimension, and K is the scaling factor referred to as pseudo-topothesy. Any sharp change in the slope of the log–log plot of the structure function vs. shift establish the corner frequency τ_c, which separates segments of different scale-invariance characteristics.

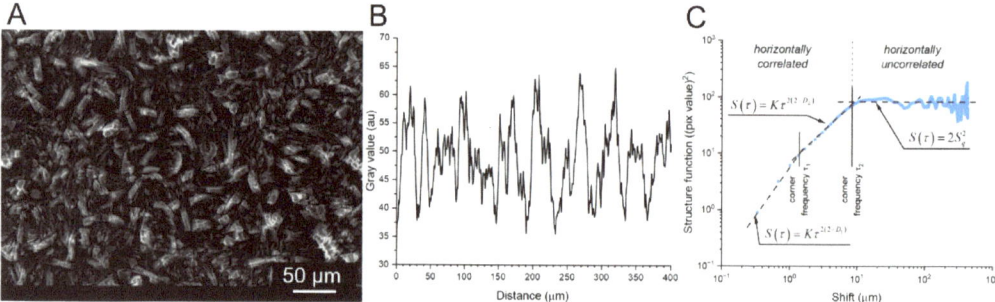

Figure 1. (**A**) SEM image of S335J2 steel under ×300 magnification and its row-averaged mean roughness profile (inset) SEI 20 kV, WD 10 mm, SS 44; (**B**) row-averaged original SEM image showing mean roughness profiles; (**C**) log–log plot of the structure function vs. horizontal shift obtained from the roughness profile (S is the RMS of pixel intensity). The corner frequency separates horizontally correlated surface bumps from the uncorrelated ones.

In the second method, surface morphology was analyzed by means of statistical shape analysis. To this end, SEM images were segmented using the watershed algorithm followed by determination of shape descriptors for planar figures: circularity, roundness, solidity, Feret's diameter and watershed diameter. Circularity is a positive fractional number that exhibits the deviation from a perfect circle. It is calculated according to the formula:

$$C = \frac{4\pi A}{P^2} \qquad (7)$$

where A—is the segment area and P—its perimeter. When the circularity decays to zero, the figure becomes increasingly elongated, and when it comes close to unity, a perfect circle appears. A similar measure is a roundness that equals the ratio of the lengths of the minor and the major semi-axes of the best fit ellipse replacing given selection area:

$$R = \frac{a_{\min}}{a_{\max}} = \frac{A}{\pi a_{\max}^2} \quad (8)$$

where $a_{\min}$, $a_{\max}$ are the minor and major semi-axes of the equivalent ellipse, respectively. In turn, solidity is the ratio of the actual area of the figure and its convex hull:

$$S = \frac{A}{A_{CH}} \quad (9)$$

For a perfectly convex figure, solidity equals one; otherwise it is less but non-zero. The last two parameters define specific size of the segments in terms of various lengths. On one hand, the watershed diameter is the diameter of the equivalent circle (same area as a given figure):

$$d_{WS} = \sqrt{\frac{4A}{\pi}} \quad (10)$$

On the other hand, Feret's diameter d_F equals the distance connecting any two points on the boundary of a segment. Among all possible d_F values, the minimum and maximum Feret's diameters are of special importance for characterization of particle shape and form.

3. Results and Discussion

The structures of the steel samples after heat treatment are shown in Figure 2. The structure of the X39Cr13 steel was tempered martensite with carbide precipitates (Figure 2A), while that of S355J2 was ferrite–pearlite (Figure 2B).

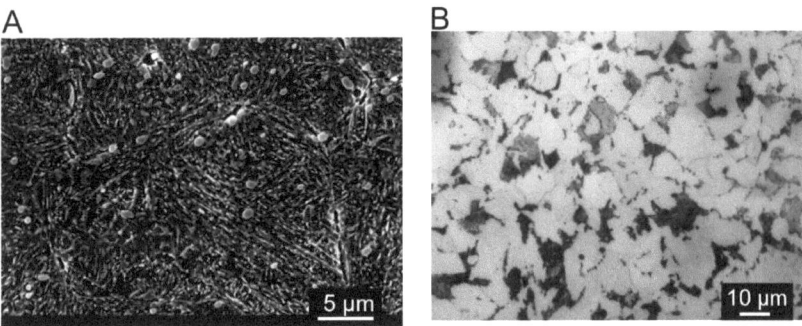

Figure 2. Structure of steel: (**A**) SEM image of X39Cr13, (**B**) optical microscope image of S355J2.

Recorded images were then used to determine the structure of the deposited phosphate coatings (Figure 3). The thickness of the coating on martensitic steel was one order smaller (~3 μm) compared to that on non-alloyed steel (~30 μm).

SEM images exhibit the crystalline structure of phosphate coatings that were made of phosphate crystals in the form of needles. A similar structure of the phosphate layer was observed by Rovani et al. [17]. Another paper reported that coatings based on zinc and phosphorus showed a predominance of needle-like crystals [33]. Some researchers identified the structure of this coating seen in SEM images as scale-like crystal structure [34]. Rossi et al. [35] demonstrated that phosphate-converted samples exhibited highly irregular surface structures, which turned out typical for this type of coating. Additionally, the thickness of the layer appeared to be between 5 and 10 μm [35]. Phosphating in an environment containing zinc and manganese ions resulted in a coating composed mainly of metal phosphates. Microscopic observations of phosphated steel surfaces showed differences in the morphology of these layers. The coating produced on X39Cr13 steel (Figure 4A) was characterized by much lower density than the coating produced on S355J2 steel (Figure 4B).

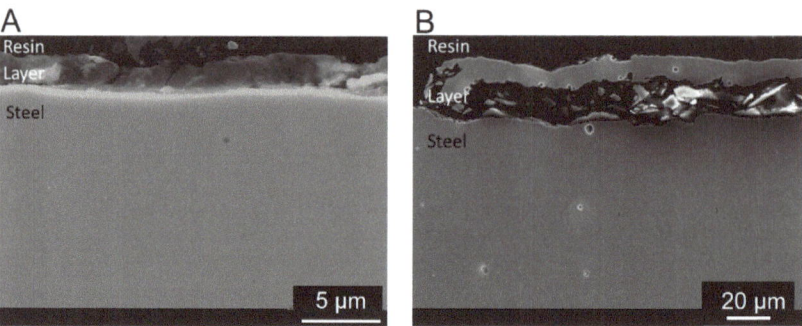

Figure 3. SEM images of phosphate layers deposited on steel samples: (**A**) X39Cr13, (**B**) S355J2.

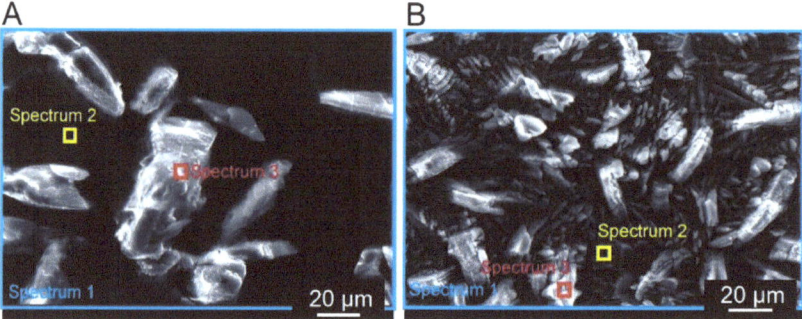

Figure 4. Steel surface after phosphating with a point and area spectrum markings: (**A**) X39Cr13 steel, SEM (SEI 30 kV, WD 10 mm, SS 46 (**B**) S355J2 steel, SEM (SEI 20 kV, WD 10 mm, SS 44).

Phosphating of non-alloyed steel resulted in the formation of a much more compact deposit compared to that on alloyed steel. The results of the chemical analysis of the surfaces of tested coatings (Table 3) confirmed the microscopic observations.

Significant differences in chemical composition of the deposited coatings were observed in X39Cr13 steel. EDS spectra taken from the needle (Figure 4A, Spectrum 3) appeared significantly different from those in the neighboring area (Figure 4A, Spectrum 2). The needles show significantly lower amounts of such elements as oxygen, phosphorus, manganese and zinc. On the other hand, an abundance of elements constituting the steel itself, such as iron and chromium, was observed in these parts of the samples. This shows that the coating made on alloyed steel is less tight than that on non-alloyed steel. The obtained XRD (Figure 5) results showed that the phosphate layer was composed of $Zn_3(PO_4)_2 \cdot 4H_2O$ and $Mn_3(PO_4)_2 \cdot 3H_2O$. The zinc-based compound predominated to a large extent. Using a phosphorus bath with a similar chemical composition, Nguyen et al. [36] showed that in coatings with a ZnO content greater than 3 g/L contains mixed phases. There are, among others, compounds such as $Mn_3(PO_4)_2 \cdot 3H_2O$ and $Zn_3(PO_4)_2 \cdot 4H_2O$. The XRD diagram presented by the researchers shows that there is only one reflection from which the $Mn(PO_4)_2 \cdot 3H_2O$ phase originates. The remaining picks come from a zinc-based compound.

Table 3. Analysis of the chemical composition (EDS) of the surface of the phosphate coating deposited on X39Cr13 and S355J2 steel samples.

Element	Weight, %					
	Spectrum 1		Spectrum 2		Spectrum 3	
	X39Cr13	S355J2	X39Cr13	S355J2	X39Cr13	S355J2
O	22.86	32.87	5.06	31.40	39.76	37.11
P	9.56	15.46	1.17	12.45	17.35	16.72
Si	-	-	0.35	-	-	-
Cr	6.43	-	13.66	-	-	-
Mn	2.75	4.24	0.59	2.10	3.58	3.33
Fe	33.91	10.09	77.90	30.10	1.73	8.80
Ni	0.83	-	-	0.36	0.70	0.42
Zn	23.66	36.90	1.27	23.02	36.88	33.19
Ca	-	0.45	-	-	-	0.24
Ti	-	-	-	-	-	0.19
Cu	-	-	-	0.58	-	-

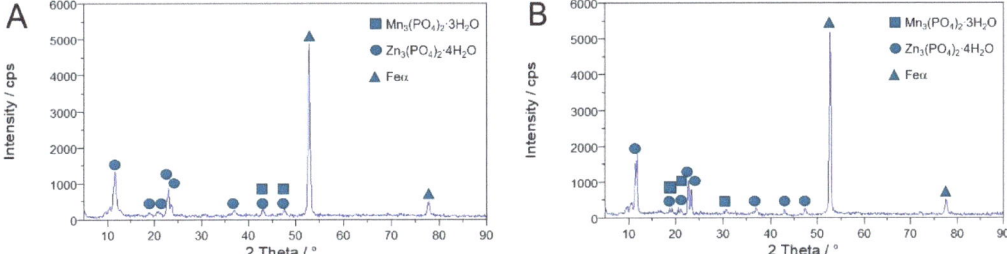

Figure 5. X-ray diffractogram: (**A**) X39Cr13 steel; (**B**) S355J2 steel.

The thinner phosphate layer on martensitic steel is probably due to its previous passivation, which in general has a beneficial effect [37]. In this case, however, passivation adversely affects the applied top layer. Due to the presence of a passive layer, the phosphating process on X39Cr13 steel was worse. According to the literature [38], the phosphating process involves dissolving a given metal in an acidic solution of soluble primary phosphates. The next process is the hydrolysis of these phosphates, which leads to the precipitation of insoluble tertiary phosphates. For the phosphating process to proceed properly, the metal should dissolve at a moderate rate, which will enable the necessary neutralization and supersaturation of the near-surface solution. Therefore, the phosphating process is deteriorated due to the presence of elements such as nickel, chromium or molybdenum in the composition of the steel. A smaller amount of precipitated phosphate is then produced. The literature states [10] that phosphating would significantly improve by the break of the chromium oxide layer. In contrast, no such large differences in the chemical composition of the coating on S355J2 steel were noted (Table 3, Figure 4B, Spectrum 2 and 3). EDS analysis also exhibits discontinuous coating layer that agrees with previous studies [17]. One possible explanation is the substrate cleaning process, which affects the nucleation and hence formation of zinc phosphate on the surface. In addition, in different morphologies of phosphate coatings applied to non-alloyed steel, substrates may be the result of the use of different phosphating baths. This also affects the different porosity of these coatings. According to researchers [33], the lowest porosity was in the coating made from a solution of zinc phosphate with ammonium niobium oxalate and benzotriazole. Significantly greater porosity was found in the zinc phosphate coating. In turn, the addition of niobium to the phosphating baths reduced the porosity of the coatings. Figure 6 presents grayscale SEM images of the investigated steel specimens: S355J2 (non-alloy quality structural steel) and X39Cr13 (martensitic stainless steel) viewed at two magnifications: ×75 and ×300.

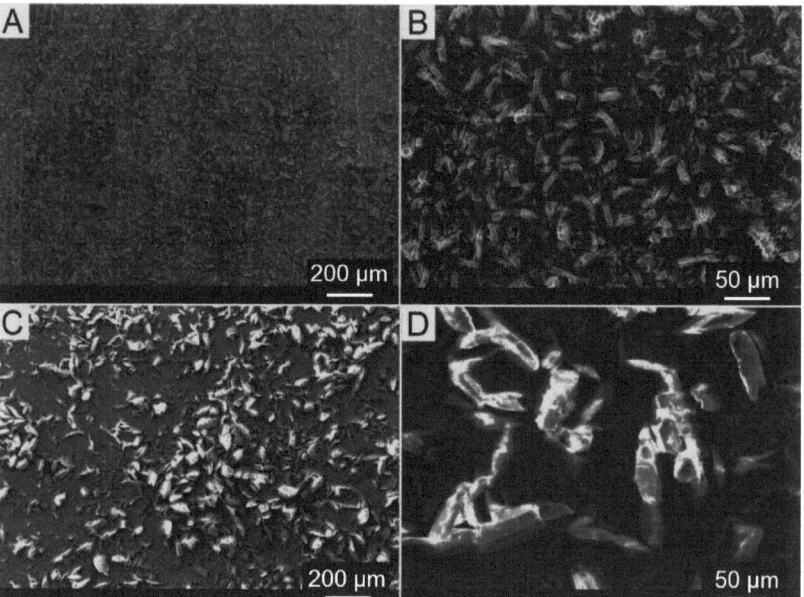

Figure 6. SEM images of steel samples under various magnifications: (**A**) S355J2 ×75 magnification, SEM (SEI 20 kV, WD 10 mm, SS 44), (**B**) S335J2 ×300 magnification, SEM (SEI 20 kV, WD 10 mm, SS 44), (**C**) X39Cr13 ×75 magnification, SEM (SEI 30 kV, WD 9 mm, SS 46), (**D**) X39Cr13 ×300 magnification, SEM (SEI 30 kV, WD 9 mm, SS 46).

Visual comparison of these images reveals the following similarities: both samples have a coarse surface covered with sharp precipitates, which are otherwise randomly oriented. Apart from that, however, the size, the shape and the alignment of these precipitates appear notably different. As a matter of fact, in the S355J2 sample shown in Figure 6A,B, the predominant geometrical forms are oblong polygons distributed evenly over the surface, the size of which ranges from a fraction of a micrometer up to few tens of micrometers, and the aspect ratio (the ratio of the shortest and the longest Feret's diameters) is between 1:3 and 1:5. Unlike that, on the surface of X39Cr13 steel sample in Figure 6C,D such precipitates can be seen that take on much more diverse and irregular shapes and might be even one order of magnitude bigger than those in the previous specimen. In addition, they are not distributed homogeneously over the surface, but instead they appear to agglomerate, forming clusters few hundreds micrometers in diameter.

Table 4 presents fractal parameters describing surface height variations in terms of allometric scaling that were derived from SEM images recorded at two magnification levels.

In the case of the S355J2 steel sample, multifractal behavior can be seen related to alignment patterns at different scale lengths. The lower scaling range, limited by the corner frequency τ_1 and defining the size of the lowest geometrical forms on the surface, extends up to 1.41 and 8.14 µm, viewed at ×300 and ×75 magnifications, respectively. On the other hand, the upper scaling ranges established by the corner frequencies τ_2 approach 8.34 and 587 µm, analyzed at ×300 and ×75 magnifications, respectively. Note, however, that when the resolutions are taken into consideration, then the frequencies τ_1 in both images correspond to ca. 5 pixels in each image regardless of the magnification, which might be a fingerprint of inevitable signal noise or high-frequency surface roughness. Note also that the scaling behaviors of the image data at both magnification levels overlap, which means that the corner frequency τ_1 in the low resolution image ($\times 75$) equals that of τ_2 in the high resolution image ($\times 300$). The same observation can be seen when comparing the values of fractal dimensions, because the fractal dimension $D_1 = 2.41$ in the low resolution image ($\times 75$) is nearly equal to $D_2 = 2.42$ in the high resolution image ($\times 300$). Such a finding might lead to a conclusion on the average size of the basic bumps on the surface of S355J2 steel, which take on elongated figures ca. 1 µm wide and ca. 8 µm long. At lower magnification, however, these bumps are found to agglomerate into clusters almost two orders of magnitude wider (600 µm in horizontal diameter). The fractal dimension was also determined by Paun et al. [24] and Kong et al. [25] using SEM images. Results for X39Cr13 steel presented in Table 3 exhibit significantly different scaling behavior, in which both corner frequencies and fractal dimensions appear similar regardless of the magnification: the corner frequency τ_1 explaining small bumps approaches ca. 6 µm, while the corner frequency τ_2 corresponding to the larger bumps is ca. 30 µm. As a result, fractal analysis reveals the appearance of oval bumps, which are around 4 times larger in their linear dimensions compared to those of the previous sample. Farias et al. [34] showed that the zinc phosphate coating had a roughness R_a of 0.47 µm and the crystals were 24.3 µm in diameter.

Table 4. Results of fractal analysis of SEM images of steel samples under investigation: D_1, D_2—fractal dimensions, τ_1, τ_2—corner frequencies.

Sample	Magnification	Image Resolution [µm/px]	D_1 [-]	D_2 [-]	τ_1 [µm]	τ_2 [µm]
S355J2	×75	4/3	2.49	2.83	8.14	587
	×300	1/3	2.19	2.42	1.41	8.34
X39Cr13	×75	4/3	2.16	2.45	7.57	32.4
	×300	1/3	2.11	2.35	5.26	29.4

To verify the results of the fractal approach, additional analysis was carried out relying on the separation of SEM images into a series of touching segments followed by statistical analysis of their form and size to determine such shape descriptors as circularity, roundness and solidity, together with specific size parameters such as Feret's diameter and watershed equivalent diameter. According to the literature [26], the Feret diameter of a particle can be defined as the distance between two parallel tangent boundaries of the object. In particular, the maximum and minimum Feret diameters, are often used for the characterization of particle sizes [27]. According to the literature [28], the minimum and maximum Feret diameters of objects are compared according to the aspect ratio, whereas the axial ratio refers to the best match between the minor and major axes of the ellipse. Detailed results are summarized in Table 5, where appropriate mean values are presented, and in Figure 6, where all data points are shown in a graphical form to reveal appropriate statistical distributions of the quantities under study together with their mean values and corresponding standard deviations. Note that SEM images with different magnification were chosen for this analysis (×300/S355J2 vs. ×75/X39Cr13) to ensure similar number of segments established in the scan area.

Comparison of obtained means proves that the predominant shape of the average segments does not vary between specimens under study. As a matter of fact, circularity equal to ca. 0.7 and roundness equal to ca. 0.6 both imply the elongated shape of such a figure, while solidity larger than 0.8 clearly points at its convex habit. Together, all three descriptors strongly suggest a nearly oval shape of the average segment. The only difference lies in the size of the segments, which agrees well with the previous findings from the fractal analysis. As a rule, both the Feret's diameter and the watershed diameter demonstrate that the size of specific bumps on the surface of S355J2 steel is roughly one order of magnitude lower than those in X39Cr13 (1.4/2.1 μm vs. 20/12 μm, respectively).

Table 5. Mean shape descriptors of the segments outlined in SEM images using the watershed algorithm.

	S355J2 (×300)	X39Cr13 (×75)
Circularity	0.67 ± 0.18	0.73 ± 0.25
Roundness	0.63 ± 0.17	0.60 ± 0.25
Solidity	0.83 ± 0.07	0.83 ± 0.13
Feret's diameter [μm]	2.1 ± 1.6	20 ± 22
Watershed diameter [μm]	1.4 ± 1.0	12 ± 14

In order to verify how the results for particular segments are distributed, Figure 7 shows half-box plots of the image data. Presented graphs confirm previous findings as to similarity of the shape descriptors: circularity, roundness and solidity and notable difference in the size of the established segments. Circularity is an important geometrical parameter in evaluation of grain shapes, as it might provide an insight into object roundness [28]. As indicated in the literature [29], surface roughness is combined with irregularity related to the level of roundness of natural grains. The plot of estimated diameters of the segments on the surface of S355J2 steel specimen exhibits quite uniform although narrow distribution of the data points with sharp edges on both sides, which might be concluded in high identity of the sizes of the segments. In contrast, the distributions of the same data on the surface of X39Cr13 steel appears strongly asymmetric, with a flat edge at the bottom, but a very long tail extending up to 100 μm at the top. This notable asymmetry is due to a large variation in the diameters of the segments, established in the image without any significant change in their habits according to the results of the shape descriptors. This allows us to refer to this surface as self-affine (self-similar within a limited range of scale lengths).

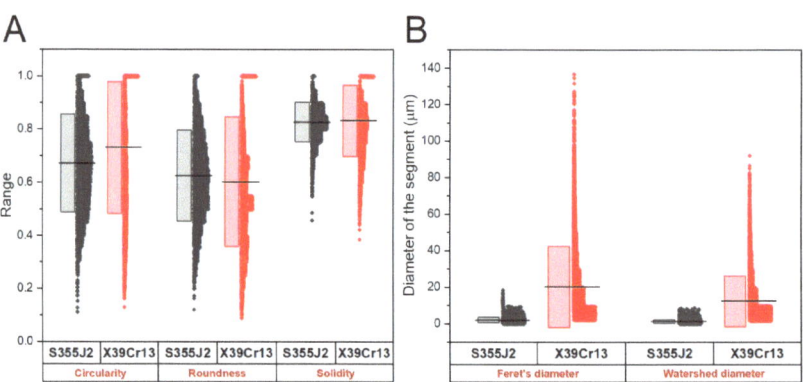

Figure 7. (**A**) Half-box plots of shape descriptors of the segments outlined in SEM images by means of the watershed algorithm: circularity (equation number 34), roundness (equation number 7) and solidity (equation number 8) and (**B**) comparison of the distributions of specific size descriptors: Feret's diameter and watershed diameter (equation number 9). Closed dots show data points and straight lines correspond to mean values, while the heights of the boxes extend to ± standard deviation.

4. Conclusions

Phosphate coatings on alloyed and unalloyed steel samples exhibit crystalline structure. This coating was composed of $Zn_3(PO_4)_2 \cdot 4H_2O$ and $Mn_3(PO_4)_2 \cdot 3H_2O$, with a significant advantage of the first phase. Much better quality of the phosphate coating was obtained on S355J2 steel. Comparative analysis by means of fractal parameters and shape descriptors reveal morphological differences between specific geometrical features of predominant forms on the surface of S355J2 and X39Cr13 steel samples. Fractal analysis uncovers aggregated structure of the surface bumps in S355J2 steel, but self-affine in X39Cr13 sample; fractal parameters reflect the appearance of oval bumps, which are around 4 times larger in their linear dimensions on the surface of S355J2 compared to X39Cr13. Statistical analysis of the shapes and sizes of the segments established in SEM images using the watershed discrimination algorithm generally confirms the findings of the fractal approach: the average segment takes on an oval shape regardless of the sample; however, its size on the surface of S355J2 steel is roughly one order of magnitude lower than that of X39Cr13. As a matter of fact, the surface of S355J2 might be referred to as clustered, while that of X39Cr13 can be regarded as self-affine. Microstructural tests combined with EDS analysis allowed for the conclusion that the phosphate coating is more tight and compact in the case of S355J2 steel. In this case, the formed phosphate layer is characterized by a strong bond with the substrate. Therefore, it can be used as a base, e.g., for a paint coating. Geometric characteristics of the structure may be desirable both to optimize production processes and to understand the effects of material deterioration.

Author Contributions: Conceptualization, M.G., M.B. and S.K.; methodology, M.G., M.B. and S.K.; formal analysis, M.G., M.B. and S.K.; investigation, M.G., M.B. and S.K.; software, M.G., M.B. and S.K.; writing—original draft preparation, M.G., M.B. and S.K.; writing—review and editing, M.G., M.B. and S.K. All authors have read and agreed to the published version of the manuscript.

Funding: This research received no external funding.

Institutional Review Board Statement: Not applicable.

Informed Consent Statement: Not applicable.

Data Availability Statement: The data presented in this study are available on request from the authors.

Conflicts of Interest: The authors declare no conflicts of interest.

References

1. Liu, D.; Huang, J.; Zhou, Y.; Ding, Y. Enhanced corrosion resistance and photocatalytic properties of Bi_2O_3/phosphate composite film prepared on AZ91D magnesium alloy by phosphating. *Int. J. Electrochem. Sci.* **2019**, *14*, 1434–1450. [CrossRef]
2. Zaludin, M.A.F.; Jamal, Z.A.Z.; Derman, M.N.; Kasmuin, M.Z. Fabrication of calcium phosphate coating on pure magnesium substrate via simple chemical conversion coating: Surface properties and corrosion performance evaluations. *J. Mater. Res. Technol.* **2019**, *8*, 981–987. [CrossRef]
3. Al-Swaidani, A.M. Inhibition effect of natural pozzolan and zinc phosphate baths on reinforcing steel corrosion. *Int. J. Corros.* **2018**, *2018*, 9078253. [CrossRef]
4. Oliveira, M.F.; de Santana, H.; Grassi, M.; Rodrigues, P.R.P.; Gallina, A.L. Comparative study of inorganic and organic phosphating of carbon steel 1008 regarding resistance to corrosion. *Matéria* **2013**, *18*, 1395–1409. [CrossRef]
5. Kanamaru, T.; Kawakami, T.; Tanaka, S.; Arai, K.; Yamamoto, M.; Mizuno, K. The structure of phosphate crystals influencing the scab corrosion-resistance. *Tetsu Hagane–J. Iron Inst. Jpn.* **1991**, *77*, 1050–1057. [CrossRef]
6. Narayanan, T.S.N.S. Surface pretreatment by phosphate conversion coatings—A review. *Rev. Adv. Mater. Sci.* **2005**, *9*, 130–177.
7. Ma, L.; Yang, X.J.; Peng, C.; Yang, Q. Rapid electrochemical phosphating at room temperature. *Asian J. Chem. Part A* **2014**, *26*, 5509–5512. [CrossRef]
8. Yan, S.; Zhao, Y.L.; Dai, Y.G.; Li, J.Z.; Shi, J.J.; Gao, X.W.; Xu, H.Y.; Yu, K.; Luo, W.B. The influence of silicon on the formation of phosphate coatings for low-carbon IF steels. *Surf. Coat. Technol.* **2022**, *441*, 128599. [CrossRef]
9. Li, G.Y.; Lian, J.S.; Niu, L.Y.; Jiang, Z.H. A zinc and manganese phosphate coating on automobile iron castings. *ISIJ Int.* **2005**, *45*, 1326–1330. [CrossRef]
10. Oskuie, A.A.; Afshar, A.; Hasannejad, H. Effect of current density on DC electrochemical phosphating of stainless steel 316. *Surf Coat. Technol.* **2010**, *205*, 2302–2306. [CrossRef]
11. Manna, M. Characterisation of phosphate coatings obtained using nitric acid free phosphate solution on three steel substrates: An option to simulate TMT rebars surfaces. *Surf. Coat. Technol.* **2009**, *203*, 1913–1918. [CrossRef]
12. Ivanova, D. Phosphating of carbon Steels in solutions containing zinc and zinc-manganese phosphates. *IOP Conf. Ser. Mater. Sci. Eng.* **2018**, *374*, 012033. [CrossRef]
13. Borko, K.; Hadzima, B.; Jackova, M.N. Corrosion resistance of Domex 700 steel after combined surface treatment in chloride environment. *Procedia Eng.* **2017**, *192*, 58–63. [CrossRef]

14. Fang, L.; Xie, L.B.; Hu, J.; Li, Y.; Zhang, W.T. Study on the growth and corrosion resistance of manganese phosphate coatings on 30CrMnMoTi alloy steel. *Phys. Procedia* **2011**, *18*, 227–233. [CrossRef]
15. Czifra, Á.; Ancza, E. Micro- and nano-roughness separation based on fractal analysis. *Materials* **2024**, *17*, 292. [CrossRef]
16. Sarul, M.; Mikulewicz, M.; Kozakiewicz, M.; Jurczyszyn, K. Surface evaluation of orthodontic brackets using texture and fractal dimension analysis. *Materials* **2022**, *15*, 2071. [CrossRef] [PubMed]
17. Rovani, A.C.; Kouketsu, F.; da Silva, C.H.; Pintaude, G. Surface characterization of three-layer organic coating applied on AISI 4130 steel. *Adv. Mater. Sci. Eng.* **2018**, *2018*, 6767245. [CrossRef]
18. Kurella, A.K.; Dahotre, N. Fractal approach to hierarchically evolved laser processed CaP coatings. *Adv. Eng. Mater.* **2010**, *12*, 517–521. [CrossRef]
19. Oshida, Y.; Tuna, E.B.; Aktören, O.; Gençay, K. Dental implant systems. *Int. J. Mol. Sci.* **2010**, *11*, 1580–1678. [CrossRef]
20. Gwoździk, M.; Nitkiewicz, Z. Topography of X39Cr13 steel surface after heat and surface treatment. *Opt. Appl.* **2009**, *39*, 853–857.
21. Pietkun-Greber, I. Comparison of resistance to damage of unalloyed carbon steels under the influence of hydrogen. *MATEC Web Conf.* **2018**, *174*, 01015. [CrossRef]
22. Polish Standard PN-EN 10088-2; Stainless Steel—Part 2: Technical Delivery Conditions for Sheet/Plate and Strip for General Purposes. 2014.
23. Polish Standard PN-EN 10025-2; Hot Rolled Products of Structural Steels—Part 2: Technical Delivery Conditions for Non-Alloy Structural Steels. 2019.
24. Paun, M.A.; Paun, V.A.; Paun, V.P. Mercury bonding to xerogel: The interface fractal dynamics of the interaction between two complex systems. *Gels* **2023**, *9*, 670. [CrossRef] [PubMed]
25. Kong, B.; Dai, C.X.; Hu, H.; Xia, J.; He, S.H. The fractal characteristics of soft soil under cyclic loading based on SEM. *Fractal Fract.* **2022**, *6*, 423. [CrossRef]
26. Vilela, F.; Bezault, A.; de Francisco, B.R.; Sauvanet, C.; Xu, X.P.; Swift, M.F.; Yao, Y.; Marrasi, F.M.; Hanein, D.; Volkmann, N. Characterization of heterogeneity in nanodisc samples using Feret signatures. *J. Struct. Biol.* **2022**, *214*, 107916. [CrossRef] [PubMed]
27. Allahverdi, Ç. Synthesis of copper nano/microparticles via thermal decomposition and their conversion to copper oxide film. *Turk. J. Chem.* **2023**, *47*, 616–632. [CrossRef] [PubMed]
28. Sinkhonde, D.; Rimbarngaye, A.; Kone, B.; Herring, T.C. Representativity of morphological measurements and 2-d shape descriptors on mineral admixtures. *Results Eng.* **2022**, *13*, 100368. [CrossRef]
29. Gresina, F.; Farkas, B.; Fábián, S.A.; Szalai, Z.; Varga, G. Morphological analysis of mineral grains from different sedimentary environments using automated static image analysis. *Sediment. Geol.* **2023**, *455*, 106479. [CrossRef]
30. Gwoździk, M.; Kulesza, S.; Bramowicz, M. Application of the fractal geometry methods for analysis of oxide layer. In Proceedings of the Metal 2017, 26th International Conference on Metallurgy and Materials, Brno, Czech Republic, 24–26 May 2017; pp. 789–794.
31. Sayles, R.S.; Thomas, T.R. The spatial representation of surface roughness by means of the structure function: A practical alternative to correlation. *Wear* **1977**, *42*, 263–276. [CrossRef]
32. Thomas, A.; Thomas, T.R. Digital analysis of very small scale surface roughness. *J. Wave-Matter Interact.* **1988**, *3*, 341–350.
33. Banczek, E.P.; Rodrigues, P.R.P.; Costa, I. Evaluation of porosity and discontinuities in zinc phosphate coating by means of voltametric anodic dissolution (VAD). *Surf. Coat. Technol.* **2009**, *203*, 1213–1219. [CrossRef]
34. Farias, M.C.M.; Santos, C.A.L.; Panossian, Z.; Sinatora, A. Friction behavior of lubricated zinc phosphate coatings. *Wear* **2009**, *266*, 873–877. [CrossRef]
35. Rossi, S.; Chini, F.; Straffelini, G.; Bonora, P.L.; Moschini, R.; Stampali, A. Corrosion protection properties of electroless Nickely-PTFE, Phosphatey MoS2 and BronzeyPTFE coatings applied to improve the wear resistance of carbon steel. *Surf. Coat. Technol.* **2003**, *173*, 235–242. [CrossRef]
36. Nguyen, T.L.; Cheng, T.C.; Yang, J.Y.; Pan, C.J.; Lin, T.H. A zinc-manganese composite phosphate conversion coating for corrosion protection of AZ91D alloy: Growth and characteristics. *J. Mater. Res. Technol.* **2022**, *19*, 2965–2980. [CrossRef]
37. López, R.; Menéndez, M.; Fernández, C.; Chmiela, A.; Bernardo-Sánchez, A. The Influence of carbon coatings on the functional properties of X39Cr13 and 316LVM steels intended for biomedical applications. *Metals* **2019**, *9*, 815. [CrossRef]
38. Flis, J.; Mańkowski, J.; Zakroczymski, T.; Bell, T. The formation of phosphate coatings on nitrided stainless steel. *Corros. Sci.* **2001**, *43*, 1711–1725. [CrossRef]

Disclaimer/Publisher's Note: The statements, opinions and data contained in all publications are solely those of the individual author(s) and contributor(s) and not of MDPI and/or the editor(s). MDPI and/or the editor(s) disclaim responsibility for any injury to people or property resulting from any ideas, methods, instructions or products referred to in the content.

Article

Influence of the Relative Displacements and the Minimum Chip Thickness on the Surface Texture in Shoulder Milling

Lukasz Nowakowski, Slawomir Blasiak and Michal Skrzyniarz *

Department of Machine Design and Manufacturing Engineering, Kielce University of Technology, al. Tysiaclecia Panstwa Polskiego 7, 25-314 Kielce, Poland; lukasn@tu.kielce.pl (L.N.); sblasiak@tu.kielce.pl (S.B.)
* Correspondence: mskrzyniarz@tu.kielce.pl; Tel.: +48-41-34-24-417

Abstract: The formation of surface texture in milling is a complex process affected by numerous factors. This paper focuses on the surface roughness of X37CrMoV51 steel machined by shoulder milling. The aim of the study was to develop a mathematical model to predict the surface roughness parameter Ra. The proposed model for predicting the surface roughness parameter Ra in shoulder milling takes into account the feed per tooth, f_z, the corner radius, r_ε, and the actual number of inserts involved in the material removal process as well as h_{min} and $D(\xi)$. The correlation coefficient between the theoretical and experimental data was high (0.96). The milling tests were carried out on a three-axis vertical milling machine using a square shoulder face mill. The geometric analysis of the face mill shows that at a feed rate of 0.04 mm/tooth, cutting was performed by three out of five inserts, and when the feed rate exceeded 0.12 mm/tooth, material was removed by all inserts. The minimum chip thickness parameter and the standard deviation of the relative displacement increased as the feed increased. Over the whole range of feeds per tooth, the displacement increased by 0.63 μm. Higher cutting speeds resulted in lower minimum chip thicknesses and the average standard deviation of the relative displacements for the whole range of cutting speeds was 2 μm.

Keywords: milling process; relative displacements; surface roughness; shoulder milling; minimum chip thickness; surface roughness; tool geometry

Citation: Nowakowski, L.; Blasiak, S.; Skrzyniarz, M. Influence of the Relative Displacements and the Minimum Chip Thickness on the Surface Texture in Shoulder Milling. *Materials* **2023**, *16*, 7661. https://doi.org/10.3390/ma16247661

Academic Editor: Francisco J. G. Silva

Received: 19 November 2023
Revised: 9 December 2023
Accepted: 13 December 2023
Published: 15 December 2023

Copyright: © 2023 by the authors. Licensee MDPI, Basel, Switzerland. This article is an open access article distributed under the terms and conditions of the Creative Commons Attribution (CC BY) license (https://creativecommons.org/licenses/by/4.0/).

1. Introduction

The manufacturing industry is facing higher and higher requirements for the quality of machine parts produced by machining. For economic reasons, it has become essential to reduce the number of finishing operations. Much of the research in this area aims to find some convenient and efficient solutions to these problems. Shoulder milling, which combines peripheral milling with face milling, may provide an answer to these expectations. These days, it is one of the most popular milling methods as it is suitable for both roughing and finishing cuts. Another approach to surface quality enhancement in machining is to use new materials for cutting tools or apply special coatings in order to improve the cutting conditions and, consequently, the process efficiency, and, ultimately, the surface quality. Changes to the tools affect the other factors responsible for the cutting process, especially the relative displacements, i.e., the displacements in the tool and workpiece system, the magnitude and distribution of cutting forces, the tool and workpiece temperatures, the friction in the cutting zone, or the minimum chip thickness. Since some of these factors largely influence the surface quality of the finished product, many studies devoted to milling or turning use them to model the surface roughness parameters, particularly Ra and Rt.

Research into the surface texture formation takes into account the contribution of various factors. For instance, a permanent change in the chip cross-section may be due to some axial and/or radial runout of the inserts employed, with this possibly resulting from insufficient geometrical or dimensional accuracy of the cutter body. The effect of the axial runout of inserts is considered in [1]; therein, the theoretical values of the surface roughness are modeled and compared with the experimental ones. A similar problem is described in [2]; the analysis concerns the influence of the tool runout on the surface quality in dynamic milling. The modeling and simulation of the surface topography in peripheral milling based on the radial and axial runout of inserts are discussed in [3]. Differences in the performance of the inserts employed resulting from their axial and radial runouts are responsible for the occurrence of displacements in the tool–workpiece system. Vibration is not a desirable phenomenon in machining because it has a negative effect on the dimensional and geometrical accuracy of the workpiece and the durability of tools and machine tool systems. The displacements in the tool–workpiece system are responsible for lower machine performance and process efficiency. It is, thus, essential to use lower values of the cutting parameters (especially the cutting speed and feed rate), which makes the process take longer. There have been many studies on the influence of vibration on the cutting process. The effects of vibration on the surface roughness parameter Ra are discussed, for example, in [4]; the research focused on the relationship between the corner radius and the relative displacements. In a further study [5], a model is proposed to predict surface topography in milling when affected by tool vibration. The method can be used to develop a formula to optimize the surface roughness parameters. Wu et al. [6] propose the use of the cutting parameters and signal features of vibration measured in milling to predict the surface roughness of S45C steel. Many studies involved using artificial neural networks or machine learning to predict the roughness parameters. For example, in [7], a model based on an artificial neural network was employed to determine surface roughness parameters. The use of machine learning methods to predict surface roughness parameters in milling was discussed in [8]. Another approach is proposed in [9], where the Long Short-Term Memory (LSTM) method was applied to predict surface roughness in milling for S45C steel. Many studies deal with the prediction of surface roughness in machining. Yan et al. [5] developed a method for predicting the surface roughness in milling based on the information about the tool vibration. The analysis of the influence of the process parameters shows that the feed per tooth has the greatest influence on surface roughness, whereas the effect of the axial depth of cut is the lowest. Many studies have aimed to determine the influence of the key machining process parameters, i.e., the feed rate, the cutting speed, and the depth of cut, on the roughness of milled surfaces [10]. Kao et al. [11] indicate that in milling, surface roughness can be predicted from the cutting forces. Their research focused on determining the relationship between the cutting force and the surface roughness of the workpiece machined on a small-size machine tool. The analysis involved employing machine learning algorithms, multi-dimensional linear regression, and a generalized neural regression network to determine the relationship between the cutting forces and surface roughness. In the machining of free-form surfaces, the major parameter responsible for surface roughness is the shape of the machining path. These days, CAM systems are used to design the shape of the tool path. Research in this area [12] reveals that tool path strategies greatly affect the actual milling time and the workpiece surface roughness. Uzun et al. [13] analyze the influence of the four most popular machining methods available in CAM systems. In micromachining, the minimum chip thickness is one of the most important and frequently considered process parameters affecting the surface roughness parameters. There have been many studies aimed to predict the minimum chip thickness by using the stagnation point [14], or to determine it through experiments [15]. Another factor responsible for the surface roughness of the workpiece is the cutting tool. Surface roughness is dependent on the tool material as well as its micro- and macrogeometry. Different materials are used for cutting tools, and the choice depends on the tool application. Recently, with the rapid development of 3D technology, additively manufactured tools are also available.

Analysis of the literature shows that much attention is given to the prediction of surface roughness parameters, particularly Ra and Rt, as these are the most popular in industry. It can be concluded that the models developed to predict surface roughness are generally used for one type of material and specific cutting conditions.

The novelty of the current article is the development of a model for predicting the surface roughness parameter Ra in shoulder milling. The parameter is determined on the basis of the following factors: the feed per tooth f_z, the corner radius, r_ε, the actual number of inserts used for cutting, the minimum chip thickness, h_{min}, and the standard deviation of the relative displacements in the tool–workpiece system, $D(\varepsilon)$. The additional novelty of the present work is the description of how to use the surface profile to determine the minimum chip thickness and how to recognize the selected areas on the profile.

2. Materials and Methods

The material selected for the experiments was X37CrMoV5 hot work tool steel, characterized by good ductility, thermal conductivity, fracture toughness under high temperature and water cooling conditions, hardenability, and resistance to tempering. The steel is suitable for die-casting molds and cores, hot scissors and guillotine cutters, press elements, light-metal forming dies, and plastic molding cores [16]. The material has been purchased from a commercial source that certifies the chemical composition of the material. The chemical composition of X37CrMoV51 steel is provided in Table 1.

Table 1. Composition of the alloying elements of the workpiece material, %.

C	Mn	Si	P	S	Cr	Ni	Mo	W	V	Co	Cu
0.32–0.42	0.2–0.5	0.8–1.2	max. 0.03	max. 0.03	4.5–5.5	max. 0.35	1.2–1.5	max. 0.3	0.3–0.5	max. 0.3	max. 0.3

The cutting was performed using a 490-050Q22-08M milling cutter (Sandvik Coromant, Sandviken, Sweden) with five 8 mm × 3.97 mm 490–08T308M–PL–1030 inserts. The inserts, with a corner radius, r_e, of 0.8 mm, are designed for light roughing to finishing operations. Although the maximum depth of cut for this type of insert is 5.5 mm, the producer recommends that it should not exceed 4 mm. The tool was mounted in the machine tool spindle using an A1B05–4022035 arbor (Sandvik Coromant, Sandviken, Sweden), which is an ISO 40 taper face mill arbor commonly used in machining centers.

The shoulder milling tests were carried out on a VMC 800 vertical machining center (FOP AVIA, Warsaw, Poland). The VMC 800 is a robust and stable machining center composed of four iron cast elements bolted together. The design of the VMC800 vertical machining center is based on a cross table moving in the X and Y axes and a horizontal spindle box moving in the Z axis along the vertical column. During the test, other factors in addition to the feed rates were controlled, namely, cutting speed, depth of cut, and workpiece position, and the relative displacement between tool and workpiece was measured.

A test rig was fitted on the machining center to measure the relative displacements between the tool and the workpiece; the major measuring device was an XL-80 laser interferometer (Renishaw plc, New Mills Woton-under-Edge Gloucestershire, Kingswood, UK). The measurement was performed for 20 s at a frequency of 500 Hz. A total of 10,000 measurement points were used in each measurement. The laser was mounted outside the machine to avoid vibration affecting its position. Before cutting, the tool was run through to measure any external factors that could affect the results. The displacement was then measured during cutting. The forward trend was then subtracted from the measured signal without removing any material.

Before the cutting tests, the insert mounting accuracy was measured using a Heidenhain TT120 tool touch probe (Dr. Johannes Heidenhain GmbH, Traunreut, Germany).

The surface topography, including roughness, was analyzed using a Taylor Hobson: Talysurf CCI—Lite Non-Contact 3D profiler (Taylor-Hobson Ltd., Leicester, UK) with a 20× magnifying lens. A measurement area of 0.8 mm × 0.8 mm provided a 1024 × 1024 pixel array. The 2D surface roughness parameters were determined using a Gaussian filter with a 0.8 cut-off wavelength.

The primary surface profile measurements for use in determining the minimum chip thickness, h_{min}, were performed using a TOPO 01P profilometer (Instytut Zaawansowanych Technologii Wytwarzania, Cracow, Poland), which is a contact-type instrument equipped with a non-sliding stylus for measuring surface roughness. The profiles were analyzed using the Topografia software running on the profilometer. The profile showed two clear peaks, which corresponded to the scratches made to act as a frame of reference (a system of coordinates) for a specimen. The distance between the scratches was 3.528 mm. In Figure 1, the plot is divided into two parts, one representing the average surface profile in milling and the other the average surface profile in grinding. The angle between the lines was measured ($\alpha = 0°23'39''$). This angle corresponded to the angle of the workpiece surface inclination to the tool during a cutting test. The profile obtained for the milled surface was smoothened in relation to the profile reported for the ground surface so that the former corresponded to zero on the axis of ordinates. The value of h_{min} (1.34 µm) was measured directly on the profile.

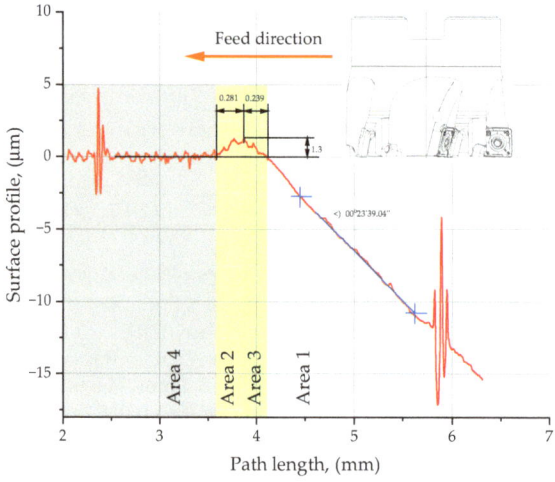

Figure 1. Profile used to determine the minimum chip thickness. Measurement with a TOPO 01P profilometer with a non-sliding stylus capturing features over a distance of 10 mm, in a range of 40 µm at a resolution of 1 µm.

Because of the principle of operation of the test rig used to measure the displacements, the machining was performed in the presence of a cutting fluid. A selective static program was used to determine the effect of each factor separately. One series of the shoulder milling tests was conducted at a constant depth of cut ($a_p = 0.2$ mm), a constant cutting speed ($v_c = 300$ m/min), and a variable feed per tooth ranging from 0.02 to 0.22, which changed every 0.02 mm/tooth. The other series of shoulder milling tests were carried out at $a_p = 0.2$ mm, $f_z = 0.1$ mm/tooth, and $v_c = 200–400$ m/min changed every 20 m/min.

3. Results

The axial and radial runouts of the tool head were calculated for each insert separately on the basis of 50 measurements. This required determining the average axial insert mounting accuracy. For simplicity and greater transparency, it was assumed that the tool length for the most axially protruding insert was 0 mm. Assumptions about tool engagement and zero tool length were used for calculations only to assess the radial and axial runout of the tool. A model of the cutting tool was developed to perform simulations and assess how many inserts were engaged in the surface texture generation at a given value of the feed per tooth. The model can be used to calculate the load each insert is subjected to, the actual feed per tooth, the instantaneous depth of cut, and the cross-sectional area of the material removed from each insert. Some of the calculation results are given in Table 2.

Table 2. Theoretical versus actual feed per tooth for each insert.

Theoretical Feed, fz, mm/tooth	Actual Feed in mm/tooth for Each Insert				
	1	2	3	4	5
0.02	0.011	0.005	0	0.004	0
0.04	0.019	0.009	0	0.012	0
0.06	0.027	0.013	0	0.019	0.001
0.08	0.035	0.017	0	0.027	0.001
0.1	0.043	0.021	0	0.034	0.002
0.12	0.048	0.025	0.001	0.041	0.005
0.14	0.052	0.029	0.003	0.046	0.01
0.16	0.056	0.033	0.007	0.05	0.014
0.18	0.061	0.037	0.01	0.054	0.018
0.2	0.065	0.041	0.014	0.058	0.022
0.22	0.069	0.045	0.017	0.062	0.027

From Table 2, it can be concluded that the insert mounting accuracy and the feed per tooth have a substantial influence on the material removal process. As can be seen from Table 2, only three out of five inserts were engaged in the cutting at a feed of 0.04 mm/tooth. When, however, the feed exceeded 0.12 mm/tooth, the material was removed by all the five inserts. The third insert at a feed rate of 0.12 mm/tool removes less material than the value of the h_{min} parameter, and at a feed rate of 0.14 mm/tool, the insert exceeds the value of the h_{min} parameter. During the cutting tests, the relative displacements in the tool–workpiece system were measured using an XL-80 laser interferometer (Renishaw plc, New Mills Woton-under-Edge Gloucestershire, Kingswood, UK). Some relative displacements in the tool–workpiece system are illustrated in Figure 2.

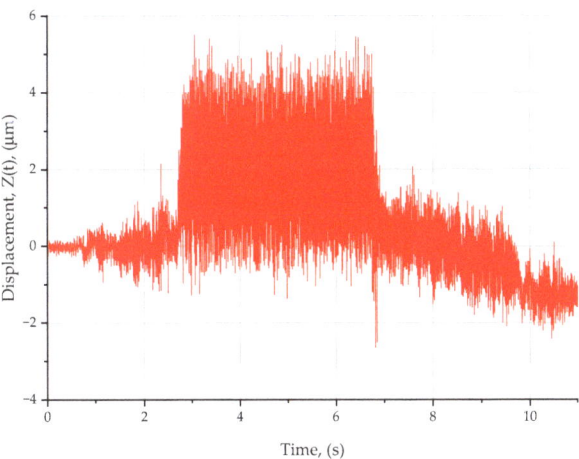

Figure 2. Relative displacements in the tool–workpiece system at v_c = 300 m/min and f_z = 0.1 mm/tooth.

Figure 2 shows a signal of the relative displacements registered in the tool–workpiece system during a test measurement under the following cutting conditions: v_c = 300 m/min,

f_z = 0.1 mm/tooth, and a_p = 0.2 mm. As can be seen, the signal varies over time. During the first phase, the displacement is stable because this represents the time when the tool approaches the workpiece and the tool is not engaged in cutting (0–2.7 s). In the second phase (2.7–6.8 s), when the tool performs shoulder milling, the displacement increases. The last phase (6.8–11 s) corresponds to the final stage of cutting, during which the tool is no longer in contact with the workpiece. From Figure 2, it can be concluded that the relative displacements increase with the depth of cut. Thus, it is clear that the displacements occurring in the first phase increase with increasing width of cut until the maximum engagement of the cutting tool is achieved, i.e., the whole diameter of the tool is used. It was then vital to determine how the cutting speed, v_c, and the feed per tooth, f_z, affected the standard deviation of the relative displacements (D(ε)) for the selected material. Table 3 shows the standard deviation of the relative displacements measured in the tool–workpiece system, D(ε), given in µm, for the cutting performed at v_c = 300 m/min, a_p = 0.2 mm, and a variable feed per tooth.

Table 3. Variation in the parameter D(ε) with feed rate and cutting speed.

v_c = 300 m/min a_p = 0.2 mm		f_z = 0.1 mm/tooth a_p = 0.2 mm	
f_z, mm/tooth	D(ε), µm	v_c, m/min	D(ε), µm
0.02	1.07	200	1.75
0.04	1.29	220	1.75
0.06	1.36	240	2.10
0.08	1.53	260	2.6
0.1	1.49	280	1.86
0.12	1.56	300	1.88
0.14	1.61	320	2.03
0.16	1.56	340	1.97
0.18	1.61	360	3.45
0.2	1.48	380	2.04
0.22	1.71	400	1.48

Table 3 shows the values of the standard deviation of the relative displacement in the tool–workpiece system, i.e., the parameter D(ε) expressed in µm, for the selected workpiece material machined at a constant cutting speed, v_c, of 300 m/min and a variable feed per tooth, and also at a constant feed of 0.1 mm/tooth and a variable cutting speed. From the data in Table 3, it is clear that increasing the feed per tooth had a negative effect on the D(ε) parameter. The higher the feed per tooth, the higher the standard deviation of the relative displacement. In the entire range of feeds per tooth, the displacement rose by 0.63 µm. When the feed varied between 0.02 mm/tooth and 0.08 mm/tooth, the parameter D(ε) first increased from 1.07 µm to 1.53 µm and then dropped slightly to 1.49 µm. At feeds of 0.1–0.16 mm/tooth, D(ε) improved from 1.49 µm to 1.61 µm to fall slightly at a feed of 0.16 mm/tooth. When the feed was 0.2 mm/tooth, there was a rapid decline in the standard deviation of the displacement to 1.48 µm, followed by a significant increase of 0.23 µm.

At feed rates of 0.10 mm/tooth and 0.16 mm/tooth, D(ε) decreases because the third and fourth inserts were involved in surface texture generation and the minimum feed rate was higher than the h_{min} parameter.

Two characteristic maximum values of the standard deviation of the relative displacements (2.6 µm and 3.45 µm) can be observed in Table 3. They were registered at cutting speeds of 260 m/min and 360 m/min, respectively. The points correspond to the two most efficient ranges of operation of the CoroMill 490 shoulder mill taking into consideration the displacements in the tool–workpiece system. The smallest displacements were reported at cutting speeds of 200–220 m/min, 280–340 m/min, and 400 m/min. The average standard deviation of the relative displacements for the whole range of cutting speeds was 2 µm.

The experiment consisted of machining 22 samples, 11 at constant feed and 11 at constant cutting speed. Table 3 shows the values of the standard deviation of the relative displacement in the tool–workpiece system for the whole range of the cutting path. Discrepancies in results for the same machining parameters may be the result of material heterogeneity. The difference in the standard deviation of relative displacements for the same machining parameters is 0.39 μm.

The next series of tests aimed to determine the influence of the cutting conditions on the minimum chip thickness (h_{min}) in shoulder milling. Table 4 shows the values of the minimum cut thickness measured at different feeds per tooth and cutting speeds.

Table 4. Variation in the values of h_{min} with feed per tooth, f_z, and cutting speed, v_c.

v_c = 300 m/min a_p = 0.2 mm		f_z = 0.1 mm/tooth a_p = 0.2 mm	
f_z, mm/tooth	h_{min}, μm	v_c, m/min	h_{min}, μm
0.02	0.706	200	0.8
0.04	0.65	220	0.695
0.06	0.276	240	0.674
0.08	1.113	260	0.465
0.1	0.95	280	0.949
0.12	0.657	300	0.42
0.14	1.312	320	0.508
0.16	1.047	340	0.431
0.18	1.313	360	0.277
0.2	0.713	380	1.07
0.22	1.71	400	1.48

Table 4 shows the influence of cutting speed and feed rate on the minimum chip thickness, h_{min}. From the data, it can be seen that in the initial phase, an increase in cutting speed (v_c = 200–260 m/min) contributed to a decrease in h_{min}. The parameter h_{min} increased when the cutting speed exceeded 280 m/min. The downward trend continued until the cutting speed reached 380 m/min. The parameter remained stable until v_c was 260 m/min. When the cutting speed exceeded this value, h_{min} increased and then decreased. The analysis of the influence of the feed per tooth, f_z, on the minimum chip thickness, h_{min}, for X37CrMoV51 steel shows that the parameter h_{min} fluctuates up and down. From the relationship between f_z and h_{min}, it can be concluded that the parameter h_{min} increases as the feed per tooth increases.

Another objective of this study was to look at the influence of the cutting process parameters in shoulder milling on the surface texture of the workpiece. The shoulder milling tests were carried out on an AVIA VMC800 vertical machining center (FOP AVIA, Warsaw, Poland). First, the tests were conducted under dry cutting conditions, i.e., without the use of cutting fluid, under specific cutting conditions assumed for a given series. The measurement results obtained for different cutting parameters were analyzed thoroughly. The data in the form of tables, plots, and maps were used to determine the 2D surface roughness parameters. Some isometric views of the surfaces of the X37CrMoV51 steel specimens are shown in Figure 3. The surface topography was analyzed using a Taylor Hobson: Talysurf CCI—Lite Non-Contact 3D profiler measurement interferometer (Taylor-Hobson Ltd., Leicester, UK).

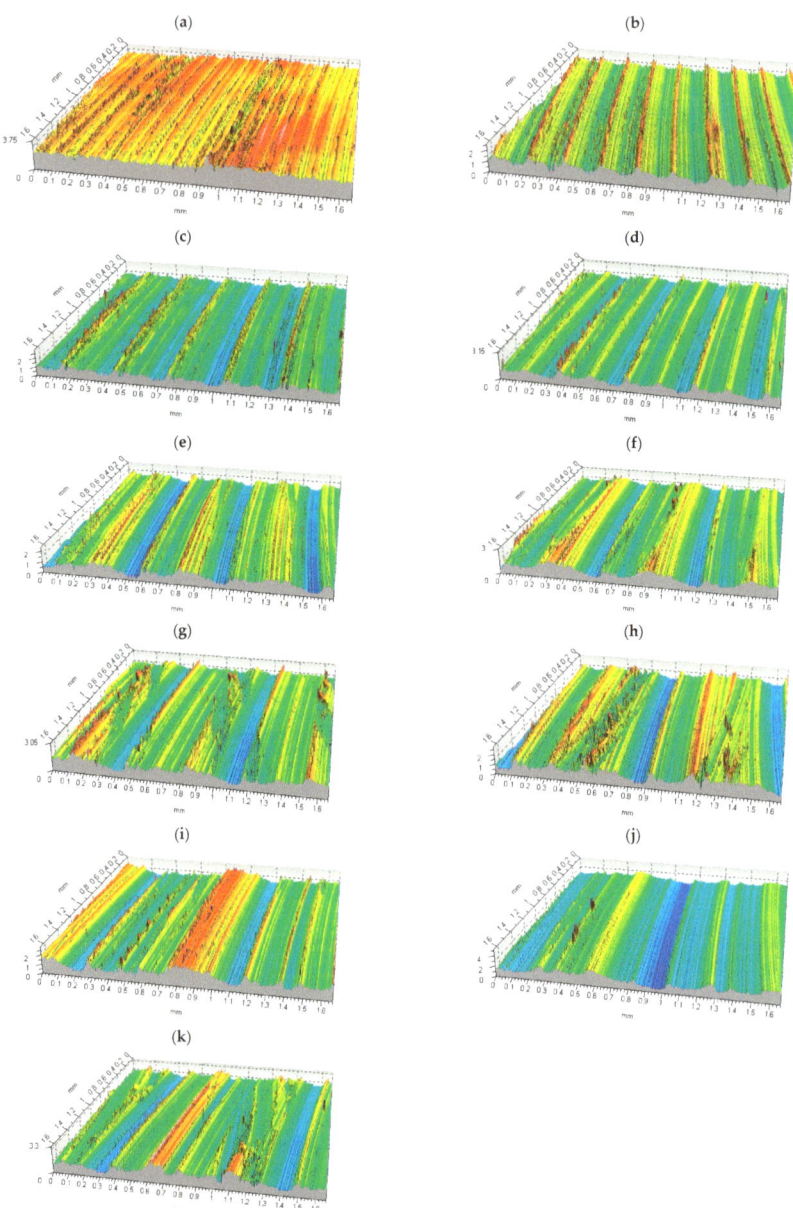

Figure 3. Isometric views of surfaces of the X37CrMoV51 steel specimens machined at different feeds per tooth. (**a**) f_z = 0.02 mm/tooth, f_n = 0.1 mm/rev, f_t = 191 mm/min; (**b**) f_z = 0.04 mm/tooth, f_n = 0.2 mm/rev, f_t = 382 mm/min; (**c**) f_z = 0.06 mm/tooth, f_n = 0.3 mm/rev, f_t = 573 mm/min; (**d**) f_z = 0.08 mm/tooth, f_n = 0.4 mm/rev, f_t = 764 mm/min; (**e**) f_z = 0.1 mm/tooth, f_n = 0.5 mm/rev, f_t = 955 mm/min; (**f**) f_z = 0.12 mm/tooth, f_n = 0.6 mm/rev, f_t = 1146 mm/min; (**g**) f_z = 0.14 mm/tooth, f_n = 0.7 mm/rev, f_t = 1338 mm/min; (**h**) f_z = 0.16 mm/tooth, f_n = 0.8 mm/rev, f_t = 1529 mm/min; (**i**) f_z = 0.18 mm/tooth, f_n = 0.9 mm/rev, f_t = 1720 mm/min; (**j**) f_z = 0.2 mm/tooth, f_n = 0.1 mm/rev, f_t = 1911 mm/min; (**k**) f_z = 0.22 mm/tooth, f_n = 1.1 mm/rev, f_t = 2102 mm/min.

As can be seen from Figure 3, at a feed of 0.02 mm/tooth, random roughness was observed. The surface roughness increased with increasing feed per tooth; the peak-to-valley distance corresponded to the distance the tool traveled during a single spindle rotation (feed per revolution). At f_z = 0.06 mm/tooth, there were no sharp peaks; burrs were observed instead. A further increase in the feed per tooth caused an increase in both the surface roughness and the surface waviness. From the isometric views, it is evident that at f_z = 0.1–0.16 mm/tooth and 0.22 mm/tooth, a two-directional cut was reported because the milling cutter milled forward and backward.

Table 5 shows the 2D surface roughness parameters for shoulder milling. The lowest value of the maximum profile height, Rz, was recorded at a feed rate of 0.1 mm/tooth, while the highest value was recorded at a feed rate of 0.22 mm/tooth. The profile characteristics, Rt and Rv, were similar to Rz. The arithmetic mean height, Ra, (Figure 4) increased steadily with increasing feed from 0.2 µm at 0.02 mm/tooth to 0.321 µm at 0.2 mm/tooth. At a feed of 0.22 mm/tooth, Ra decreased slightly to 0.304 µm. The root mean square deviation, Rq, and Rp, the maximum profile peak height, were reported to be identical to Ra over the whole range of feeds. The mean height of the profile elements, Rc, decreased with increasing feed. After the feed exceeded 0.06 mm/tooth, Rc fluctuated around 0.55 µm and then increased when the feed was 0.1–0.18 mm/tooth. At feed rates higher than 0.18 mm/tooth, the parameter Rc decreased. The mean width of the profile elements, RSm, first increased and then decreased with increasing feed per tooth. At f_z = 0.06 mm/tooth, RSm decreased to 0.03 mm. Then, at feed rates of 0.08–0.12 mm/tooth, RSm increased to 0.048 mm and then decreased to 0.034 mm. A further increase in feed to 0.2 mm/tooth caused the parameter RSm to reach a maximum of 0.058 mm and then drop to 0.04 mm.

Table 5. Variation in the values of the 2D surface roughness parameters in the shoulder milling of X37CrMoV51 steel with the feed per tooth.

Parameter	Unit	Feed per Tooth f_z, mm/tooth										
		0.02	0.04	0.06	0.08	0.1	0.12	0.14	0.16	0.18	0.2	0.22
Rp	µm	0.961	0.981	1.106	1.135	1.032	0.992	1.164	1.202	1.004	1.225	1.376
Rv	µm	1.112	0.759	0.691	0.801	0.661	0.768	0.784	0.891	0.847	0.869	0.970
Rz	µm	2.074	1.740	1.797	1.936	1.693	1.760	1.948	2.093	1.851	2.095	2.346
Rc	µm	0.646	0.620	0.509	0.521	0.530	0.685	0.675	0.774	0.809	0.751	0.717
Rt	µm	2.078	1.748	1.798	1.952	1.696	1.782	2.035	2.114	1.915	2.111	2.352
Ra	µm	0.200	0.221	0.213	0.222	0.227	0.260	0.249	0.279	0.290	0.321	0.304
Rq	µm	0.272	0.279	0.278	0.285	0.286	0.314	0.315	0.348	0.349	0.402	0.384
RSm	mm	0.024	0.032	0.030	0.030	0.036	0.048	0.034	0.043	0.050	0.058	0.040

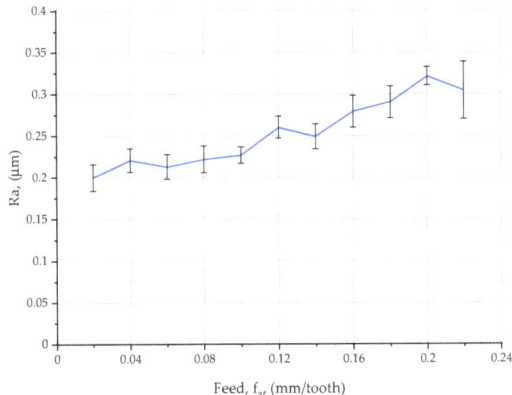

Figure 4. Relationship between the feed per tooth and the surface roughness parameter Ra in the shoulder milling of X37CrMoV51 steel.

The next stage of the study was to develop a model for predicting the surface roughness parameter, Ra, in shoulder milling. First, relationship (1) was formulated as a linear function to link the minimum chip thickness with the feed per tooth. The coefficients used in the formula were selected in such a way that the feed could be expressed in mm. As there were substantial fluctuations in h_{min}, the model took into consideration only the minimum chip thickness, which increased with increasing feed per tooth.

$$h_{min} = 2.6536 \cdot f_z + 0.58 \tag{1}$$

Then, relationship (2) was formulated in the form of a power function to calculate the parameter D(ξ) using the value of the feed per tooth. The coefficient of correlation between the theoretical results determined from relationship (2) and the experimental data obtained through tests was 0.89.

$$D(\xi) = 2.1461 \cdot f_z^{0.1636} \tag{2}$$

The last stage of the study was to formulate relationship (3), which is the model for predicting the surface roughness parameter, Ra, on the basis of the number of inserts engaged, the corner radius, the minimum chip thickness, and the relative displacements. It was assumed that the relative displacements combined with the geometry of the insert (wiper for finishing) contribute to lower roughness as they help remove some of the machining debris.

$$Ra = \frac{f_{za}^2}{8 \cdot r_\varepsilon} + \frac{k \cdot h_{min}}{D(\xi)} \tag{3}$$

where f_{za} is the actual value of the feed per tooth dependent on the actual number of inserts taking part in the cutting process, z_a.

$$f_{za} = \frac{f_z \cdot z}{z_a} \tag{4}$$

k is the material coefficient describing the contribution of the minimum chip thickness to the surface roughness formation. The material coefficient, k, is the ratio of the average peak-to-valley distance on a surface profile to the minimum chip thickness.

$$k = \frac{\left(\frac{\sum_{i=1}^n a_n}{n}\right)}{h_{min}} = \frac{\left(\frac{0.30+0.25+0.29+0.29}{4}\right)}{0.75} = 0.373 \tag{5}$$

The analysis to determine the effect of the selected factors required finding the value of the material coefficient, k. The average value of this coefficient ($k = 0.4$) was calculated using the data from measurements of 11 X37CrMoV51 steel specimens.

The model for predicting the surface roughness parameter, Ra, in shoulder milling was verified using relationship (3) by substituting the actual feed per tooth with relationship (4), the material coefficient, k, with relationship (5), and the parameters h_{min} and D(ξ) with relationships (1) and (2), respectively. The predicted surface roughness Ra was thus dependent on the feed per tooth, f_z, the corner radius, r_ε, the actual number of inserts taking place in the material removal process, and the parameters h_{min} and D(ξ).

Table 6 shows the data used to verify the proposed model. As can be seen, the difference between the measured values and the calculated values is very small.

Figure 5 compares the plots obtained from the experiments with those from the calculations.

The red line in Figure 5 represents the predicted values of the surface roughness parameter Ra in shoulder milling, which were calculated using the proposed model based on relationship (3). The blue line shows the relationship between the actual surface roughness and the feed per tooth in shoulder milling at a cutting speed of 300 m/min and a depth of cut of 0.2 mm. The X37CrMoV51 steel specimens were measured with a Taylor Hobson Talysurf CCI–Lite Non-Contact 3D Profiler (Taylor-Hobson Ltd., Leicester, UK).

Table 6. Data used to verify the proposed model.

X37CrMoV51 Steel	Actual Number of Inserts	Predicted Roughness	Actual Roughness	Standard Deviation
Feed, f_z, mm/tooth	z_{rz}, Number	Ra, μm	Ra, μm	σ, μm
0.02	3	0.224	0.2	0.016
0.04	3	0.217	0.221	0.014
0.06	4	0.219	0.213	0.015
0.08	4	0.225	0.222	0.016
0.1	4	0.232	0.227	0.01
0.12	5	0.239	0.26	0.013
0.14	5	0.248	0.249	0.015
0.16	5	0.257	0.279	0.019
0.18	5	0.266	0.29	0.019
0.2	5	0.276	0.321	0.011
0.22	5	0.285	0.304	0.034
		Coefficient of correlation $R^2 = 0.96$		

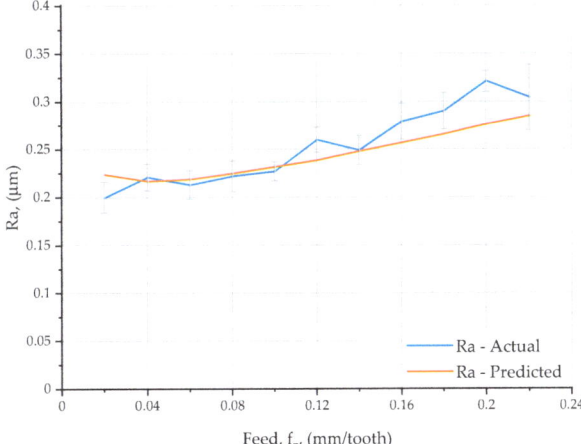

Figure 5. Actual and predicted surface roughness of face milled X37CrMoV51steel versus feed per tooth.

From Table 6 and Figure 5, it is evident that for feed rates ranging from 0.02 to 0.18 mm/tooth, the predicted surface roughness Ra determined using Equation (3) is similar to the actual surface roughness. At a feed rate of 0.2 mm/tooth, however, the predicted values are slightly different from the actual ones. The coefficient of correlation between the theoretical and experimental data was 0.96. This value confirms that the developed model is correct.

4. Conclusions

The major conclusions drawn from the shoulder milling experiments for X37CrMoV51 steel are as follows:

1. The material removal process was largely affected by the feed per tooth and the number of inserts engaged in the cutting process. The analysis of the experimental data reveals that at a feed of 0.04 mm/tooth, the cutting was performed by three out of five inserts. However, when the feed exceeded 0.12 mm/tooth, the material was removed by all the inserts. At feed rates of 0.10 mm/tooth and 0.16 mm/tooth, D(ξ)

decreases because the third and fourth inserts were involved in surface texturing and the minimum feed rate was higher than the value of the h_{min} parameter.

2. The average standard deviation of the displacement in the tool–workpiece system for the whole range of cutting speeds was 2 μm. The smallest relative displacements were recorded at cutting speeds of 200–220 m/min, 280–340 m/min, and 400 m/min, so these parameters guarantee the most stable conditions.
3. Higher cutting speeds resulted in lower minimum chip thicknesses, h_{min}. However, the parameter h_{min} increased with increasing feed. Under variable feed conditions, the lowest and highest values of h_{min} were 0.276 μm and 1.71 μm, respectively. Under variable cutting speed conditions, however, the lowest value of h_{min} was 0.277 μm, while the highest was 1.48 μm.
4. The parameter Ra showed a gradual increase over the entire range of feed rates: from a minimum of 0.2 μm at a feed of 0.02 mm/tooth to a maximum of 0.321 μm at 0.2 mm/tooth. At f_z = 0.1–0.16 mm/tooth and 0.22 mm/tooth, a two-directional cut was reported because the milling cutter milled forward and backward.
5. The most favorable machining conditions, based on the lowest values of h_{min}, standard deviation of the displacement, and the number of inserts involved in the cutting process, is a feed rate of 0.10 mm/tooth.

Author Contributions: Methodology, L.N.; software, S.B.; validation, L.N., M.S. and S.B.; formal analysis, L.N.; investigation, L.N. and M.S.; resources, S.B.; data curation, L.N.; writing—original draft preparation, L.N.; writing—review and editing, M.S.; visualization, S.B.; supervision, S.B. All authors have read and agreed to the published version of the manuscript.

Funding: This research was funded by the National Science Foundation, Poland, grant number 2021/05/X/ST8/01470.

Institutional Review Board Statement: Not applicable.

Informed Consent Statement: Not applicable.

Data Availability Statement: The data presented in this study are available on request from the corresponding author.

Conflicts of Interest: The authors declare no conflict of interest.

References

1. Felhő, C.; Kundrák, J. Effects of Setting Errors (Insert Run-Outs) on Surface Roughness in Face Milling When Using Circular Inserts. *Machines* **2018**, *6*, 14. [CrossRef]
2. Weinert, K.; Surmann, T.; Enk, D.; Webber, O. The effect of runout on the milling tool vibration and surface quality. *Prod. Eng. Res. Devel.* **2007**, *1*, 265–270. [CrossRef]
3. Chen, H.-Q.; Wang, Q.-H. Modelling and simulation of surface topography machined by peripheral milling considering tool radial runout and axial drift. *Proc. Inst. Mech. Eng. Part B J. Eng. Manuf.* **2019**, *233*, 2227–2240. [CrossRef]
4. Skrzyniarz, M.; Nowakowski, L.; Miko, E.; Borkowski, K. Influence of Relative Displacement on Surface Roughness in Longitudinal Turning of X37CrMoV5-1 Steel. *Materials* **2021**, *14*, 1317. [CrossRef] [PubMed]
5. Yan, B.; Zhu, L.; Liu, C. Prediction model of peripheral milling surface geometry considering cutting force and vibration. *Int. J. Adv. Manuf. Technol.* **2020**, *110*, 1429–1443. [CrossRef]
6. Wu, T.Y.; Lei, K.W. Prediction of surface roughness in milling process using vibration signal analysis and artificial neural network. *Int. J. Adv. Manuf. Technol.* **2019**, *102*, 305–314. [CrossRef]
7. Zain, A.M.; Haron, H.; Sharif, S. Prediction of surface roughness in the end milling machining using Artificial Neural Network. *Expert Syst. Appl.* **2010**, *37*, 1755–1768. [CrossRef]
8. Sizemore, N.E.; Nogueira, M.L.; Greis, N.P.; Davies, M.A. Application of Machine Learning to the Prediction of Surface Roughness in Diamond Machining. *Procedia Manuf.* **2020**, *48*, 1029–1040. [CrossRef]
9. Manjunath, K.; Tewary, S.; Khatri, N. Surface roughness prediction in milling using long-short term memory modelling. *Mater. Today Proc.* **2022**, *64*, 1300–1304. [CrossRef]
10. Nowakowski, L.; Bartoszuk, M.; Skrzyniarz, M.; Blasiak, S.; Vasileva, D. Influence of the Milling Conditions of Aluminium Alloy 2017A on the Surface Roughness. *Materials* **2022**, *15*, 3626. [CrossRef] [PubMed]
11. Kao, Y.C.; Chen, S.J.; Vi, T.K.; Feng, G.H.; Tsai, S.Y. Study of milling machining roughness prediction based on cutting force. *IOP Conf. Ser. Mater. Sci. Eng.* **2021**, *1009*, 12027. [CrossRef]

12. de Souza, A.F.; Machado, A.; Beckert, S.F.; Diniz, A.E. Evaluating the Roughness According to the Tool Path Strategy When Milling Free Form Surfaces for Mold Application. *Procedia CIRP* **2014**, *14*, 188–193. [CrossRef]
13. Uzun, M.; Usca, Ü.A.; Kuntoğlu, M.; Gupta, M.K. Influence of tool path strategies on machining time, tool wear, and surface roughness during milling of AISI X210Cr12 steel. *Int. J. Adv. Manuf. Technol.* **2022**, *119*, 2709–2720. [CrossRef]
14. Son, S.M.; Lim, H.S.; Ahn, J.H. Effects of the friction coefficient on the minimum cutting thickness in micro cutting. *Int. J. Mach. Tools Manuf.* **2005**, *45*, 529–535. [CrossRef]
15. Rezaei, H.; Sadeghi, M.H.; Budak, E. Determination of minimum uncut chip thickness under various machining conditions during micro-milling of Ti-6Al-4V. *Int. J. Adv. Manuf. Technol.* **2018**, *95*, 1617–1634. [CrossRef]
16. Stal Narzędziowa do Pracy na Gorąco: Oferta | Techmet-Opoczno. Available online: http://www.stalnarzedziowa.pl/stal-narzedziowa/do-pracy-na-goraco (accessed on 4 December 2023).

Disclaimer/Publisher's Note: The statements, opinions and data contained in all publications are solely those of the individual author(s) and contributor(s) and not of MDPI and/or the editor(s). MDPI and/or the editor(s) disclaim responsibility for any injury to people or property resulting from any ideas, methods, instructions or products referred to in the content.

Article

Identification of the Production of Small Holes and Threads Using Progressive Technologies in Austenite Stainless Steel 1.4301

Dana Stančeková [1,*], Filip Turian [1], Michal Šajgalík [1], Mário Drbúl [1], Nataša Náprstková [2], Anna Rudawská [3] and Miroslav Špiriak [1]

1. Faculty of Mechanical Engineering, University of Zilina, Univerzitná 8215/1, 01026 Žilina, Slovakia; filip_turian@elmaxzilina.sk (F.T.); michal.sajgalik@fstroj.uniza.sk (M.Š.); mario.drbul@fstroj.uniza.sk (M.D.); spirimro@schaeffler.com (M.Š.)
2. Faculty of Mechanical Engineering, University J. E. Purkyně in Ústí nad Labem, Pasteurova 3544/1, 400 96 Ústí nad Labem, Czech Republic; natasa.naprstkova@ujep.cz
3. Faculty of Mechanical Engineering, Lublin University of Technology, 20-618 Lublin, Poland; a.rudawska@pollub.pl
* Correspondence: dana.stancekova@fstroj.uniza.sk

Citation: Stančeková, D.; Turian, F.; Šajgalík, M.; Drbúl, M.; Náprstková, N.; Rudawská, A.; Špiriak, M. Identification of the Production of Small Holes and Threads Using Progressive Technologies in Austenite Stainless Steel 1.4301. *Materials* 2023, 16, 6538. https://doi.org/10.3390/ma16196538

Academic Editors: Przemysław Podulka and Andrea Di Schino

Received: 4 August 2023
Revised: 21 September 2023
Accepted: 29 September 2023
Published: 3 October 2023

Copyright: © 2023 by the authors. Licensee MDPI, Basel, Switzerland. This article is an open access article distributed under the terms and conditions of the Creative Commons Attribution (CC BY) license (https://creativecommons.org/licenses/by/4.0/).

Abstract: This article focuses on the technologies used by a manufacturing company to produce threads in chrome–nickel steel 1.4301 at specific sheet thicknesses. To enhance production quality, two specific technologies were chosen for hole formation, considering the requirements of the company. Both conventional drilling and nonconventional laser cutting methods were evaluated as potential techniques for hole production. Conventional thread-cutting technology and progressive forming technology were employed to create metric internal threads. The aim of integrating these diverse technologies is to identify the optimal solution for a specific sheet thickness in order to prevent the occurrence of defective threads that could not fulfil the intended purpose. The evaluation of the threads and holes relies on the examination of surface characteristics, such as the quality of the surface, as well as the lack of any signs of damage, cracks, or burrs. Furthermore, residual stresses in the surface layer were monitored because these stresses have the potential to cause cracking. Additionally, extensive monitoring was performed to guarantee that the form and size of the manufactured threads were correct to ensure smooth assembly and optimal functionality.

Keywords: thread; hole; cutting; laser; quality; shape accuracy

1. Introduction

As numerous authors in the field of engineering have noted, designers and production engineers frequently face the challenge of producing highly durable connections capable of facilitating the assembly and disassembly of components made of various materials. To fulfil this task, fasteners and internal threads represent the prevailing choices, as references [1,2] indicate. Threaded connections are one of the most prevalent solutions for assembling mechanical components because they result in assemblies with high strength and stiffness, as substantiated by references [1,3]. While screw connections find applications in technical structures, it is worth emphasizing that a substantial proportion of mechanical, aerospace, medical, and dental products incorporate at least one threaded component, as documented in references [4–6]. This underlines the versatility and practicality of threaded connections, further enhanced by their notably rapid component assembly and disassembly, as denoted in reference [7].

Historically, the use of nuts was the most common approach for ensuring the stability of mechanical components. However, in other circumstances, the direct integration of internal threads onto the joint component is used as an alternative method of fastening.

According to the data reported in reference [5], the evaluation of mechanical stresses experienced by machine components demonstrates that internal threads are continuously subjected to tensile forces. According to Liu [8], the continued use of threads is bound to cause gradual deterioration, which may grow in intensity and demonstrate a correlation with the overall quality of the thread connections. It is crucial to note that the failure of fastening mechanisms, both permanent and non-permanent, can have significant consequences for the entire operation of mechanical systems, as cited in reference [9].

In their work, Val A.G.D. and colleagues [10] emphasise that the process of thread cutting is a multifaceted operation focused on the creation of internal openings for screws. This complexity stems from the pivotal need for precise synchronization between the rotational motion of the tap and the vertical movement, as well as for substantial contact between the tool and the workpiece. This synchronization and contact are especially critical when threading is performed under demanding cutting conditions. The production of internal threads typically involves one of three methods: cutting, cold forming, or milling. Thread cutting and thread milling represent conventional metal cutting techniques, whereas the cold forming of threads provides an alternative method that eliminates chip generation [11].

During the process of thread cutting, the cutting edges of the tap progressively shape the internal thread, which is a continual process. An integrated thread is produced by a single, complete thread-cutting process [12]. The requirements of thread cutting are high, primarily due to the critical need for precise synchronization between the rotational and vertical movements [1].

Threads formed through the thread-forming process are the result of plastic deformation and material displacement. In their study, Masmoudi, N. et al. [13] expound on the numerous advantages offered by the cold-forming thread process in comparison to conventional cutting techniques. However, it is essential to consider specific process characteristics and material properties to ensure the consistent and reliable production of internal threads. This aspect was explored by De Oliveira et al. [14], who delved into the impact of tooling and process parameters on the quality of internally formed threads. They observed that the optimization of thread quality can be achieved by selecting appropriate parameters such as forming speed, tap surface finish, and tap geometry. Moreover, in a separate investigation, Maciel et al. [15] ascertained the optimal revolutions for both forming and cutting external threads in a titanium alloy, relying on hardness testing and thread profile control.

In engineering, the setup of milling operations can be quite intricate due to the multitude of cutting parameters and tool geometrical aspects that demand consideration. Thread milling, when viewed from a geometric standpoint, represents a multifaceted 3D machining configuration, incorporating elements such as the toolpath, tool geometry, and the intricacies of the cutting process [16]. This particular technology carries a greater financial burden and necessitates access to suitable infrastructure and a high level of expertise for its effective application.

Over the past two decades, preliminary studies have focused on evaluating thread profile quality based on geometric errors [5]. In their work, Dogra and colleagues [17] delved into the evaluation of deviations between the tap's feed rate and the thread pitch. Meanwhile, Freitas et al. [18] evaluated thread profile quality, employing the fundamental profile specifications of the ISO metric thread as outlined in ISO-68-1 and ISO 68-2, particularly during the threading of test samples.

A crucial factor in achieving a high-quality thread is a proper machined hole. The selection of hole manufacturing technology must consider not only its diameter but also factors encompassing material properties, depth, and the requisite levels of precision and functionality. Vigilant oversight of the resulting geometric and dimensional accuracy, alongside scrutiny of surface quality (including the identification of potential cracks or burrs) [19,20] proves indispensable, for it is impossible to produce a top-tier thread within an inadequately machined hole.

A specialized threaded joint involves the insertion of a screw into a thread directly within the thin-walled sheet of a manufactured structure. Nonetheless, during the manufacturing process, deviations in terms of thread shape and dimensions, or even damage, may manifest, ultimately resulting in suboptimal connections. Therefore, the appropriate selection of technology combinations for hole and thread production assumes vital importance. This forms the core focus of the experimental validations carried out within this study, as there is limited availability regarding information on this particular matter.

Analysis of the Current Situation in Practice

This study examines the prevalent practices at ELMAX ŽILINA Slovakia., a company primarily employed in the production of electrical distribution boards and various sheet metal products, including those intended for use in the food industry. In the context of food sector products, it is common practice to employ corrosion-resistant steels, with a prevalent preference for austenitic chrome–nickel steel 1.4301, which is available in a range of thicknesses.

In the mentioned company, an analysis was carried out to solve a problem with making internal metric threads in semi-finished sheet metal with different thicknesses made of corrosion-resistant steel, as shown in Figure 1. While manufacturing the most frequently employed metric threads, evident deformations were observed after the production process, as depicted in Figure 2. These deformations emerge as cracks, rendering the threads partially non-functional and preventing them from meeting the standard-defined parameters.

Figure 1. Electrical distribution cabinets monitored internal threads.

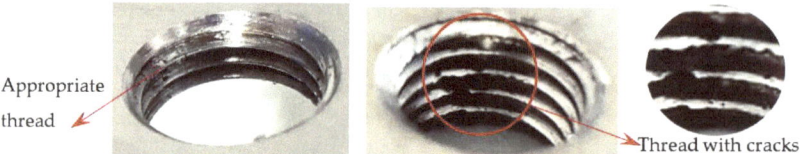

Figure 2. Internal thread M10 on the left without cracks in 5 mm thick sheet metal, and, on the right, with cracks in 10 mm thick sheet metal.

The key issue lies in the frequent occurrence of potential thread damage, which poses a significant risk to both individual components and, in certain cases, the entire product itself. Frequently, these defective products are beyond repair, necessitating their removal from the manufacturing process. The need to remanufacture these products naturally results in higher production costs, longer production times, and a reduction in the actual profit margin of the product.

In the initial phase of our research, we discovered that holes created through laser machining displayed distinct defects (see Figure 3), leading to imprecise and irregular hole shapes. Furthermore, visible deformations within these holes were observed, posing a potential risk of imperfections in the internal threads manufactured within them.

  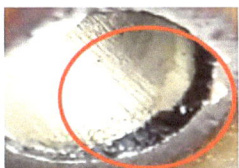

| Deformation of the functional surface of the hole with the remainder of the unseparated melt. | Deformation of the functional surface of the hole with the melted area and with the remainder of the unseparated melt. | Deformation of the functional surface of the hole with loss of material on the lower edge. |

Figure 3. Holes made using LASER.

The rationale behind employing laser machining for hole production in the aforementioned company, despite it encountering minor deformations, lies in its capacity to encompass all the outlines of the product, including the holes, in a single operation. This eliminates the need for additional drilling operations.

However, it is crucial to find the point where these deformations are still within acceptable limits and can be used to make high-quality, crack-free threads inside the holes that meet geometric standards.

When choosing a suitable advanced technology, it becomes especially crucial to analyze the problem of cracks and deformations in metric threads. These issues significantly diminish thread functionality, result in non-compliance with standards, and frequently render the product unusable, ultimately leading to its removal from the manufacturing process.

2. Materials and Methods

The mentioned company works with material 1.4301, which belongs to the category of materials characterized by specific properties acquired during the machining process. The chemical composition is shown in Table 1.

Table 1. Chemical composition of material 1.4301 according to EN 10088.

Chemical Composition [%]						
C	Cr	Mn	Ni	P	S	Si
0.07	17–20	2	9–11.5	0.045	0.03	1
Mechanical Properties						
Yield strength R_e [MPa]		Tensile strength R_m [MPa]		Hardness [HB]		Ductility A [%]
min 230		540–750		160–210		45

Steel 1.4301, known for its exceptional corrosion resistance, demonstrates resilience against water, steam, airborne moisture, and even edible acids. However, during the machining process, chromium–nickel steels themselves tend to undergo work hardening. This effect is predominantly undesirable and problematic, as it diminishes machinability and adversely impacts the final quality of the product [21].

The focus of this study is to examine the effects of modifications in manufacturing methods on the characteristics of threads manufactured from a particular material. The selection of utilized technologies is customized to meet the specific requirements of the company, ELMAX, inc. As a result, the choice of employed technologies is tailored to suit their requests. Threads were manufactured on sheets with varying thicknesses of 3, 5, 8, and 10 mm. Two distinct technologies were employed to create holes: progressive laser cutting technology and conventional drilling technology. Subsequently, the holes were utilized to create threads using two different methodologies: standard cutting technology and progressive forming technology.

The experimental measurements were set up following the specifications provided in Table 2. This encompassed the initial stages of preparing holes for threads and then inserting the threads within these holes. All necessary arrangements were made to facilitate subsequent comparisons and measurements.

Table 2. Sample production process table.

	Product	Production Process	Machine	Tool
Exp. 1	Hole	Laser	TRULASER 3030	Laser Beam
	Thread	Cutting	Thread Cutter MOSQUITO 300–600	HSS-E-PM 6H TiN DIN 371
Exp. 2	Hole	Drilling	HURCO VMX30t	HSS TiN DIN 338
	Thread	Cutting	Thread Cutter MOSQUITO 300–600	HSS-E-PM 6H TiN DIN 371
Exp. 3	Hole	Laser	TRULASER 3030	Laser Beam
	Thread	Forming	Thread Cutter MOSQUITO 300–600	HSS-E 6HX TiN DIN 2174
Exp. 4	Hole	Drilling	HURCO VMX30t	HSS TiN DIN 338
	Thread	Forming	Thread Cutter MOSQUITO 300–600	HSS-E 6HX TiN DIN 2174

Samples, as shown in Figure 4, were prepared using a TruLaser 3030 machine, along with pre-cut openings of varying diameters, according to Table 3. In the course of the testing process, threads matching the dimensions commonly employed by the company were subsequently generated within these pre-prepared openings.

Figure 4. Laser production of one of the samples.

Table 3. Laser production of one of the samples.

Thread Size	Hole Diameter for Thread Made by Cutting [mm]	Hole Diameter for Thread Made by Forming [mm]
M4	ø 3.5	ø 3.7
M5	ø 4.2	ø 4.6
M6	ø 5.0	ø 5.5
M8	ø 7.0	ø 7.4
M10	ø 8.5	ø 9.3

For tools, we employed DIN 338 drill bits made from high-speed steel, featuring a TiN coating and an angle of 118°. According to industry standards, these drill bits are suitable for drilling this specific type of steel. Cooling was facilitated through the use of Zubor 65 H Extra cooling emulsion, sourced from the German company ZEELER + GMELIN. The cutting parameters adhered closely to the recommendations of the manufacturer, maintaining a cutting speed (v_c) of 14 m.min^{-1}, while the feed rate varied in accordance with the diameter of the drill bit.

Following the production of the holes, the production of metric threads through cutting was initiated. The MOSQUITO 300–600 threading machine was employed for this purpose. We utilized specially designed machine taps, namely DIN 371 HSS-E-PM-6H TiN taps. These taps are made from high-speed steel, enriched with a 5% cobalt alloy, and manufactured as sintered carbide with a coating, as illustrated in Figure 5a. For thread forming, DIN 2174 HSS-E-6HX TiN taps were chosen. These taps are composed of high-speed steel with a 5% cobalt addition and feature a coating, with one of them depicted in Figure 5b. To optimise the phase of the cutting efficiency, prevent potential tap binding, and reduce wear, we applied ARIANA cutting paste for threading.

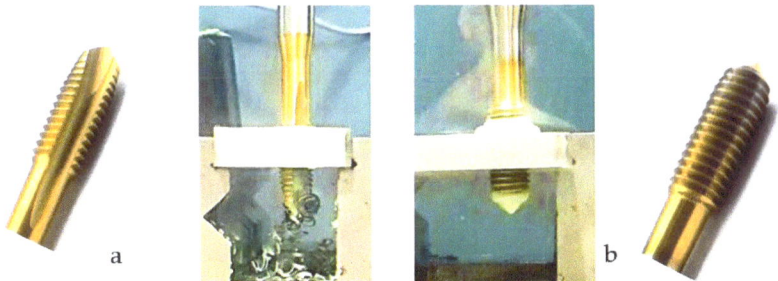

Figure 5. Manufacturing the M10 thread in an 8 mm thick sheet metal, (**a**)—by cutting and (**b**)—by forming.

Certain holes and, consequently, their corresponding threads, could not be produced due to technological limitations. Moreover, there were cases in which holes formed with deviations in their shape, resulting in negative consequences for the subsequent process of producing threads.

3. Results and Discussion

3.1. Visual Comparison of Functional Hole and Thread Surfaces

We conducted a visual inspection and quality evaluation for all the holes and threads. The functional surfaces of the holes are presented in Table 4, whereas those of the threads are detailed in Tables 5–8.

Upon initial examination, several threads manufactured internally displayed evident visible signs of damage and lacked the appropriate thread profile. These thread profiles suffered from deformation, cracking, and insufficiency. Notably, these issues were most pronounced in threads produced using thread-forming technology within holes produced using laser cutting, regardless of sample thickness. This predicament arises from the thermally affected zone on the functional surfaces of the holes, a byproduct of the laser-cutting hole production process. Consequently, it can be deduced that employing the forming process within holes created using laser cutting is an unfavourable choice, given the considerably larger-than-expected thermally affected zone. Furthermore, imperfections in the thread profile were only apparent in threads produced by cutting within holes produced using laser cutting, specifically in samples with thicknesses of 8 mm and 10 mm and for M8 and M10 threads. This inconsistency can be attributed to the suboptimal geometry of holes created by laser cutting in thicker samples and those with ø 8.5 and ø 7 diameters.

Table 4. Functional hole surfaces in the cut for various sheet metal thicknesses and hole diameters.

			Hole for Thread				
			M4	M5	M6	M8	M10
Sheet thickness	3 mm	Laser					
		Drilling					
	5 mm	Laser					
		Drilling					
	8 mm	Laser					
		Drilling					
	10 mm	Laser					
		Drilling					
	Technology						

Table 5. Functional thread surfaces in the cut for a sheet metal thickness of 3 mm.

					Thread				
					M4	M5	M6	M8	M10
Sheet thickness 3 mm	Technology of hole production	Laser	The technology of thread production	Cut					
				Formed					
		Drilling		Cut					
				Formed					

Table 6. Functional thread surfaces in the cut for a sheet metal thickness of 5 mm.

Sheet thickness 5 mm	Technology of hole production		The technology of thread production	Thread				
				M4	M5	M6	M8	M10
		Laser	Cut					
			Formed					
		Drilling	Cut					
			Formed					

Table 7. Functional thread surfaces in the cut for a sheet metal thickness of 8 mm.

Sheet thickness 8 mm	Technology of hole production		The technology of thread production	Thread				
				M4	M5	M6	M8	M10
		Laser	Cut					
			Formed					
		Drilling	Cut					
			Formed					

Table 8. Functional thread surfaces in the cut for a sheet metal thickness of 10 mm.

Sheet thickness 10 mm	Technology of hole production		The technology of thread production	Thread				
				M4	M5	M6	M8	M10
		Laser	Cut					
			Formed					
		Drilling	Cut					
			Formed					

3.2. Measurement of Residual Stresses

The process of measuring residual stresses through X-ray diffraction can be explained with the Proto iXRD device (Figure 6). This equipment enables the measurement of residual stresses in all crystalline materials and, due to its design, is suitable for use both in laboratory settings and in real manufacturing processes [22,23]. Residual stress evaluation for the produced holes and threads was carried out utilizing X-ray diffraction on the Proto iXRD measurement device.

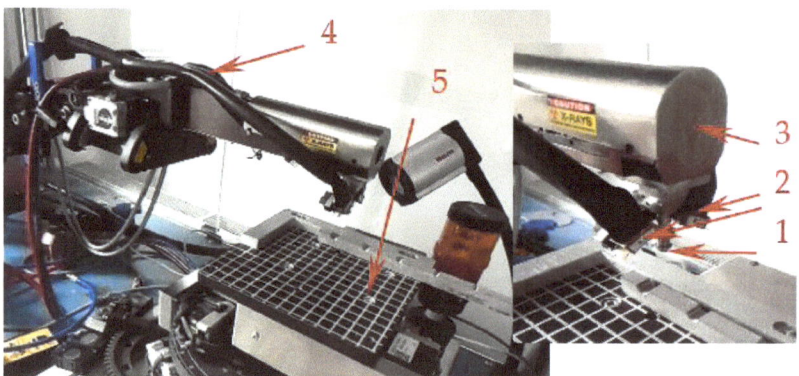

Figure 6. Device for measuring residual stresses Proto iXRD (laboratory setup): 1—collimator directing the X-ray beam; 2—a pair of detectors capturing the diffraction cone; 3—X-ray tube with Cr target; 4—Cobralink ® flexible measuring arm; 5—adjustable and rotating table.

The same measurement parameters were used when measuring residual stresses in individual holes and threads: a fixed number of measurements (angles) at one position, ±30° with 15 angular positions, X-ray tube: Mn_K(α), Collimator: 1mm diameter, filter: Cr, voltage: 20 KV, current: 4mA, and beta oscillation: 3°.

Normal and shear stresses were measured in all holes and threads. The recorded values were used to calculate nominal (average) values, and graphical representations were generated for better visualization, as depicted in Figures 7–12.

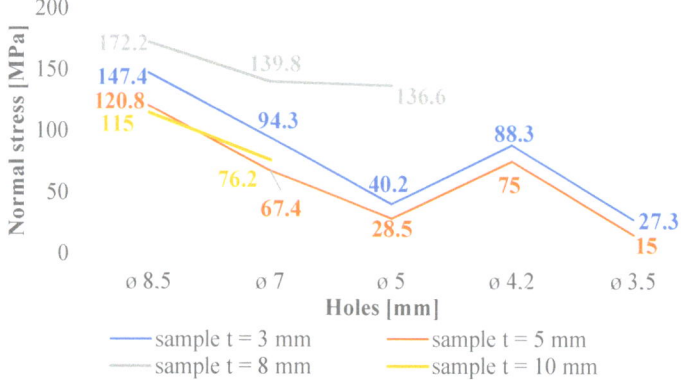

Figure 7. Normal stress in subsurface layers of holes produced using laser cutting.

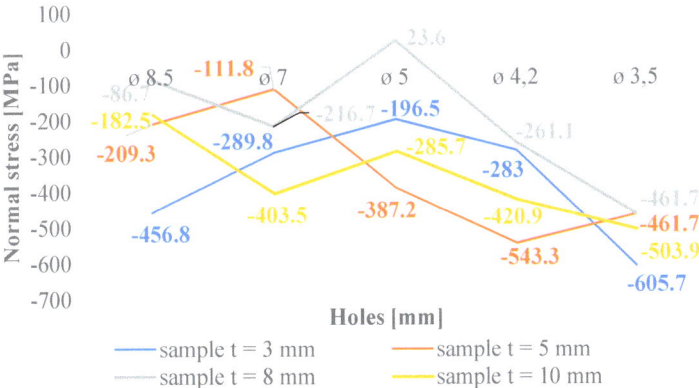

Figure 8. Normal stress in subsurface layers of holes produced by drilling.

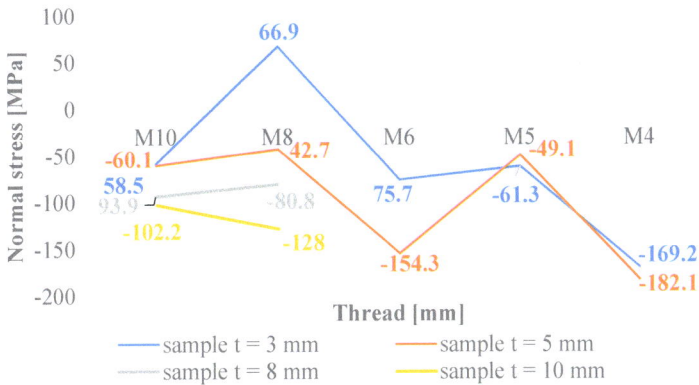

Figure 9. Normal stress in subsurface layers of threads produced by cutting in the holes produced using laser cutting.

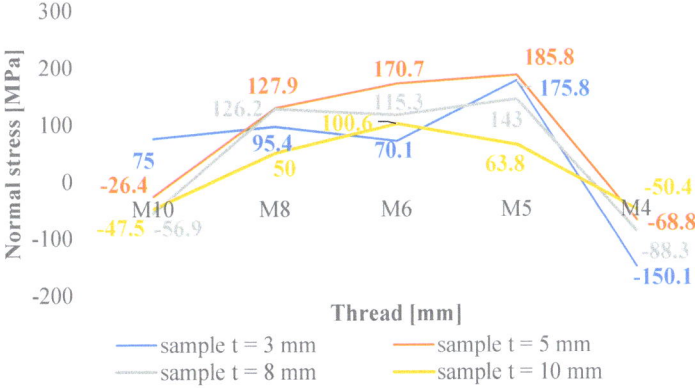

Figure 10. Normal stress in subsurface layers of threads produced by cutting in the holes produced by drilling.

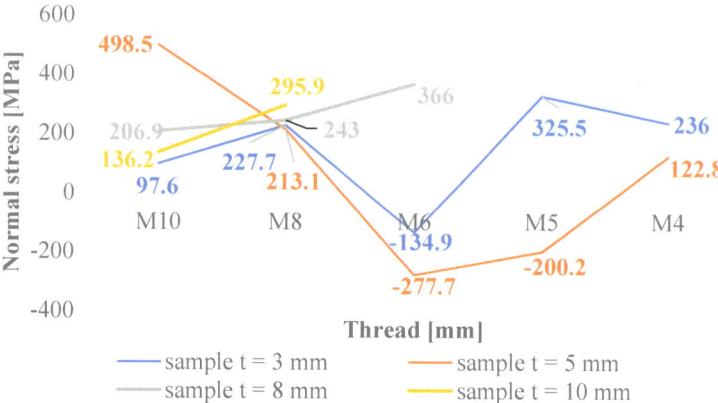

Figure 11. Normal stress in subsurface layers of threads produced by forming in the holes produced using laser cutting.

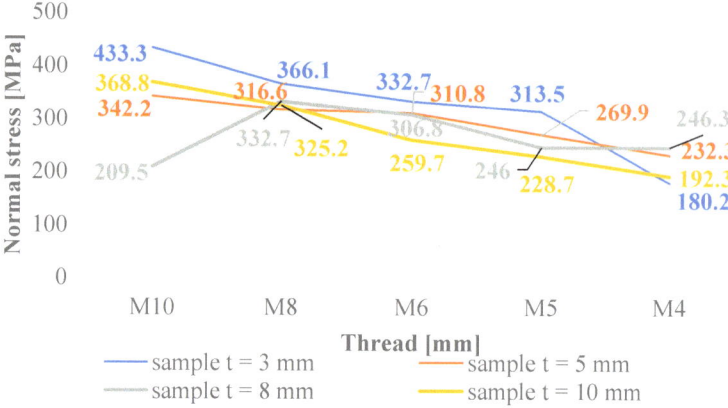

Figure 12. Normal stress in subsurface layers of threads produced by forming in the holes produced by drilling.

The measured values in the holes created using laser cutting (Figure 7) show low values above the zero axis, confirming minimal normal tensile stresses close to the material's equilibrium state. Figure 8 reveals that holes created by drilling are below the zero axis, indicating significant normal compressive stresses, implying material strengthening in subsurface layers.

Figure 9 provides an opportunity for observation, revealing that the production of threads through the process of cutting into holes generated using laser cutting typically results in minor normal compressive stresses. However, it is worth noting that an exception is observed in the case of the M8 thread. On the other hand, it can be observed that threads formed within drilled holes (as depicted in Figure 10) experience normal compressive stresses in the case of M4 and M10 threads. While the M8, M6, and M5 threads demonstrate normal tensile stresses.

One could argue that the production of threads by cutting into pre-drilled holes induces larger normal compressive stresses in these holes, thereby creating a strengthened layer. However, when threads are cut through this layer, it loses its integrity. This means that most threads have opposing normal tensile stresses that are low and close to the equilibrium state of the material.

For threads made by cutting into holes made using laser cutting, where there were initially small normal tensile stresses, the process of cutting the threads changed these values so that most of the stresses in the threads were normal compressive stresses. Nonetheless, these stress levels are closely aligned with the equilibrium state of the material, thereby avoiding substantial modifications in the subsurface layers and maintaining the material free from extreme internal stresses.

In Figure 11, the graph reveals that the majority of measured values indicate the presence of normal compressive stresses in the subsurface layers of threads produced by forming holes created by laser cutting. These stress values exhibit some variability, which can be attributed to the thermally affected zone generated during the laser-cutting process. Conversely, when examining the measured values of threads produced by forming holes created by drilling, as depicted in Figure 12, we observe that these values are closely grouped. This clustering unequivocally illustrates that only normal compressive stresses were induced in the subsurface layers of these threads, with a less pronounced thermal influence during hole production.

3.3. Measurement of Thread Parameters

The thread profiles of all manufactured threads were assessed using the CONTOURECORD 1700SD3 measuring device, employing a probing tip with a 25 μm radius for the measurements. The instrument recorded the profiles of all threads, allowing for the evaluation of two fundamental thread parameters. The values of these parameters, following the STN ISO 262:2000 (01 4010) standard, are detailed in Table 9.

Table 9. Table of nominal values of two measured parameters on manufactured threads.

Thread size [mm]	M4	M5	M6	M8	M10
Spacing P [mm]	0.7	0.8	1	1.25	1.5
Thread profile angle α [°]			60		

Due to the large volume of data generated for evaluating measurable thread parameters, we have chosen to present only selected measurements and summarize the results in a table for better comprehension.

3.4. Results of Measurements of Thread Parameters Produced by Cutting

Figures 13 and 14 provide a basis for comparing two sets of measurements for internally threaded holes with identical dimensions. These holes were created by cutting into the same sample thickness but using different hole production technologies. This comparison leads to the conclusion that the threads produced by cutting into holes created by laser cutting (Figure 13) are deemed unsatisfactory. This is evident because the two thread profiles do not conform to the required specifications, and it was not feasible to measure the parameters on these profiles. Additionally, a visual inspection reveals that the thread heights, and, consequently, the major diameter of the thread, are smaller when compared to threads produced by cutting into holes created by drilling (Figure 14). Hence, despite the measured parameter values falling within acceptable limits, the threads produced by cutting into holes created using laser cutting are considered unsatisfactory.

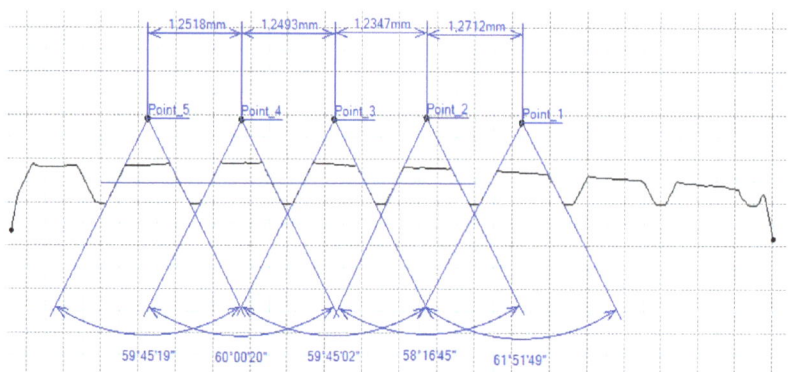

Figure 13. The M8 thread is produced by cutting into a hole made by a laser in sheet metal with a thickness of 10 mm.

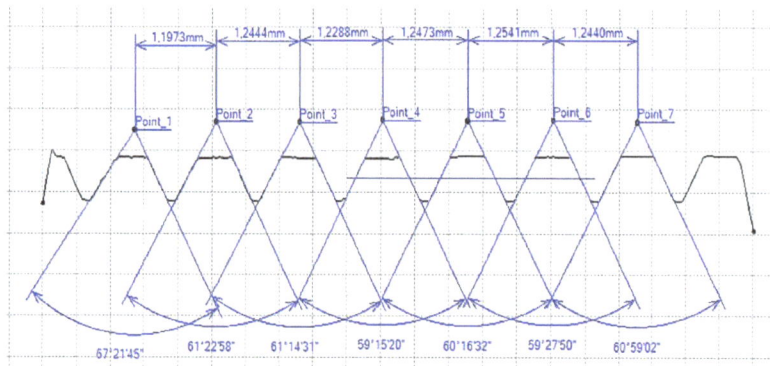

Figure 14. The M8 thread is produced by cutting into a hole made by drilling in sheet metal with a thickness of 10 mm.

3.5. Results of Measurements of the Parameters of Threads Produced by Forming

In Figures 15 and 16, we can compare two sets of contour measurements for internally threaded holes with identical dimensions. These threads were created within the same 10 mm thick sample but using different hole production technologies. This comparison clearly shows that the shape and parameters of the threads produced by forming within holes created using laser cutting (Figure 15) are unsatisfactory, unlike the threads produced by forming within holes created by drilling (Figure 16).

Measurements revealed that all internally threaded holes made by forming within holes made by laser cutting were unsatisfactory, except for one case. Upon evaluation, we identified deficiencies such as imperfect shapes with deformations or unsatisfactory parameters. Hence, it is evident that manufacturing using these technologies for both the hole and thread lacks precision, and it is not feasible to produce high-quality threads in this manner. Consequently, this approach cannot be employed in practical applications at all.

Using data from the contour measuring device, we assessed all threads based solely on their profile contours, specifically examining their shape, potential cracks, deformations, and the completeness of the thread profile. Threads with noticeable shape imperfections were categorized as unsatisfactory, while those with acceptable shapes were considered satisfactory, as indicated in Table 10.

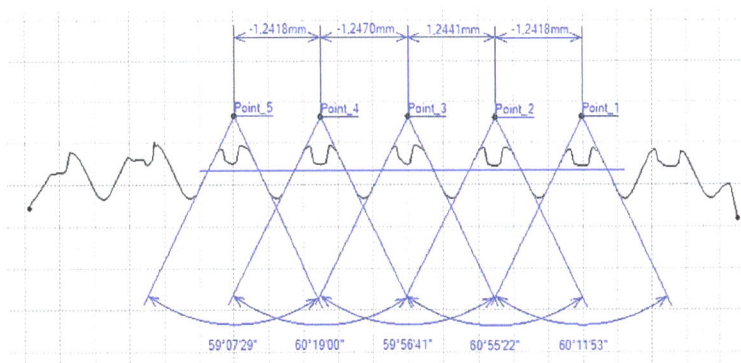

Figure 15. The M8 thread is produced by forming into hole made by a laser, a sheet metal thickness of 10 mm.

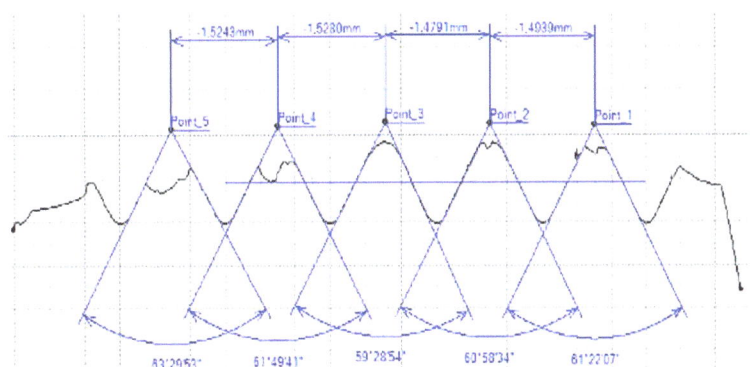

Figure 16. The M8 thread is produced by forming into hole made by drilling, a sheet metal thickness of 10 mm.

Table 10. Evaluation of the measured outputs of the thread profile.

Sheet Thickness	Production Technology		Thread				
	Hole	Thread	M4	M5	M6	M8	M10
3 mm	Laser	Cut	✔	✔	✔	✔	✔
		Formed	✘	✘	✘	✘	✘
	Drilling	Cut	✔	✔	✔	✔	✔
		Formed	✘	✘	✘	✘	✘
5 mm	Laser	Cut	✔	✔	✔	✔	✔
		Formed	✔	✘	✘	✘	✘
	Drilling	Cut	✔	✔	✔	✔	✔
		Formed	✘	✘	✘	✘	✘
8 mm	Laser	Cut	-	-	-	✘	✘
		Formed	-	-	✘	✘	✘
	Drilling	Cut	✔	✔	✔	✔	✔
		Formed	✔	✔	✔	✔	✔
10 mm	Laser	Cut	-	-	-	✘	✘
		Formed	-	-	-	✘	✘
	Drilling	Cut	✔	✔	✔	✔	✔
		Formed	✔	✔	✔	✔	✔

4. Conclusions and Benefits

All the threads that were produced and measured, and subsequently assessed for thread quality, residual stress levels within the threads, thread profile shape, and measured parameters (thread profile angle α and thread pitch P), were processed following the guidelines outlined in Table 11.

Table 11. Comprehensive evaluation of form-fitting threads.

Thread Marking	Thread M10 into Drilled Hole, t = 10 mm		Thread M8 into Drilled Hole, t = 10 mm	
	Řezaný	Tvárněný	Řezaný	Tvárněný
Normal stresses in the hole [MPa]	−182.5 ± 19.4	-	−403.5 ± 24.1	-
Shear stress in the hole [MPa]	−32.9 ÷ 10	-	−1.3 ± 12.4	-
Normal stresses in the thread [MPa]	−47.5 ± 35.8	368.5 ± 106.5	50 ± 30.7	332.7 ± 37.1
Shear stresses in the thread [MPa]	−11 ± 17.2	−40.8 ± 51.2	−17.6 ± 14.7	−11 ± 17.8
Arithmetic diameter of thread profile angles α [°]	61° 16	59° 30′	61° 26′	60° 7′
Arithmetic diameter of thread pitches P [mm]	1.48	1.49	1.24	1.24
Evaluation	good	very good	good	very good

Subsequently, individual threads were evaluated as either good ✔, very good ✔✔, or poor ✘ (see Table 12).

Table 12. Evaluation of the quality of individual produced threads.

	M4	M5	M6	M8	M10
Internal threads produced by cutting into holes made by drilling					
t = 3 mm	✔	✔	✔	✘	✘
t = 5 mm	✔✔	✔✔	✔	✘	✘
t = 8 mm	✔	✔	✔✔	✔	✔
t = 10 mm	✔	✔✔	✔✔	✔	✔
Internal threads produced by cutting into holes made by laser cutting					
t = 3 mm	✔✔	✔✔	✔✔	✘	✘
t = 5 mm	✔✔	✔✔	✔	✘	✘
t = 8 mm	✘	✘	✘	✘	✘
t = 10 mm	✘	✘	✘	✘	✘
Internal threads produced by forming into holes made by drilling					
t = 3 mm	✘	✘	✘	✘	✘
t = 5 mm	✘	✘	✘	✘	✘
t = 8 mm	✘	✔✔	✔✔	✔	✔✔
t = 10 mm	✔✔	✔✔	✔✔	✔✔	✔✔
Internal threads produced by forming into holes made by laser cutting					
t = 3 mm	✘	✘	✘	✘	✘
t = 5 mm	✔	✘	✘	✘	✘
t = 8 mm	✘	✘	✘	✘	✘
t = 10 mm	✘	✘	✘	✘	✘

The finalized results of the conducted experimental verifications and their evaluations are presented in Table 12. This table clearly illustrates that the most satisfactory internal threads can be achieved by employing thread-cutting technology within holes created by drilling. However, it is unviable to produce larger threads like M8 and M10 in thinner sheets with thicknesses of t = 3 mm and t = 5 mm using this approach. Furthermore, manufacturing a small M4 thread with a thickness of 10 mm is also unfeasible.

The feasibility of producing threads within holes created using laser cutting, including the suitable thread sizes, material thicknesses, and production methods, has also been investigated. Based on the findings, it is feasible to create high-quality threads by cutting into holes created by laser cutting, particularly for the M4, M5, and M6 threads with thicknesses of t = 3 mm and t = 5 mm. Nevertheless, considering the continually unsatisfactory outcomes of all previous threads generated using this technique, it is implausible to consistently deem the threads formed within holes created using laser cutting as acceptable.

The threads produced by forming within holes created by drilling were satisfactory only for thicknesses of t = 8 mm and t = 10 mm.

The primary contribution of this study lies in the obtained results, which serve as valuable recommendations for selecting suitable manufacturing technologies for the production of the analyzed metric internal threads (M4, M5, M6, M8, M10) within the material 1.4301, considering various thicknesses ranging from 3 mm to 10 mm, specifically tailored to the needs of the aforementioned company.

These recommendations and insights will be put into practice at ELMAX ŽILINA inc. and subjected to further validation. However, it is important to note that the results have broader applicability and can be useful in a wide range of practical scenarios where the manufacturing technologies tested in the conducted verifications are employed.

Author Contributions: Conceptualization, F.T. and D.S.; methodology, D.S., F.T. and M.Š. (Michal Šajgalík); validation, D.S., M.D. and F.T.; formal analysis, D.S., A.R. and M.Š. (Miroslav Špiriak); investigation, D.S., M.Š. (Michal Šajgalík) and M.D.; resources, F.T., N.N. and A.R.; data curation, N.N. and A.R.; writing—original draft preparation, F.T. and D.S.; writing—review and editing, D.S., N.N. and M.Š. (Miroslav Špiriak); visualization, D.S., A.R. and M.D.; supervision, M.Š. (Michal Šajgalík) and D.S.; project administration, D.S. and M.Š. (Miroslav Špiriak); funding acquisition, D.S. and M.D. All authors have read and agreed to the published version of the manuscript.

Funding: The article was supported by grant project KEGA 015ŽU-4/2023, Modernization of teaching chip technologies with elements of information technologies based on networked virtual laboratories.

Institutional Review Board Statement: Not applicable.

Informed Consent Statement: Not applicable.

Data Availability Statement: The data are available in a publicly accessible repository.

Conflicts of Interest: The authors declare no conflict of interest.

References

1. Val, A.G.D.; Veiga, F.; Penalva, M.; Arizmendi, M. Oversizing Thread Diagnosis in Tapping Operation. *Metals* **2021**, *11*, 537. [CrossRef]
2. Croccoloa, D.; de Agostinisa, M.; Finia, S.; Olmia, G.; Robustoa, F.; Vincenzib, N. Steel screws on aluminium nuts: Different engagement ratio tapped threads compared to threaded inserts with a proper tolerance choice. *Tribol. Int.* **2019**, *138*, 297–306. [CrossRef]
3. Fromentin, G.; Poulachon, G.; Moisan, A.; Julien, B.; Giessler, J. Precision and surface integrity of threads obtained by form tapping. *CIRP Ann. Manuf. Technol.* **2006**, *541*, 519–522. [CrossRef]
4. Chakhari, J.; Daidié, A.; Chaib, Z.; Guillot, J. Numerical model for two-bolted joints subjected to compressive loading. *Finite Elem. Anal. Des.* **2008**, *44*, 162–173. [CrossRef]
5. Brandão, G.L.; Carmo Silva, P.M.; Freitas, S.A.; Pereira, R.B.D.; Lauro, C.H.; Brandão, L.C. State of the art on internal thread manufacturing. *Int. J. Adv. Manuf. Technol.* **2020**, *110*, 3445–3465. [CrossRef]
6. Frydrýšek, K.; Šír, M.; Pleva, L.; Szeliga, J.; Stránský, J.; Čepica, D.; Kratochvíl, J.; Koutecký, J.; Madeja, R.; Dedková, K.P.; et al. Stochastic Strength Analyses of Screws for Femoral Neck Fractures. *Appl. Sci.* **2022**, *12*, 1015. [CrossRef]
7. Heiler, R.; Tischkau, P. *Hochleistungsgewindefertigung in Titanlegierungen*; Forschungsprojekt Hochschulbasierte Weiterbildung für Betriebe an der HTW: Berlin, Germany, 2015.
8. Liu, Y.; Fan, L.; Wang, W.; Gao, Y.; He, J. Failure Analysis of Damaged High-Strength Bolts under Seismic Action Based on Finite Element Method. *Buildings* **2023**, *13*, 776. [CrossRef]

9. Monka, P.; Monkova, K.; Modrak, V.; Hric, S.; Pastucha, P. Study of a tap failure at the internal threadsmachining. *Eng. Fail. Anal.* **2019**, *100*, 25–36. [CrossRef]
10. Suárez, A.; Arizmendi, M. Thread Quality Control in High-Speed Tapping Cycles. *Manuf. Mater. Process.* **2020**, *4*, 9. [CrossRef]
11. Heiler, R. Cold thread Forming—The chipless alternative for high resistant internal threads. *MATEC Web Conf.* **2018**, *251*, 02046. [CrossRef]
12. Anna, C.A.; Guillaume, F.; Gérard, P. Analytical and Experimental Investigations on Thread Milling Forces in Titanium Alloy. *Int. J. Mach. Tools Manuf.* **2013**, *67*, 28–34.
13. Masmoudi, N.; Soussi, H.; Krichen, A. Determination of an adequate geometry of the flanged hole to perform formed threads. *Int. J. Adv. Manuf. Technol.* **2017**, *92*, 547–560. [CrossRef]
14. de Oliveira, J.A.; Filho, S.L.M.R.; Brandao, L.C. Investigation of the influence of coating and the tapered entry in the internal forming tapping process. *Int. J. Adv. Manuf. Technol.* **2019**, *101*, 1051–1063. [CrossRef]
15. Maciel, D.T.; Filho, S.L.M.R.; Lauro, C.H.; Brandão, L.C. Characteristics of machined and formed external threads in titanium alloy. *Int. J. Adv. Manuf. Technol.* **2015**, *79*, 779–792. [CrossRef]
16. Fromentin, G.; Araujo, A.C.; Poulachon, G.; Paire, Y. Modélisation géometrique du filetage à la fraise. *Mécanique Ind.* **2011**, *12*, 469–477. [CrossRef]
17. Dogra, A.P.S.; de Vor, E.R.; Kapoor, S.G. Analysis of feed errors in tapping by contact stress model. *J. Manuf. Sci. Eng.* **2002**, *124*, 248–257. [CrossRef]
18. De Freitas, S.A.; Vieira, J.T.; Filho, S.L.M.R.; Cardoso, L. Experimental investigation of tapping in CFRP with analysis of torque-tension resistance. *Int. J. Adv. Manuf. Technol.* **2019**, *104*, 757–766. [CrossRef]
19. Lekkala, R.; Bajpai, V.; Singh, R.K.; Joshi, S.S. Characterization and modeling of burr formation in micro-end milling. *Precis. Eng.* **2011**, *35*, 625–637. [CrossRef]
20. Lauderbaugh, K. Analysis of the effects of process parameters on exit burrs in drilling using a combined simulation and experimental approach. *J. Mater. Process Technol.* **2009**, *209*, 1909–1919. [CrossRef]
21. Moganapriya, C.; Rajasekar, R.; Santhosh, R.; Saran, S.; Santhosh, S.; Gobinath, V.K.; Sathish Kumar, P. Sustainable Hard Machining of AISI 304 Stainless Steel Through TiAlN, AlTiN, and TiAlSiN Coating and Multi-Criteria Decision Making Using Grey Fuzzy Coupled Taguchi Method. *J. Mater. Eng. Perform.* **2022**, *16*, 7302–7314. [CrossRef]
22. Drbúl, M.; Šajgalík, M.; Šemcer, J.; Czánová, T.; Petrkovská, L. *Strojárska Metrológia a Kvalita Povrchov Vytvorených Technológiami Obrábania*; EDIS: Žilina, Slovakia, 2014.
23. Kalincová, D.; Ťavodová, M.; Kapustová, M.; Novák, M. Evaluation of residual stress in coinage tools. In *TEAM 2011: Proceedings of the 3rd International Scientific and Expert Conference with Simultaneously Organised 17th International Scientific Conference CO-MAT-TECH 2011*; University of Applied Sciences of Slavonski Brod: Slavonski Brod, Croatia, 2011; pp. 208–211. ISBN 978-953-55970-5-6.

Disclaimer/Publisher's Note: The statements, opinions and data contained in all publications are solely those of the individual author(s) and contributor(s) and not of MDPI and/or the editor(s). MDPI and/or the editor(s) disclaim responsibility for any injury to people or property resulting from any ideas, methods, instructions or products referred to in the content.

MDPI AG
Grosspeteranlage 5
4052 Basel
Switzerland
Tel.: +41 61 683 77 34

Materials Editorial Office
E-mail: materials@mdpi.com
www.mdpi.com/journal/materials

Disclaimer/Publisher's Note: The title and front matter of this reprint are at the discretion of the Guest Editor. The publisher is not responsible for their content or any associated concerns. The statements, opinions and data contained in all individual articles are solely those of the individual Editor and contributors and not of MDPI. MDPI disclaims responsibility for any injury to people or property resulting from any ideas, methods, instructions or products referred to in the content.

www.ingramcontent.com/pod-product-compliance
Lightning Source LLC
LaVergne TN
LVHW072338090526
838202LV00019B/2440